Energy Systems Analysis and Modelling towards Decarbonisation

Energy Systems Analysis and Modelling towards Decarbonisation

Editors

Panagiotis Fragkos
Pelopidas Siskos

MDPI • Basel • Beijing • Wuhan • Barcelona • Belgrade • Manchester • Tokyo • Cluj • Tianjin

Editors

Panagiotis Fragkos
National Technical University of Athens
Greece

Pelopidas Siskos
National Technical University of Athens
Greece

Editorial Office
MDPI
St. Alban-Anlage 66
4052 Basel, Switzerland

This is a reprint of articles from the Special Issue published online in the open access journal *Energies* (ISSN 1996-1073) (available at: https://www.mdpi.com/journal/energies/special_issues/ESA_MD).

For citation purposes, cite each article independently as indicated on the article page online and as indicated below:

LastName, A.A.; LastName, B.B.; LastName, C.C. Article Title. *Journal Name* **Year**, *Volume Number*, Page Range.

ISBN 978-3-0365-3885-3 (Hbk)
ISBN 978-3-0365-3886-0 (PDF)

Contents

About the Editors

Panagiotis Fragkos is a manager and senior researcher in E3-Modelling. Dr. Panagiotis holds a Ph.D. from the National Technical University of Athens (NTUA), a M.Sc and a B.Sc. in Electrical and Energy Engineering. He is co-leading the development of energy-economy modelling tools, including Prometheus, MENA-EDS, and PRIMES-Gas. His main research areas include climate change and energy economics, energy systems modelling, risk and uncertainty and sustainable development. He has co-authored over 40 articles in peer-reviewed scientific journals.

Pelopidas Siskos has a great deal of experience in energy and transport projects in relation to their transition to low- and zero-carbon solutions. His experience also involves managing multi-disciplinary teams and developing and applying bespoke quantitative tools or complex models for strategic analysis. Some of his most important milestones involve his horizontal engagement in the quantitative analysis of all mobility- and energy-related initiatives of the EU Fit for 55 policy package. Pelopidas Siskos holds an engineering (BSc and MSc) degree from the School of Mechanical Engineering of NTUA and a PhD in energy economics. As part of his doctoral thesis, Pelopidas developed the PRIMES-TREMOVE, a large-scale energy economic model for the simulation of the transport sector. The model has been widely used as a tool for medium and long-term strategic planning and policy-making purposes for the EU, during the 2011–2021 time period. He is also a member of the Editorial Board of the journal *Energy Systems* (Springer).

energies

Editorial

Energy Systems Analysis and Modelling towards Decarbonisation

Panagiotis Fragkos [1,2,*] and Pelopidas Siskos [1,2]

1 E3MLab, School of Electrical and Computer Engineering, National Technical University of Athens, 9 Iroon Polytechniou Street, Zografou, 15773 Athens, Greece; psiskos@e3modelling.com
2 E3-Modelling SA, Panormou 70-72, 11524 Athens, Greece
* Correspondence: fragkos@e3modelling.com; Tel.: +30-2106775696

1. Introduction

The Paris Agreement establishes a process to combine Nationally Determined Contributions with the long-term goal of limiting global warming to well below 2 °C and even to 1.5 °C. Responding to this challenge, national and regional low-emission strategies are prepared by both EU and non-EU countries, outlining clean energy transition pathways.

The aim of this Special Issue is to provide rigorous quantitative assessment of the challenges, impacts and opportunities induced by ambitious low-emission pathways. It aims to explore how deep emission reductions can be achieved in all energy demand and supply sectors, exploring the interplay between mitigation options, including energy efficiency, renewable energy uptake and electrification to decarbonise inflexible end-uses such as mobility and heating. The high expansion of renewable energy poses high technical and economic challenges with regard to system configuration and market organisation, requiring the development of new options, such as batteries, prosumers, grid expansion, chemical storage through power-to-X and new tariff-setting methods. The uptake of disruptive mitigation options (hydrogen, CCUS, clean e-fuels), as well as carbon dioxide removal (BECCS, direct air capture, others) may also be required in the case of net zero emission targets but raises market, regulatory and financial challenges.

This Special Issue assesses low-emission strategies at the national and global level and their implications for energy system development, technology uptake, energy system costs as well as the socioeconomic and industrial impacts of low-emission transitions.

Citation: Fragkos, P.; Siskos, P. Energy Systems Analysis and Modelling towards Decarbonisation. *Energies* **2022**, *15*, 1971. https://doi.org/10.3390/en15061971

Received: 25 February 2022
Accepted: 6 March 2022
Published: 8 March 2022

2. Scientific Contribution of this Special Issue: A Brief Overview

Yang et al. [1] examine greenhouse gas emission reduction possibilities from the electricity sector. The authors develop an agent-based model of the electricity system with heterogeneous agents who invest in power generating capacity under uncertainty. The heterogeneity is characterised by the hurdle rates the agents employ (to manage risk) and by their expectations of the future carbon prices. The results show that under an increasing CO_2 tax scenario, the agents start investing heavily in wind, followed by nuclear and to some extent in natural gas fired power plants both with and without carbon capture and storage, as well as biogas fired power plants. However, the degree to which different technologies are used depend strongly on the carbon tax expectations and the hurdle rate employed by the agents. Comparing to the case with homogeneous agents, the introduction of heterogeneity among the agents leads to a faster CO_2 reduction.

The Green Deal of the European Union defines extremely ambitious climate targets for 2030 (−55% emissions compared to 1990) and 2050 (−100%). Kattelmann et al. [2] focus on how the emission reduction targets shall be distributed across EU countries. The authors analyse the necessary burden sharing within the EU from both an energy system and an overall macroeconomic perspective. For this purpose, they use the energy system model TIMES PanEU and the computational general equilibrium model NEWAGE. The results show that excessively strong targets for the Emission Trading System (ETS) in 2030 are not system-optimal for achieving the 55% overall target. Economically weaker regions would have to reduce their CO_2 emissions until 2030 by up to 33% on top of the currently decided targets in the Effort Sharing Regulation, which leads to higher energy system costs, as well as losses in gross domestic product (GDP). Depending on the policy scenario applied, GDP losses in the range of −0.79% to −1.95% relative to baseline can be found for single EU regions. In the long-term, an equally strict mitigation regime for all countries in 2050 is not optimal from a system perspective; the total system costs would be higher by 1.5%.

The achievement of climate neutrality in the European Union by 2050 will not be possible solely through a reduction in fossil fuels and the development of energy generation from renewable sources. Large-scale implementation of various technologies is necessary, including bioenergy with carbon capture and storage (BECCS), carbon capture and storage (CCS), and carbon capture and utilisation (CCU), as well as industrial electrification, the use of hydrogen, the expansion of electromobility, low-emission agricultural practices, and afforestation. Tatarewicz et al. [3] carry out an analysis of BECCS as a negative emissions technology (NET) and the assessment of its implementation impact upon the possibility of achieving climate neutrality in the EU. The modelling approach utilises tools developed within the LIFE Climate CAKE PL project and includes the MEESA energy model and the d-PLACE CGE economic model. The authors identify the scope of the required investment in generation capacity and the amount of electricity production from BECCS necessary to meet the greenhouse gas (GHG) emission reduction targets in the EU, examining the technology's impact on the overall system costs and marginal abatement costs (MACs). The modelling results confirm the key role of BECCS technology in achieving EU climate goals by 2050.

The promising power-to-gas (P2G) technology makes it possible for wind farms to absorb carbon and trade in multiple energy markets. Considering the remoteness of wind farms equipped with P2G systems and the isolation of different energy markets, the scheduling process may suffer from inefficient coordination and unstable information. An automated scheduling approach is proposed in the work of Ji et al. [4]. Firstly, an automated scheduling framework enabled by smart contract is established for reliable coordination between wind farms and multiple energy markets. Considering the limited logic complexity and insufficient calculation of smart contracts, an off-chain procedure as a workaround is proposed to avoid complex on-chain solutions. The scheduling strategy takes into account not only the revenues from multiple energy trades but also the penalties for violating contract items in smart contracts. Then, the implementation of smart contracts under a blockchain environment is presented with multiple participants, including voting in an agreed scheduling result as the plan.

The COVID-19 pandemic left a remarkable impact on how people and businesses behave. At the same time, there is still high uncertainty as to how the impacts will evolve in the future. Emissions pathways after COVID-19 will be shaped by how governments' economic responses translate into infrastructure expansion, energy use, investment planning and societal changes. As a response to the COVID-19 crisis, most governments worldwide launched recovery packages aiming to boost their economies, support employment, and enhance their competitiveness. Climate action is pledged to be embedded in most of these packages but with sharp differences across countries. The work of Rochedo et al. [5] provides novel evidence on the energy system and greenhouse gas (GHG) emissions implications of post-COVID-19 recovery packages by assessing the gap between pledged recovery packages and the actual investment needs of the energy transition to reach the Paris Agreement goals. Using two well-established Integrated Assessment Models (IAMs) and analysing various scenarios combining recovery packages and climate policies, the authors conclude that currently planned recovery from COVID-19 is not enough to enhance societal responses to climate urgency and that it should be significantly upscaled and prolonged to ensure compatibility with the Paris Agreement goals.

Scaling down the big picture at the level of businesses, more and more companies choose to disclose carbon information, respond to the national policy of carbon emission reduction and focus on the sustainable development of enterprises. The paper of Lu et al. [6] investigates the impact of carbon disclosure on financial performance based on the 2011–2018 CDP report, taking the Fortune 500 companies as a sample. The study finds that for carbon-intensive industries, carbon disclosure cannot significantly contribute to the improvement of financial performance in the current period, but for carbon-non-intensive industries, carbon disclosure can significantly contribute to the improvement of financial performance in the current period, and the positive impact of carbon disclosure on financial performance in the current period can be extended to the next period. Finally, based on the findings of the empirical study, this paper puts forward policy recommendations for the construction of China's carbon disclosure system.

Jamróz et al. [7] contribute to this SI from a technical perspective on improving the efficiency of power plants to contributing to the energy transition and reduction of emissions from power generation. Combined cycle power plants are characterized by high efficiency, now exceeding 60%. The economic analysis of the authors revealed that the difference between the annual revenue from the sale of electricity and the annual fuel cost is considerably higher for power plants set to supercritical parameters, reaching approx. USD 14 million per annum. It is proposed that investments in adapting components of the steam part to supercritical parameters may be balanced out by a higher profit.

It is widely accepted that the market uptake of electric vehicles is essential for the decarbonisation of transport. However, scaling up the roll out of electric vehicles (EV) is challenging considering the lack of charging infrastructure. The latter is, currently, developing in an uneven way across the EU countries. A charging infrastructure with wide coverage addresses range limitations but requires high investment with uncertain returns during the early years of deployment. The aim of the work of Statharas et al. [8] is to assess how different policy options affect EV penetration and the involvement of private sector in infrastructure deployment. The authors propose a mathematical programming model of the decision problem and the interaction between the actors of EV charging ecosystem and apply it to the case of Greece from the time period until 2030. Greece represents a typical example of a country with ambitious targets for EV penetration by 2030 (10% of the total stock) with limited effort made until now. The results indicate that it is challenging to engage private investors in the early years, even using subsidies; thus, publicly financed infrastructure deployment is important for the first years. In the mid-term, subsidization on the costs of charging points is necessary to positively influence the uptake of private investments. These are mainly attracted from 2025 onwards, after a critical mass of EVs and infrastructure has been deployed.

Carbon leakage features prominently in the climate policy debate in economies implementing climate policies, especially in the EU. The imposition of carbon pricing impacts negatively the competitiveness of energy-intensive industries, inducing their relocation to countries with weaker environmental regulation. Unilateral climate policy may complement domestic emissions pricing with border carbon adjustment to reduce leakage and protect the competitiveness of domestic manufacturing. Fragkos et al. [9] use an enhanced version of GEM-E3-FIT model to assess the macro-economic impacts when the EU unilaterally implements the EU Green Deal goals, leading to a leakage of 25% over 2020 to 2050. The size and composition, in terms of GHG and energy intensities, of the countries undertaking emission reductions matter for carbon leakage, which is significantly reduced when China joins the mitigation effort, as a result of its large market size and the high carbon intensity of its production. Chemicals and metals face the stronger risks for relocation to non-abating countries. The Border Carbon Adjustment can largely reduce leakage and the negative activity impacts on energy-intensive and trade-exposed industries of regulating countries by shifting the emission reduction to non-abating countries through implicit changes in product prices.

Research and Innovation (R&I) are a key part of the EU's strategy towards stronger growth and the creation of more and better jobs while respecting social and climate objectives. In the last decades, improvements in costs and performance of low-carbon technologies triggered by R&I expenditures and learning-by-doing effects have increased their competitiveness compared to fossil fuel options. So, in the context of ambitious climate policies as described in the EU Green Deal, increased R&I expenditures can increase productivity and boost EU economic growth and competitiveness, especially in countries with large innovation and low-carbon manufacturing base. The analysis of Fragkiadakis et al. [10] captures the different nature of public and private R&I, with the latter having more positive economic implications and higher efficiency as it is closer to industrial activities. Public R&D commonly focuses on immature highly uncertain technologies, which are also needed to achieve the climate neutrality target of the EU. The model-based assessment shows that a policy portfolio using part of carbon revenues for public and private R&D and development of the required skills can effectively alleviate decarbonisation costs, while promoting high value-added products and exports (e.g., low-carbon technologies), creating more high-quality jobs, and contributing to climate change mitigation.

Author Contributions: P.F. and P.S. contributed equally to this work. All authors have read and agreed to the published version of the manuscript.

Funding: This research received no external funding.

Acknowledgments: The authors are grateful to the MDPI Publisher for the invitation to act as guest editors of this special issue and are indebted to the editorial staff of "Energies" for the kind co-operation, patience and committed engagement.

Conflicts of Interest: The authors declare no conflict of interest.

References

1. Yang, J.; Azar, C.; Lindgren, K. Modelling the Transition towards a Carbon-Neutral Electricity System—Investment Decisions and Heterogeneity. *Energies* **2022**, *15*, 84. [CrossRef]
2. Kattelmann, F.; Siegle, J.; Cunha Montenegro, R.; Sehn, V.; Blesl, M.; Fahl, U. How to Reach the New Green Deal Targets: Analysing the Necessary Burden Sharing within the EU Using a Multi-Model Approach. *Energies* **2021**, *14*, 7971. [CrossRef]
3. Tatarewicz, I.; Lewarski, M.; Skwierz, S.; Krupin, V.; Jeszke, R.; Pyrka, M.; Szczepański, K.; Sekuła, M. The Role of BECCS in Achieving Climate Neutrality in the European Union. *Energies* **2021**, *14*, 7842. [CrossRef]
4. Ji, Z.; Guo, Z.; Li, H.; Wang, Q. Automated Scheduling Approach under Smart Contract for Remote Wind Farms with Power-to-Gas Systems in Multiple Energy Markets. *Energies* **2021**, *14*, 6781. [CrossRef]
5. Rochedo, P.R.R.; Fragkos, P.; Garaffa, R.; Couto, L.C.; Baptista, L.B.; Cunha, B.S.L.; Schaeffer, R.; Szklo, A. Is Green Recovery Enough? Analysing the Impacts of Post-COVID-19 Economic Packages. *Energies* **2021**, *14*, 5567. [CrossRef]
6. Lu, W.; Zhu, N.; Zhang, J. The Impact of Carbon Disclosure on Financial Performance under Low Carbon Constraints. *Energies* **2021**, *14*, 4126. [CrossRef]
7. Jamróz, M.; Piwowarski, M.; Ziemiański, P.; Pawlak, G. Technical and Economic Analysis of the Supercritical Combined Gas-Steam Cycle. *Energies* **2021**, *14*, 2985. [CrossRef]
8. Statharas, S.; Moysoglou, Y.; Siskos, P.; Capros, P. Simulating the Evolution of Business Models for Electricity Recharging Infrastructure Development by 2030: A Case Study for Greece. *Energies* **2021**, *14*, 2345. [CrossRef]
9. Fragkos, P.; Fragkiadakis, K.; Paroussos, L. Reducing the Decarbonisation Cost Burden for EU Energy-Intensive Industries. *Energies* **2021**, *14*, 236. [CrossRef]
10. Fragkiadakis, K.; Fragkos, P.; Paroussos, L. Low-Carbon R&D Can Boost EU Growth and Competitiveness. *Energies* **2020**, *13*, 5236. [CrossRef]

energies

MDPI

Article

Modelling the Transition towards a Carbon-Neutral Electricity System—Investment Decisions and Heterogeneity

Jinxi Yang *, Christian Azar and Kristian Lindgren

Department of Space, Earth and Environment, Chalmers University of Technology, SE-412 96 Gothenburg, Sweden; christian.azar@chalmers.se (C.A.); kristian.lindgren@chalmers.se (K.L.)
* Correspondence: jinxi.yang@chalmers.se

Abstract: To achieve the climate goals of the Paris Agreement, greenhouse gas emissions from the electricity sector must be substantially reduced. We develop an agent-based model of the electricity system with heterogeneous agents who invest in power generating capacity under uncertainty. The heterogeneity is characterised by the hurdle rates the agents employ (to manage risk) and by their expectations of the future carbon prices. We analyse the impact of the heterogeneity on the transition to a low carbon electricity system. Results show that under an increasing CO_2 tax scenario, the agents start investing heavily in wind, followed by nuclear and to some extent in natural gas fired power plants both with and without carbon capture and storage as well as biogas fired power plants. However, the degree to which different technologies are used depend strongly on the carbon tax expectations and the hurdle rate employed by the agents. Comparing to the case with homogeneous agents, the introduction of heterogeneity among the agents leads to a faster CO_2 reduction. We also estimate the so called "cannibalisation effect" for wind and find that the absolute value of wind does not drop in response to higher deployment levels, but the relative value does decline.

Keywords: agent-based modelling; low carbon electricity system; investment decisions; heterogeneous agents; value factor of wind

Citation: Yang, J.; Azar, C.; Lindgren, K. Modelling the Transition towards a Carbon-Neutral Electricity System—Investment Decisions and Heterogeneity. *Energies* **2022**, *15*, 84. https://doi.org/10.3390/en15010084

Academic Editors: Panagiotis Fragkos and Pelopidas Siskos

Received: 25 October 2021
Accepted: 15 December 2021
Published: 23 December 2021

1. Introduction

The Paris Agreement [1] implies that emissions may have to drop by as much as half over the next decade or two, and then reach approximately zero by the middle of the century [2,3] (The required rate of emission reduction depends on the climate sensitivity, the temperature target as well as the assumed pathway towards the ultimate target (i.e., the degree to which an overshoot is accepted or not [3]).

In order to achieve the climate objectives of the Paris Agreement, the power sector must undergo a rapid transition towards carbon-neutral energy technologies over the next couple of decades. Since investments in power plants today will shape the power system for decades to come, understanding investment decisions in the context of current climate ambitions is critical.

Understanding how this transition may take place requires energy-system models. Typically, these are optimisation models that assume perfect foresight and one central (rational) optimiser making decisions for the entire economy in pursuit of the least-cost path that meets climate or emission constraints [4–8].

We believe such models can provide important insights. However, there is also a need to analyse the transition through a perspective that more closely resembles the workings of a real economy, where companies compete against one another and make investments in the power sector based on their risk aversion and capital costs as well as the available information, including expected electricity, fuel and carbon prices. An agent-based model (ABM) may provide such a perspective and relevant insights into complex energy-transition dynamics [9].

An ABM consists of a system of interacting agents, and it includes an explicit description of agents' decisions based on their (possibly limited) information about the system (including other agents). An ABM is often referred to as a simulation technique that enables modelling individual, heterogeneous and autonomous agents, where each agent individually assesses its situation and makes decisions on the basis of a set of rules [10]. In the context of the energy-system transition, ABMs can provide a complementary perspective to optimisation models.

Several ABMs have been used to study the capacity investments in the energy system under climate policies. For instance, Chappin, et al. [11] have developed an ABM—the Energy Modelling Laboratory (EMLab) — and modelled power companies' investments in generation capacity under climate policy. Agents in EMLab use past years data to forecast future fuel, electricity and CO_2 prices, and agents' heterogeneity is represented by the length of years that an agent used for making the forecast. Jonson et al. [12] compared the capacity investments from an ABM with a corresponding optimization model, and found results were similar, but that due to the limited foresight of the agent in the ABM, there are underinvestments and overinvestments capacity in the ABM compared to the results from the optimisation model. Kraan, et al. [13] studied energy transition pathways by simulating decision-making in investing in new power plants. Heterogeneity in this study is represented by the agents' discount rates (which range from 4% to 20% in this study). The results show that under a constant carbon-price scenario, the capacity mix of the system stays the same over time when agents use the same discount rate (homogeneous agents), while in the heterogenous agents case, a range of possible transition pathways are taken. Barazza and Strachan [14] explored the impacts of agent heterogeneity on investment decisions about power generation technologies. Heterogeneity in this study is represented by the agent's aim, technology preference, foresight and costs of capital. They modelled the transitions of the UK, German, and Italian electricity sectors and illustrated the impact of agents' heterogeneous characteristics on the number of investments made, which may accelerate or hamper efforts to reach decarbonisation goals.

These ABM studies find that at the aggregated system level, results relating to the overall capacity mix and electricity output differ from cases where models only include homogeneous agents or a central optimiser. This demonstrates that there may be significant value to considering the fact that agents in real world electricity systems tend to be heterogeneous, and therefore, it may be beneficial to use ABM with heterogeneous characteristics for exploring the energy transition.

In our previous ABM study [12], we identified how a bottom up analysis of companies' investment decisions may provide a different perspective or results compared to an optimisation modelling approach, and in that way having a model that more directly reflects decisions on liberalised electricity markets in the real world (where there is no central optimiser making decisions about the development of the electricity system).

The purpose of with the present paper is to further develop and investigate reasonable investment mechanisms in a simulation model of the electricity system development. In this paper, we extend the model [12] by introducing heterogeneous agents that use individual investment hurdle rates or individual expectations about the future carbon price.

This study also contributes to the literature by looking at one characteristic of the agents at a time and addressing issues such as how the investment in different technologies and the dynamics of the system are affected by heterogeneity. We also analyse the competition among agents with different characteristics, and look at if there are certain characteristics that come to dominate the investments.

Furthermore, this study contributes to the literature by looking at how the investment decisions and the heterogeneity of the agents affect the electricity prices and output variabilities, and how fast carbon emission declines towards carbon neutrality. We also pay attention to how the value of wind is affected by the so-called "cannibalisation effect" in a non-equilibrium model, i.e., that more wind power tends to undercut the value of additional investments in wind.

In this study, we focus on two characteristics of the agents (hurdle rates and expectations of future carbon price), but we think other heterogeneous characteristics of the agents are also important for their investment decisions, and we wish to analyse those other characteristics in future studies. The reason that we focus on agent heterogeneity in hurdle rates and the future carbon tax is that these two factors are central for power companies when they make investment decisions.

When evaluating the Net Present Value (NPV) of future revenue streams of different generation technologies, investors need to select a discount rate (here referred to as their hurdle rate). This choice has a strong impact on the profitability of different investment options [12,15,16]. In general, a high hurdle rate favours options with a low investment cost such as gas-fired power plants, whereas a low hurdle rate tends to favour capital-intensive energy technologies such as nuclear power plants and wind. However, we will in this paper notice that systems effects may complicate this pattern.

When comparing future revenue streams of different technologies, investors also need to estimate the future carbon price since the level of the carbon price will impact the running cost of the fossil technologies, the merit-order and therefore impact the profits of different technologies. A recent survey shows that companies worldwide are using (or planning to use) the internal carbon price for their investment and business operations, and power and fossil fuel industries have the highest proportion of companies currently using (or planning to use) an internal carbon price [17]. (An internal price is a monetary value on greenhouse gas emissions a company considers when making investment decisions and operating businesses). However, there are uncertainties in future carbon price levels (i.e., see the carbon price in the EU Emission Trading System [18,19]), and companies expect different future carbon prices [20]. Studies show that different carbon price assumptions and scenarios can lead to very different investment decisions (see, e.g., [21–23]). Therefore, it is important to consider how the competition between companies with different expectations for the future carbon tax may play out.

In Section 2 we present the model and our scenario assumptions, in Section 3 we describe three separate cases designs (one homogeneous agent case and two heterogeneous agents cases), and in Section 4 we present and discuss our results. In Section 5 we provide our main conclusions.

2. Methodology

2.1. Model Description

2.1.1. Overall Structure

We develop an agent-based model of the electricity system. The agents are individual power companies that (i) operate their power plants in an ideal electricity market and (ii) invest in new power-generating capacity. The agents choose investments to maximise their estimated profits (based on their expectations about the future).

2.1.2. Technologies

There are six power generation technologies: coal-fired, natural gas combined cycle (NGCC), NGCC with carbon capture and storage (NGCC with CCS), nuclear, solar PV, and wind power. The plants are characterised by their capacities, fuel costs, investment costs, lifetimes, and CO_2 emission intensities. The NGCC plant can be fuelled by natural gas or biogas, a decision made by the plant owner based on which has the lower operating cost when including the carbon tax. (For details on the parameter settings, see Table S1 in Supplementary Materials).

2.1.3. Time Slices

When modelling the power system in which variable renewable power sources may provide significant shares of the power supply, the model needs to address the variability of both supply and demand over time. In our model, we divide each year into 64 time slices individually characterised by electricity demand and the availability of variable renewable supply from wind and solar (for details, see Supplementary Materials).

2.1.4. The Electricity Market

We assume an ideal electricity market and that the agents do not engage in strategic behaviour. Instead, they sell all the electricity they can generate so long as the electricity price is greater than (or equal to) the operating cost of the corresponding plant, following the merit order. This defines the electricity supply function at each time slice, τ, of the year. An iso-elastic demand function determines the price, p_τ, that is reached when the electricity produced meets the demand,

$$q_\tau = q_{0,\tau} \times \left(\frac{p_\tau}{p_0}\right)^{\epsilon} \quad (1)$$

where q_τ is the electricity produced and consumed in time slice τ. $\varepsilon = -0.05$ is the elasticity, and the reference demand $q_{0,\tau}$ reflects the varying demand over the year (see the Supplementary Materials) and gives the demand for electricity when the price of electricity equals the reference level p_0. The reference demand $q_{0,\tau}$ is chosen so that the time slices correspond to the varying electricity demand in Germany in 2011, with $p_0 = 3.25$ EUR ct/kWh, following Jonson et al. [12] (In the sensitivity analysis, we have tested a case where the electricity demand is growing over time. See Section S5.4 in the Supplementary Materials). The carbon tax is part of the operating cost for fossil-fuel-based technologies and thus influences the electricity supply function.

2.1.5. The Investment Decision Process

The decision process follows the following steps. First, power plants that reached the end of their lifespan are removed from the current installed capacity mix each year. If multiple plants are retiring in the same year, the model removes one at a time. Following any plant removal, the agents take turns, in a randomly set order, making investment decisions for new plants. An agent adds a hypothetical power plant (testing the range of available technologies) to the existing capacity mix. The agent can calculate the electricity price (see Equation (1)), production profile, and the expected annual profit of this newly added plant for next year.

The agent knows the CO_2 tax rate for the next year. Agents thus have a good estimate of revenues and costs for the first year after an investment, even though subsequent investment decisions may alter the capacity mix and thereby change the estimated profitability. However, the agents have limited information about the CO_2 tax and the electricity prices for subsequent periods. In the simplest setting, agents assume that both the CO_2 tax and the estimated annual profit remain constant throughout the life of the power plant. (We also expand the model to allow for more advanced agents that have different expectations on the CO_2 tax ten years into the future, see Section 3.) The agent chooses the investment option with the highest positive *profitability index*, i.e., the highest profit per invested EUR corrected for the lifetime of the plant.

$$profitability\ index = \frac{NPV}{I} \times CRF \quad (2)$$

$$CRF = \frac{r}{1-(1+r)^{-T}} \quad (3)$$

where *NPV* is the Net Present Value of all future revenues and costs, *T* is a plant's lifespan, *I* is a plant's investment cost, and *r* is the hurdle rate used by the agent. By multiplying the *NPV* with the capital recovery factor (*CRF*), investment projects with different lifespans can be compared [24,25]. (We correct for the life of the investment in order to enable a fair comparison across technologies. However, it turns out that differences in plant lifespans in the relevant range (say, 20 to 40 years) do not make a major difference to the results since revenues 20 to 40 years into the future are discounted rather heavily.)

Following an agent's decision, all agents are informed and take the information into account in subsequent decisions. These decision rounds continue as long as agents are willing to invest. Then, the next plant is removed, and a new round of decisions takes place, until all plants to be retired in that year have been removed from the capacity mix. The newly invested plants are added to the capacity mix and start producing electricity the following year.

As the process of "out with the old, in with the new" repeats every year, the capacity mix keeps changing over time. In the present version of the model, agents do not take their current plant portfolio into account when evaluating an investment. (One could argue that a company with lots of, say, coal-fired power plants should not invest in wind if that investment would reduce electricity prices and induce lower profits or even losses for the coal-fired power plants. However, if the wind plant is profitable, it is likely that another company would carry out the investment anyway, and therefore electricity prices would drop by as much, but the profits from the plant would just accrue to some other company. That said, incumbent energy companies may still be less willing to invest in new technologies. In future work, we plan to consider how strategies that optimise the profitability of a company's entire portfolio may affect the overall development of the power system and the performance of each company).

2.2. *Carbon Tax*

In the scenario investigated, we assume a rising carbon tax that is zero during the first 10 years, after which it grows linearly to 100 EUR /ton CO_2 in year 50 and then stays there.

3. Case Design: Homogeneous and Heterogeneous Agents

We investigate three cases for the model set-up, a base case with homogeneous agents and two cases with heterogeneous agents: in one, the agents apply different hurdle rates; in the other, they instead make different assumptions about the rate of the future CO_2 tax.

3.1. *Homogeneous Case*

In the homogeneous case, all agents are identical. They use the same hurdle rate (8%/year), employ the same investment decision criterion, and have the same expectations about the future. They *know* the carbon price for the next year. They also know the present and already planned power generation capacity. They *assume* that the carbon price, power capacity, and electricity prices that are estimated for the subsequent year remain constant thereafter.

3.2. *Heterogeneous Hurdle Rates (HHR) Case*

In the second case, 25 agents use hurdle rate values between 5% and 11% per year (with steps of 0.25%). We want to investigate how the market develops if companies exhibit different degrees of risk aversiveness (and we assume that this is expressed by using different hurdle rates). We will refer to this case as the heterogeneous hurdle rate (HHR) case. In this run, all agents have otherwise the same expectations for the future as in the homogeneous case.

Here, we assume a hurdle rate of 8%/year in the base case, but in the heterogeneous cases, we have a range from 5–11%/yr. The base case value and the range are based on studies by IEA [26] who reported a weighted average cost of capital (WACC) for major power companies of around 8%/year in 2006 and a drop to around 5%/year in 2018. IRENA [27] reports a real WACC equal to 7.5%/year for OECD countries and China, and 10%/year for the rest of the world. The National Energy Modelling System (NEMS) of the US Energy Information Administration estimates a WACC of roughly 7%/year (see Figure 8 in Appendix 3. C in [28]), and a recent EU study reported that WACC varied across the EU Member States from 3.5%/year in Germany to 12%/year in Greece for onshore wind projects in 2014 [29].

3.3. Heterogeneous Foresight (HF) Case

In the third case, there are 16 agents with different expectations about the future growth rate of the carbon tax. We will refer to this case as the heterogeneous foresight (HF) case. In this case, an agent uses the future CO_2 tax rate (as per its expectation), along with the current capacity mix (including already-made decisions), to estimate the cash flow 10 years ahead. The agents expect future CO_2 tax levels $E_{t+10}(\beta)$ that increase linearly from the known level T_{t+1} in the next year (t+1) to a future level 10 years ahead (t+10), as described by the following expression,

$$E_{t+10}(\beta) = T_{t+1} + \beta(T_{t+10} - T_{t+1}) \quad (4)$$

where t is the current year, and β reflects the expected growth rate of the CO_2 tax, ranging uniformly from 0 to 1.5 (with steps of 0.1). This means that agent 1 (with $\beta = 0$) believes that the CO_2 tax will not increase, and agents 2 to 10 underestimate the growth of the tax, while agent 11 (with $\beta = 1$) expects the true future tax, T_{t+10}, and agents 12 to 16 overestimate the tax. Note that the agents calculate the future electricity prices based on their expected CO_2 tax. However, all agents assume that the installed capacity mix is unchanged over the next 10 years, apart from the new investment decisions that have already been taken. Linear interpolation is used to estimate the profitability of the different options over time. In the HF case, all agents use an 8%/year hurdle rate.

In all cases, the model starts from the stationary state at year 0 and runs for 80 years, when the system is close to a stationary state again. Initially, there are only coal and gas power plants, with a total capacity of 64 GW and 2 GW, respectively. The initial plants are owned and operated by a separate company that does not take part in further investments. These coal and gas plants have varying remaining lifetimes and are gradually removed from the system over time. (We also test a case where there are external PV investments from the households. See Section S5.3 in the Supplementary Materials).

4. Results and Discussion

We present our results from both the system-level perspective and the agent perspective. For the system level, we present the installed capacity, electricity price and production, price variations, CO_2 emission trajectories, and the ex-post evaluated internal rate of return (IRR) for each investment. For the agent level, we present the individual agents' capacity investments and economic performances. We also compare our model results with results from the literature or real-world examples in order to corroborate the model.

4.1. Installed Capacity

In all three cases, coal-fired power plants are gradually phased out from the capacity mix as a result of the increasing carbon tax. Meanwhile, the installed capacities of all other technologies increase, but to varying degrees.

This pattern can also be observed in the real world, for example, the increasing carbon price in the EU ETS has led to a rather rapid reduction in coal used for electricity generation (in particular over the last years), and an increasing use in variable renewable energy [30,31]. The growth in solar and wind has historically primarily been driven by various support schemes, but with current EU ETS prices, investments in these technologies become profitable even without other support schemes.

4.1.1. Homogeneous Case

In the homogeneous case (Figure 1a), the installed capacity of coal starts to decline around year 15. At that time, investments in wind starts to grow, but the installed capacity of wind begin to decline some twenty years later due to competition from nuclear, which then expands significantly (Figures 1a and 2a). It is interesting to note that this happens for purely economic reasons despite the fact that the assumed levelised cost of wind is significantly lower than that of nuclear in the model.

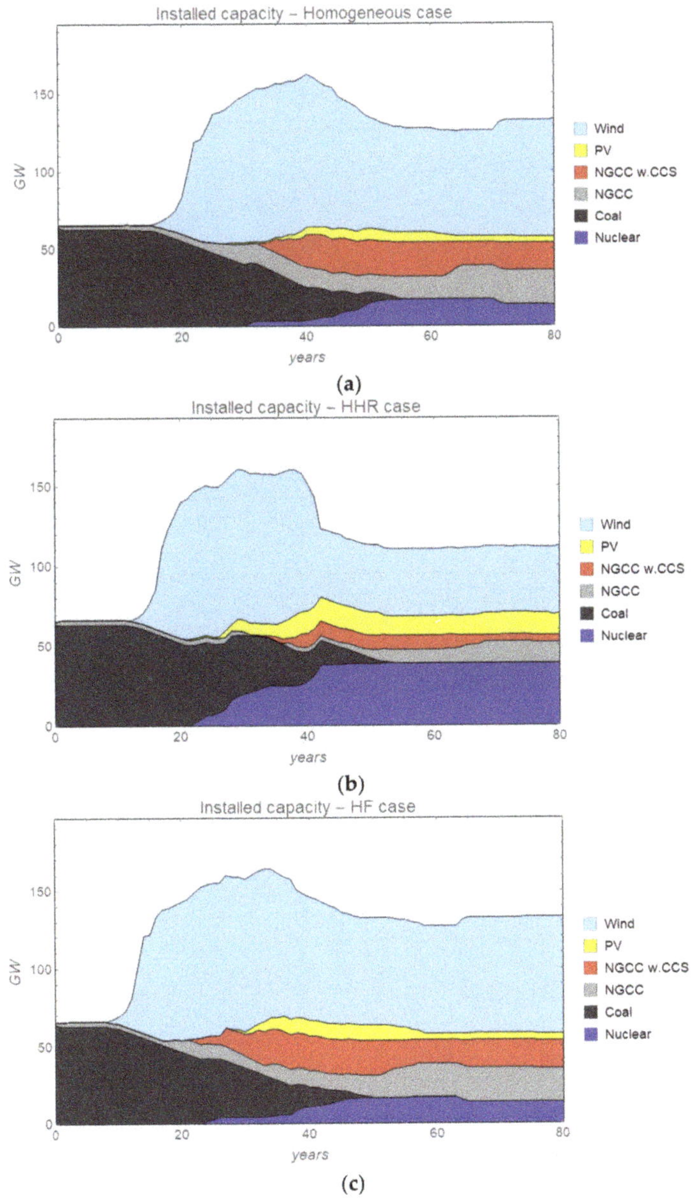

Figure 1. Installed capacity in the (**a**) homogeneous case, (**b**) HHR case, and (**c**) HF case. In all three cases, the capacity mix goes from coal-based to low-carbon, but the transition speed and final mix vary.

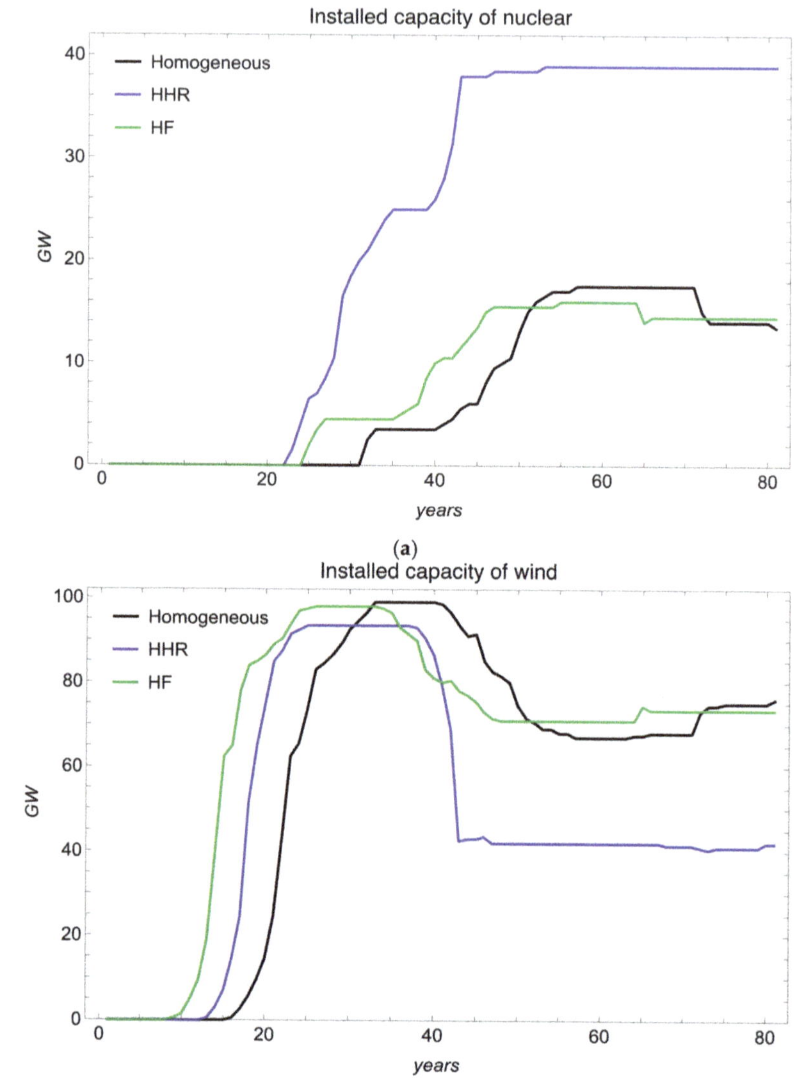

Figure 2. Installed capacity by technology in homogeneous, HHR, and HF cases. (**a**) Nuclear, (**b**) wind. In the HHR case, there is significantly more nuclear and less wind. The earliest wind expansion is observed in the HF case.

This illustrates why the levelised cost of electricity (LCOE) metric is inappropriate when comparing the competitiveness of, in particular, dispatchable technologies such as nuclear and gas-fired power plants with that of intermittent technologies such as wind and solar [32]. The fundamental reason for this is that the LCOE metric does not consider that the price of electricity varies during the course of a year and hence that different generation technologies will receive different average revenues per kWh of electricity produced. Therefore, in this case, when the installed capacity of wind grows to sufficiently high levels, wind eventually begins to receive lower revenues per kWh than does nuclear (because they produce power at different points in time). (See more about the revenue and value factor of wind in Section 4.2.2.)

4.1.2. Heterogeneous Hurdle Rate (HHR) Case

In the HHR case, nuclear starts to grow earlier and at a higher rate compared to the homogeneous case (see Figure 2a). By the end of the period we have 40 GW nuclear in the HHR case compared to some 15 GW in the homogeneous case. We also see an earlier introduction of wind in the HHR case than in the homogeneous case. Interestingly, towards the end of the period, the installed wind capacity is significantly lower in the HHR case (75 GW in the homogeneous case, but 42 GW in the HHR case, see Figure 2b). The impact of agent hurdle rate heterogeneity in the long run is thus different for wind and nuclear.

The explanation for this has two steps. First, when the model is run with heterogeneous hurdle rates, the investments are unevenly distributed among the agents (Figure 3), and the agent with the lowest hurdle rate (5%/year) does almost all the investing. With a low hurdle rate among those who carry out almost all the investments, the levelised costs of both nuclear and wind drop more than the cost of coal does, so wind and nuclear are introduced earlier than in the homogeneous case.

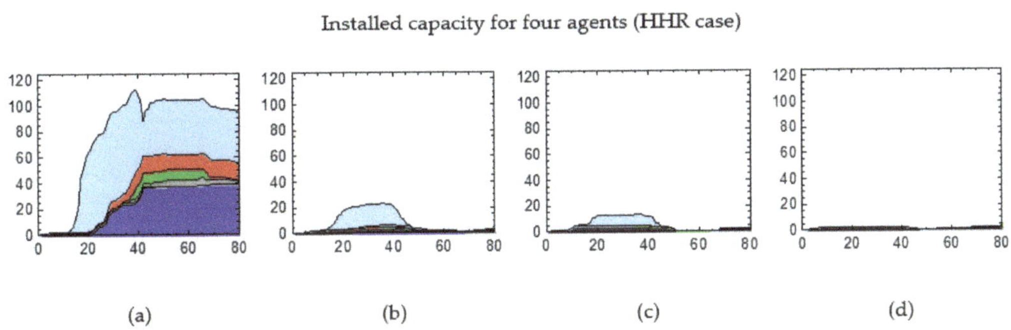

Figure 3. Illustration of the installed capacity for some agents in HHR case, (**a**) r = 5.0%, (**b**) r = 5.25%, (**c**) r = 5.5%, (**d**) r = 5.75%. The agent with r = 5% dominates the investing, while agents with r > 6.75% make no investments.

Second, the lower hurdle rate makes the cost of nuclear drop more than for its dispatchable competitors—NGCC—and natural gas with CCS, and therefore nuclear takes market shares from these technologies. Now, since wind and gas-fired power plants tend to be more complementary to, than competitive with, each other [12,33], the installed wind capacity actually drops as a consequence of the decline of gas-fired power plants (both NGCC and natural gas with CCS). Therefore, towards the end of the period, there is significantly more nuclear and less wind in the HHR case than in the homogeneous case. (We conducted a sensitivity analysis of the impact of the hurdle rate on the installed capacity of wind and how it is related to the availability of gas-fired power plants. See Supplementary Materials for details).

When looking at the HHR case from the agent's investment perspective, we see that while the agent with the lowest hurdle rates (5%/year) dominates investing, agents with hurdle rates over 6.75%/year do not make any investments at all (see Figure 3a–d). This is because agents with lower hurdle rates are more willing to invest than other agents, as they have a lower requirement on future net revenues, which in turn lowers the electricity price. This reduces returns on investment so much that agents requiring a higher return never invest and are largely outcompeted.

Our finding here also supports the results in [13,14], where they find that with agents applying various hurdle rates, the transition to a low-carbon system is accelerated due to the investments from agents that apply lower hurdle rates, and those agents take a major share of the investments [34].

4.1.3. Heterogeneous Foresight (HF) Case

The expansion of wind and nuclear in the HF case is similar to the expansion in the homogeneous case, and the installed capacity in these two technologies towards the end is roughly the same as in the homogeneous case, but investments in wind and nuclear start a decade earlier than in the homogeneous case.

However, one interesting difference is the faster phase-out of coal plants (compared to in the other two cases) and the expansion of gas-fired power plants is much faster, and this in turn depends on two factors: (i) the agents have 10-year foresight, so many of them expect that the carbon tax will be higher in the future, and that prompts them to invest in gas with CCS earlier; and (ii) some agents expect a carbon price that is even higher than the one that eventually materialises, and this creates an even stronger incentive to act early. These findings can be also observed in the actions taken by companies in the energy sector. For example, the Dutch company Shell expects the carbon cost to increase to more than $100 per tonne of GHG emissions by 2050, and this high price expectation influences Shell's decisions to invest in carbon capture technology (as well as investments in natural gas and biofuels) [35].

Agents that anticipate a high carbon price dominate the overall investment activity (but only prior to the stabilisation of the carbon price). Figure 4 shows that the agent that expects the highest carbon price makes most of the investments during the transient phase (years 10 to 50). This agent (falsely) believes the electricity price will be higher than the other agents think, due to overestimating the future carbon price. However, during the final 20 years, when the carbon price has stabilised, all agents foresee the same future carbon price, and they all invest and gradually approach equal installed capacities.

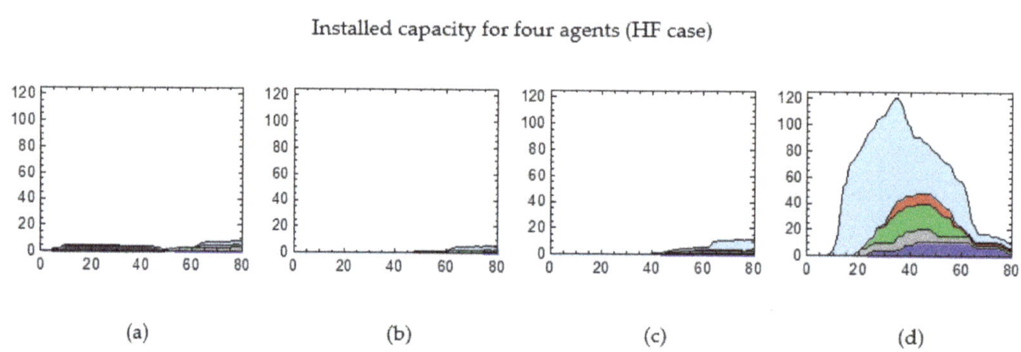

Figure 4. Illustration of the installed capacity for some agents in HF case. (**a**) $\beta = 0.0$, (**b**) $\beta = 1.0$, (**c**) $\beta = 1.2$, (**d**) $\beta = 1.5$. (β is the multiplication factor of expectation of CO_2 tax). The agent with $\beta = 1.5$ dominates the investing in the first 50 years; afterwards, all agents contribute to investing in new capacities.

4.2. Electricity Price and Production

4.2.1. Average Electricity Price

The annual average electricity price increases over time in all three cases, but to different degrees and at different speeds (Figure 5). The electricity price increases because the growing carbon price results in higher marginal costs of the fossil plants, which sets the market price. Once essentially all CO_2 emissions are phased out, the electricity price is nevertheless higher than in the beginning of the modelling period, since the cost of the technologies that replaced coal are higher than the cost of coal.

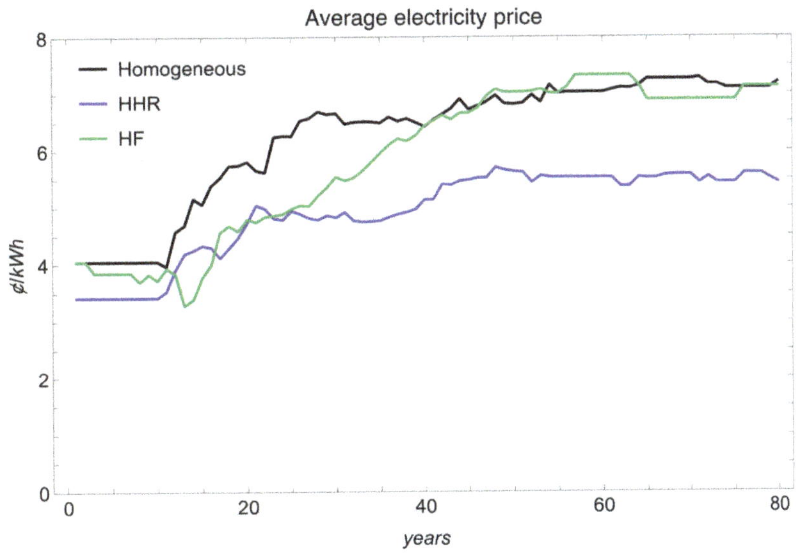

Figure 5. Average electricity prices for the homogeneous case, HHR case, and HF case.

In the long run, the electricity price is the same in the HF case as in the homogeneous case, because the CO_2 tax stabilises (and the hurdle rate is the same). However, in the near term, the electricity price is lower in the HF case, since agents invest more heavily in the wind as some agents believe that carbon prices will be higher than they really turn out to be. This increases available capacity and hence lowers electricity prices during the transient phase.

In the HHR case, due to the lower hurdle rate applied by the dominating agent, the price increase is smaller. In this case, the agents use lower hurdle rates, so with a lower cost of capital it becomes less costly to produce electricity, and therefore the price of electricity drops in a competitive market.

4.2.2. Average Revenues per kWh Received by Wind—The So-Called "Cannibalisation Effect"

Several studies have found that as the installed capacity of wind increases, the value of wind (the average electricity price received by wind producers) tends to drop [36–38]. The average revenues per kWh received by wind producers are expected to drop in absolute terms if the wind is introduced through, for instance, an investment subsidy (this is what is sometimes referred to as the cannibalisation effect). However, here we find that when wind is introduced as a result of an increasing carbon tax, the revenue received by wind generators will not drop in absolute terms but in relative terms (see Figure 6a,b). Our results here corroborate the findings by Brown and Reichenberg [39] and Jonson et al. [12]. The reason why the absolute value of wind does not drop is that on a competitive market, agents will not invest in wind if the average revenues received per kWh are lower than the levelised cost of wind. The reason why the relative value (the wind value factor) tends to drop is that while the average revenues per KWh received by wind operators remain roughly constant, the average electricity price increases in response to the increasing carbon tax. Hence, the relative value of wind drops.

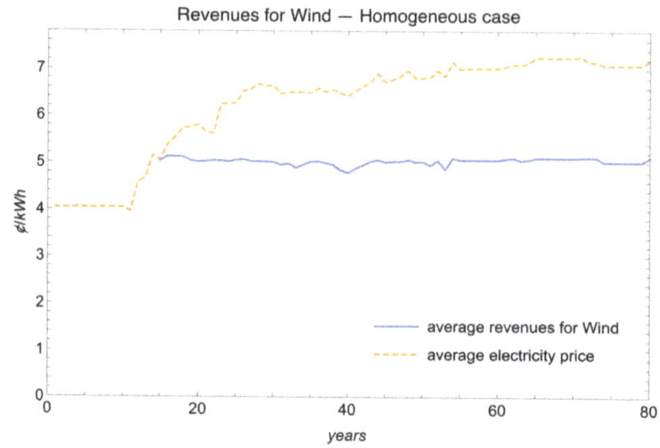

(**a**)

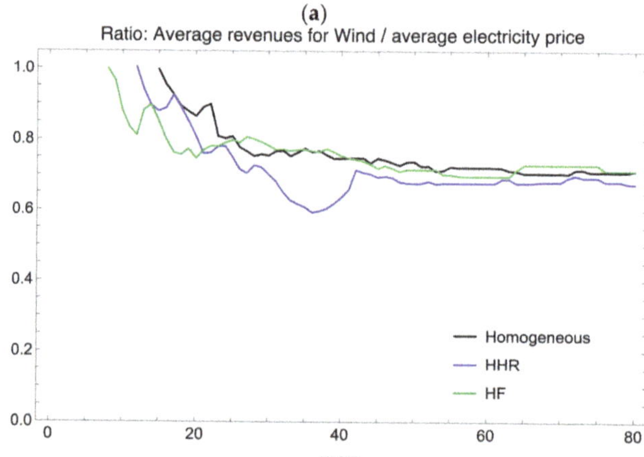

(**b**)

Figure 6. Revenues received by wind: (**a**) in absolute term (homogeneous case), (**b**) in relative term (all three cases).

4.2.3. Electricity Price Variations

Another important question for a transition towards a greater share of variable renewable energy is whether and to what extent price variations are more pronounced over the course of the year. Studies, e.g., [40–42], have shown that the higher share of renewable energy has increased the variance of wholesale electricity prices. Our results here show a similar pattern.

Figure 7 shows the amount of electricity produced at different prices for the homogeneous case. In year 5, while coal still dominates production, we see that most electricity is produced and sold at 3–4 EUR ct/kWh, but there is also some electricity produced at around 20 EUR ct/kWh, associated with ~500 h of the highest reference demand. In year 25, most electricity produced, from both wind and coal, is sold at 5–6 EUR ct/kWh. Meanwhile, the high wind capacity can result in very low prices, ~1 EUR ct/kWh, as well as very high prices, 25 EUR ct/kWh or above, for some hours of the year (a similar observation has been made by Auer and Hass [43]). When agents have invested heavily in wind, there is less incentive to invest in other technologies (because prices have been depressed), and when demand is high and wind output is low during certain hours of the year, prices may become very high. By year 40, the expansion of wind has led to a situation with many hours having a price level of less than 1 EUR ct/kWh. However, the high price spikes are damped by the increased presence of biogas, natural gas with CCS, and nuclear power (see Figure 7 for year 80). The price variation is also presented by the 10th and 90th percentiles in the Supplementary Materials, see Figure S6a–c.

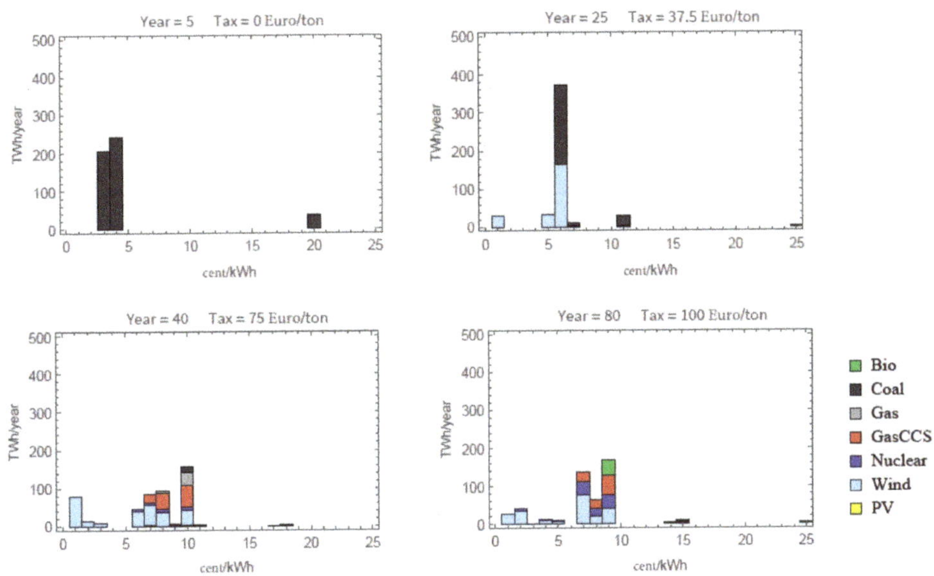

Figure 7. Power output at different price levels for four years (years 5, 25, 40, and 80) in the homogeneous case. Each column represents all production sold at prices in an interval [p–1, p] EUR cent/kWh, for integer values of p. The 25 EUR cent level also collects all production from higher prices. Each column shows produced amounts in merit order from bottom to top.

4.3. CO_2 Emissions

As the electricity system shifts from a coal-based mix to low-carbon energy sources, CO_2 emissions drop from around 480 megatons per year to 5 megatons in the homogeneous and HF cases, and to 1 megaton in the HHR case (Figure 8). The earliest emission decline happens in the HF case.

The HF case illustrates that the uncertainty about future carbon prices may lead to a faster transition. However, this result does not hold if "uncertainty" is interpreted as if companies are uncertain whether the tax will increase at all and the true subsequent increase is higher than what all the companies expected (see e.g., Figure S9 in Supplementary Materials, where we compare the rate of emission reductions in a case where agents have a one-year correct foresight for the carbon tax and a ten-year correct foresight).

What we have found (and estimated the impact of) is a case where the companies are uncertain about how much the carbon tax will increase and some companies expect that carbon tax will increase more than it eventually does, while others expect it will increase less.

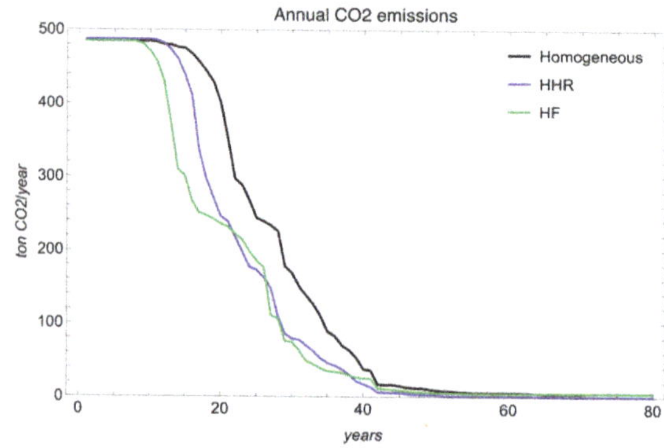

Figure 8. Annual CO_2 emissions trajectory for the homogeneous case, HHR case, and HF case.

4.4. Economic Performances—Ex-Post Analysis of the Investments

4.4.1. Internal Rate of Return of Investments

Figure 9a–c show ex-post analyses of the profitability of the investment projects over time. We compare the internal rate of return (IRR) for all investments done in the three cases. Each dot represents an investment that was made in a particular year.

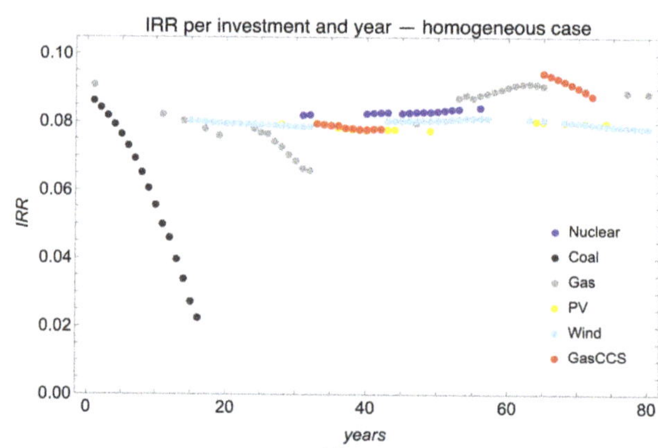

(a)

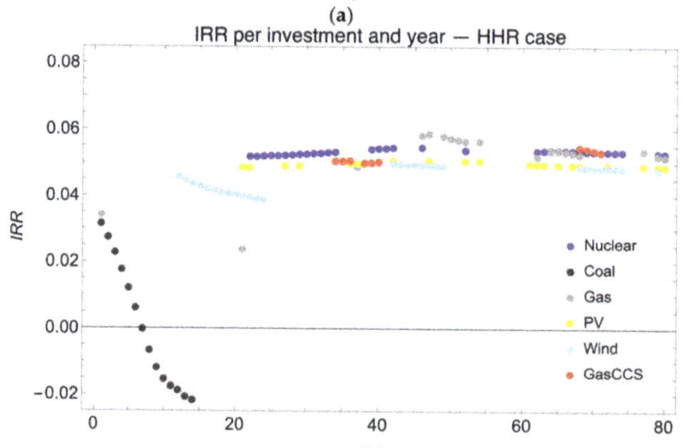

(b)

Figure 9. *Cont.*

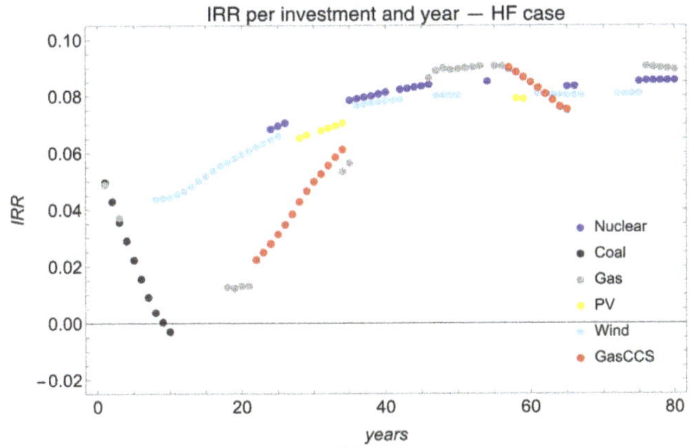

(c)

Figure 9. Internal rate of return (IRR) of investments in the (**a**) homogeneous case, (**b**) HHR case, and (**c**) HF case. Except for coal, the IRRs are mainly close to the level of the hurdle rates used by agents who make the investments.

In the homogeneous case, the IRR of coal drops rapidly from above 8% to 2% per year. This is because the increasing carbon tax makes investments in coal-fired power plants increasingly unprofitable. The reason why investments are nevertheless made in this technology is that the agents do not foresee the increase in the tax. For gas plants, the IRR declines during the first 30 years due to the carbon tax, but after switching to biogas around year 42, the IRR goes up and exceeds 9% per year at its highest. Meanwhile, all other investments have an IRR around 8% per year, which indicates that agents in general are close to the target set by their hurdle rate, despite their limited foresights (the reason we do not get exactly 8%/year is that (i) we have limited foresight so all investments are not perfect, and (ii) we have discrete name plate capacities, which means we cannot invest to get the exactly "optimal" solution).

The declining profitability of coal power can be observed in Germany and the EU. Analyses have shown that the coal plants have been operating at a loss in recent years due to the carbon price in the EU ETS and competition from renewable energies [44,45].

In the HHR case, IRRs of most investments are around 5%/year, which is the hurdle rate applied by the agent that made almost all the investments. The IRR of coal is even lower in this case, due to a lower electricity price.

In the HF case, during the first half of the simulation, the IRRs are much lower than in the homogeneous case. This is due to overinvestment by the agent that expects the highest carbon tax, which in turn yields low electricity prices during the transient phase (Figure 5). In the second half period, when the carbon tax begins to stabilise, all agents have roughly correct foresight, and the IRRs of wind and gas grow back to around 8%/year, similarly to the homogeneous case.

4.4.2. Economic Performance of Individual Agents

To measure an individual agent's economic performance, we use a *performance index* which is defined as the agent's annual profit divided by the total investment costs of all plants it owns. Here we assume that an agent buys a plant by borrowing money from the bank at a 4% interest rate. The annual profit is then calculated as the revenue minus the operating costs and the annuitised capital costs, where the capital cost is calculated with the 4% bank rate. This means that the varying hurdle rates assumed by the agents in our model (5% to 11%) reflect different risk premiums or margins compared to the bank rate. A positive performance index in a certain year thus means that an agent's net revenues cover more than the annuitised capital cost of its plants.

In the homogeneous case (Figure 10a), the index starts at around 5%, but falls to 1% around year 20, due to the dropping revenue from coal investments and large expenditures on wind investments. However, the index gradually grows back to around 3.5%.

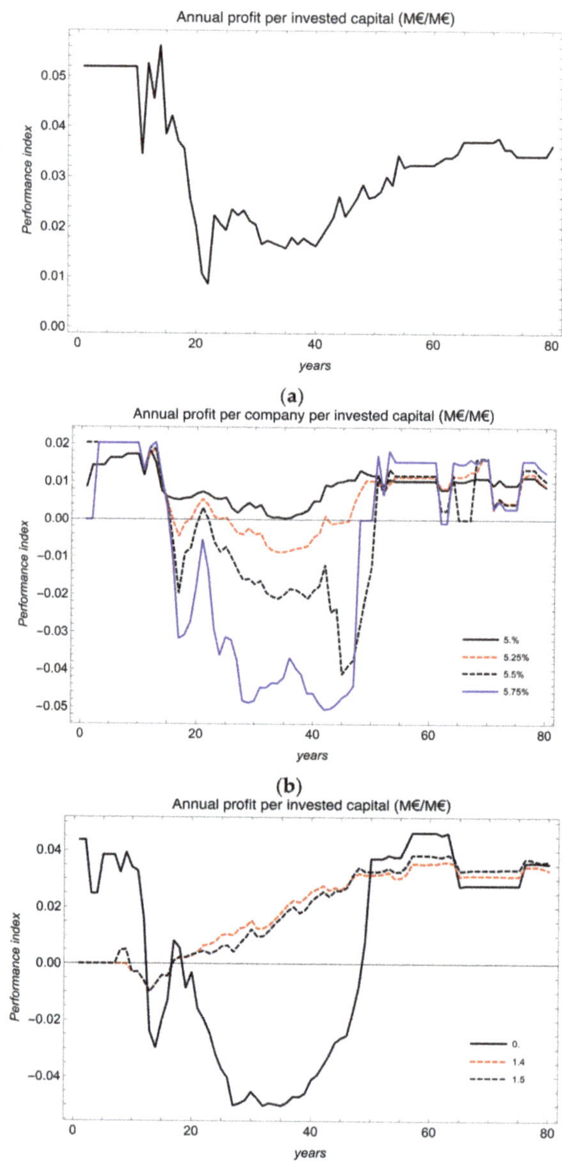

Figure 10. Individual agents' economic performance in the (**a**) homogeneous case, (**b**) HHR case, and (**c**) HF case. In the homogeneous case, the index stays positive during the whole model run, while in the two heterogeneous cases, the index varies among agents due to their different investment behaviours.

In the HHR case (Figure 10b), the performance varies among agents with different hurdle rates. The agent with the lowest hurdle rate performs well as the index stays above 0 during the entire 80 years. The performance indexes of agents with hurdle rates from 5.25% to 6.5% per year show a different pattern. Their indexes are positive during the first 15 years, then drop to negative, and stay negative for most of the period between years 20 and 50. These agents invest in coal early on, and that turns out to be a bad investment. The indexes return to positive during the final 30 years, thanks to the profits from subsequent investments in low-carbon technologies. Agents with hurdle rates over 6.5%/year made no investments, so their indexes are 0.

In the HF case (Figure 10c), it is helpful to divide the discussion of our results into two separate periods—before and after year 50 (when the CO_2 tax stabilises). During the first period, agents that expect lower CO_2 prices (β smaller than 0.5) perform poorly due to losses in coal plant investments. However, these losses are actually rather small and largely stem from the fact that these agents did not make additional investments until year 50. Meanwhile, agents with β values between 0.6 and 1.3 either did not invest or invested in several gas plants (see Figure S4 in Supplementary Materials). During that period, agents with high β values (i.e., β = 1.4 and 1.5) had already started investing in wind, and they performed relatively well. In the second period, all agents perform equally well because after the carbon tax stabilises, agents foresee the same CO_2 tax and make the same investment decisions.

One important and perhaps counterintuitive result is that the two agents that expect the highest future CO_2 tax exhibit the best overall performance, even though their assumptions were incorrect. These agents invest heavily in wind, and that turns out to be profitable (see IRR in Figure 9c).

5. Impacts of Assumptions and Simplifications

In this section, we discuss some limitations of our model and the impact they may have on our results. We also present some additional sensitivity analysis with respect to these assumptions.

5.1. Type of Agents

All agents in the study are assumed to be power companies, and investments from other types of agents are not included, for example, distributed solar PV invested by households.

We have tested a case where there are "external" investments made in solar PV to resemble solar PV investments from households. We find that when significant (around 70 GW) exogenous PV is installed (by the households), power companies invest less in nuclear capacity, but actually more in wind and gas, compared to the case where there is no exogenous PV investment. This leads to less electricity generation from nuclear plants but more from solar, wind and gas power plants. Detailed results are presented in the Supplementary Materials, see Supplementary Section S5.3.

5.2. Electricity Demand

The electricity demand in this study is assumed to vary during the course of the year (seasons and day and night variations) but this demand pattern is also assumed to remain the same from one year to the other. We have tested a case where there is a 1%/year increasing demand, and find that the installed capacity of each type of technology increases compared to our base case scenario, but the overall pattern does not change. Detailed results are presented in the Supplementary Materials, see Supplementary Section S5.4.

5.3. Size of the Plant

Here we tested a case with a uniform plant size of 50 MW instead of 500 MW, which is used in this study. We find that with smaller plant sizes, there are slightly more nuclear and less wind and gas capacities, but the overall result is similar to the case with 500 MW plants. The reason why there is more nuclear is because nuclear has a rather high investment cost per MW, and a smaller plant size makes it possible to invest in nuclear incrementally. Detailed results are presented in the Supplementary Materials, see Supplementary Section S5.5.

6. Conclusions

In this paper, we have developed an agent-based model of the electricity system with heterogeneous agents. Results from this study can be useful for both modellers in the energy system community as well as scholars who investigate the low-carbon energy system transition. Results pertaining to the dynamics of the electricity system transition could also be in the interests of policy-makers.

We primarily analysed how investments in new generating capacity are affected by heterogeneity and uncertainty in terms of how agents form their capital cost (presented by the hurdle rate in this study), expectations about the future carbon price, and the competition with other agents. We then looked at how investments in the power sector change over time, and how this affects the capacity mix, electricity output, electricity prices (the average and the variability), carbon emissions and profits for the various agents.

In optimisation models, there is a central decision maker that ensures that all necessary investment decisions are made. In liberalised electricity markets, those decisions are left for the different agents to make. One interesting feature with the type of simulation model employed here is that it more closely mimics the workings of a decentralized electricity market and the fact that the system properties emerges from decisions made by individual agents that compete against one another (bottom up rather than top down). Another feature is that we may with this approach investigate how different successful companies with different investment decisions strategies eventually turn out to be and which investment decision criteria that lead to either losses or a low activity.

At the systems level, we find two distinct phases of the transition:

- The *first phase* takes place when wind starts to increase rapidly in response to the increasing carbon tax. This tax makes the electricity price grow, which makes investment in wind profitable. However, as the tax continues to grow, the price of electricity tends to increase most during those time periods of the year when wind output is low (because during those time periods dispatchable coal-based power generation is still used in combination with natural gas fired power plants).
- This takes us into the *second phase* (around the years 30–40) when changes of the profitability of future investment start to occur. During this phase it becomes more profitable to invest in technologies that can generate electricity also during time periods when wind output is low (and the electricity price is high). Thus, in this phase we see investments in nuclear, gas-fired power plants with biogas and natural gas with CCS, but also solar PV, and during this period wind capacity actually drops.

The technology mix that eventually comes to dominate the electricity system (during phase two and thereafter, see Figure 1a–c) depends on the characteristics of agents that make most of the investments.

In the heterogenous hurdle rate case, most investments are made by the agents with the lowest hurdle rate. One interesting dynamic here is that once these agents start to invest, electricity prices become lower (because of their lower cost of capital), and this then tends to outcompete the agents with a higher cost of capital. Furthermore, the lower hurdle rate—it turns out—leads to that nuclear becomes a much bigger player (twice as big as in the homogenous and the heterogenous forecast case, see Figure 2a). Interestingly, it also means that wind ends up playing a less important role—despite the fact that a low hurdle rate is typically considered advantageous for wind (given that wind is a capital-intensive technology). The reason for this has to do with the systems interactions that take place between nuclear, wind and gas-fired power plants (see the discussions in Section 4.1.2). Thus, one result from our paper is that the impact of the hurdle rate on the technology mix is significantly more complex than what is often thought.

In the heterogenous forecast case, it is the agents with the highest expectation for the carbon tax that makes most investments, and that means the transition starts a bit earlier and happens at a slightly faster rate (see the discussions in Section 4.1.3.) (see Figures 2 and 10). However, it should also be observed that when the agents believe that the carbon tax will be significantly higher than it really turns out to be, they tend to overinvest in carbon neutral technologies which in turn leads to depressed electricity prices and lower profit margins (Figures 9c and 10c).

We can hence conclude that, in the present modelling effort, the heterogeneity in terms of variations in hurdle rates and tax expectations results in a "selection" of companies that dominate the investment decisions. This leads to more investments in low carbon technologies, as compared to a homogenous agents situation where all agents have a middle value in the range of possible hurdle rates and tax expectations. This in turn speeds up the transition towards lower CO_2 emissions. In the case with homogeneous agents, emissions drop by more than 50% once the carbon tax reaches around 50 EUR /ton CO_2. However, when we analyse the heterogeneous cases, emissions drop much faster. When the carbon tax reaches 50 EUR /ton CO_2, emissions are down by roughly 80% (see Figure 8).

Moreover, we have estimated the value factor for wind (i.e., the revenues received by wind per kWh compared to the average electricity price, see [37]) during a non-equilibrium transient phase. We find that although wind expands to significant shares of the electricity system, the absolute revenues for wind per kWh do not drop (which is in line with Brown and Reichenberg [39] who performed this in an equilibrium setting). However, the relative value of wind drops. The relative value decline for wind and the increasing price variations calls for policy interventions to deal with the increasing share of the variability of wind and solar. It could include state support for an expansion of the grid, expansion of storage technologies, and various demand side management strategies [46–48].

Supplementary Materials: The following are available online at: https://www.mdpi.com/article/10.3390/en15010084/s1, Table S1: Assumed power plants parameters – costs and efficiencies; Table S2: Number of hours in each of the time slice; Table S3: Values of the demand reference, solar availability and wind availability; Table S4: Installed capacity of wind in year 80 in different set-ups in terms of hurdle rates and gas plants availability; Figure S1a–c: Annual electricity production by technologies in the (a) homogeneous case, (b) HHR case, and (c) HF case; Figure S2: The reference demand level (dashed) and the dispatched power (solid) across the time slices of the year, for four different years (years 5, 25, 40, and 80) in the homogeneous case; Figure S3: Illustration from one simulation showing the installed capacity of agents with hurdle rates from 5%/year (top figure) to 6.75%/year (bottom figure) in the HHR case; Figure S4: Individual agent's financial performance (measured by Performance Index) in the HHR case; Figure S5: Individual agents´ installed capacity (GW) over the time period of 80 years in the HF case; Figure S6: Individual agent's financial performance (measured by Performance Index) in the HF case; Figure S7a–c: Electricity-price spread over time illustrated by 10th and 90th percentiles, as well as the average and median prices for the (a) homogeneous case, (b) HHR case, and (c) HF case; Figure S8: Standard deviation of electricity price for the homogeneous case, HHR case, and HF case; Figure S9: Annual emissions in the homogeneous case, HF case and homogeneous 10-year foresight case; Figure S10: The system installed capacity in the HHR case; Figure S11: Annual electricity production by technology in the HHR case; Figure S12: Installed capacity in the HHR case in the two demand scenarios; Figure S13: Installed capacity of individual technologies in the HHR case.

Author Contributions: Conceptualization, K.L. and C.A.; methodology, K.L. and C.A.; software, K.L and J.Y.; validation, K.L., C.A. and J.Y; formal analysis, K.L., C.A. and J.Y.; investigation, K.L., C.A. and J.Y.; writing—original draft preparation, J.Y.; writing—review and editing, K.L. and C.A.; visualization, K.L. and J.Y.; supervision, K.L. and C.A. All authors have read and agreed to the published version of the manuscript.

Funding: J.Y. received funding from ENSYSTRA (with funding through the EURpean Union's Horizon 2020 research and in-novation programme under the Marie Skłodowska-Curie grant agreement No 765515). CA and KL received funding from Carl Bennet AB, MISTRA Electrification and AFRY.

Data Availability Statement: The model code, data and documentation are available online at https://github.com/happiABM/HAPPI.

Acknowledgments: We would like to thank ENSYSTRA (with funding through the European Union's Horizon 2020 research and innovation programme under the Marie Skłodowska-Curie grant agreement No 765515), Carl Bennet AB, MISTRA Electrification and AFRY for financial support. We would also like to thank Daniel Johansson, Lina Reichenberg, Paulina Essunger, Markus Wråke, Göran Carstedt, Stefan Park, Sabine Fuss and Cookie Belfrage for valuable discussions and suggestions.

Conflicts of Interest: The authors declare no conflict of interest.

References

1. United Nations. *Paris Agreement*; United Nations: Paris, France, 2015; pp. 1–27.
2. Rogelj, J.; Shindell, D.; Jiang, K.; Fifita, S.; Forster, P.; Ginzburg, V.; Handa, C.; Kheshgi, H.; Kobayashi, S.; Kriegler, E.; et al. *Mitigation Pathways Compatible with 1.5 °C in the Context of Sustainable Development*; Intergovernmental Panel on Climate Change: Geneva, Switzerland, 2018.
3. Johansson, D.J.A.; Azar, C.; Lehtveer, M.; Peters, G.P. The role of negative carbon emissions in reaching the Paris climate targets: The impact of target formulation in integrated assessment models. *Environ. Res. Lett.* **2020**, *15*, 124024. [CrossRef]
4. Azar, C.; Lindgren, K.; Larson, E.; Möllersten, K. Carbon Capture and Storage from Fossil Fuels and Biomass—Costs and Potential Role in Stabilizing the Atmosphere. *Clim. Chang.* **2006**, *74*, 47–79. [CrossRef]
5. Braff, W.A.; Mueller, J.M.; Trancik, J.M.M.J.E. Value of storage technologies for wind and solar energy. *Nat. Clim. Chang.* **2016**, *6*, 964–969. [CrossRef]
6. Capros, P.; Paroussos, L.; Fragkos, P.; Tsani, S.; Boitier, B.; Wagner, F.; Busch, S.; Resch, G.; Blesl, M.; Bollen, J. EURpean decarbonisation pathways under alternative technological and policy choices: A multi-model analysis. *Energy Strat. Rev.* **2014**, *2*, 231–245. [CrossRef]
7. Capros, P.; Paroussos, L.; Fragkos, P.; Tsani, S.; Boitier, B.; Wagner, F.; Busch, S.; Resch, G.; Blesl, M.; Bollen, J. Description of models and scenarios used to assess EURpean decarbonisation pathways. *Energy Strat. Rev.* **2014**, *2*, 220–230. [CrossRef]
8. Reichenberg, L.; Hedenus, F.; Odenberger, M.; Johnsson, F. The marginal system LCOE of variable renewables—Evaluating high penetration levels of wind and solar in EURpe. *Energy* **2018**, *152*, 914–924. [CrossRef]
9. Hansen, P.; Liu, X.; Morrison, G.M. Agent-based modelling and socio-technical energy transitions: A systematic literature review. *Energy Res. Soc. Sci.* **2019**, *49*, 41–52. [CrossRef]
10. Bonabeau, E. Agent-based modeling: Methods and techniques for simulating human systems. *Proc. Natl. Acad. Sci. USA* **2002**, *99*, 7280–7287. [CrossRef] [PubMed]
11. Chappin, E.J.; de Vries, L.J.; Richstein, J.C.; Bhagwat, P.; Iychettira, K.; Khan, S. Simulating climate and energy policy with agent-based modelling: The Energy Modelling Laboratory (EMLab). *Environ. Model. Softw.* **2017**, *96*, 421–431. [CrossRef]
12. Jonson, E.; Azar, C.; Lindgren, K.; Lundberg, L. Exploring the competition between variable renewable electricity and a carbon-neutral baseload technology. *Energy Syst.* **2018**, *11*, 21–44. [CrossRef]
13. Kraan, O.; Kramer, G.; Nikolic, I. Investment in the future electricity system—An agent-based modelling approach. *Energy* **2018**, *151*, 569–580. [CrossRef]
14. Barazza, E.; Strachan, N. The impact of heterogeneous market players with bounded-rationality on the electricity sector low-carbon transition. *Energy Policy* **2020**, *138*, 111274. [CrossRef]
15. Egli, F.; Steffen, B.; Schmidt, T.S. Bias in energy system models with uniform cost of capital assumption. *Nat. Commun.* **2019**, *10*, 4588. [CrossRef]
16. Hirth, L.; Steckel, J.C. The role of capital costs in decarbonizing the electricity sector. *Environ. Res. Lett.* **2016**, *11*, 114010. [CrossRef]
17. Bartlet, N.; Coleman, T.; Schmid, S. *Putting a Price on Carbon—The State of Internal Carbon Pricing by Corporates Globally*; CDP: Brussels, Belgium, 2021; 24p, Available online: https://cdn.cdp.net/cdp-production/cms/reports/documents/000/005/651/original/CDP_Global_Carbon_Price_report_2021.pdf?1618938446 (accessed on 21 October 2021).
18. Trading Economics. EU Carbon Permits. 2021. Available online: https://tradingeconomics.com/commodity/carbon (accessed on 20 May 2021).
19. EURpean Commission. EU Emissions Trading System (EU ETS). 2021. Available online: https://ec.EURpa.eu/clima/eu-action/eu-emissions-trading-system-eu-ets_en (accessed on 17 November 2021).
20. Kruger, J. Hedging an Uncertain Future: Internal Carbon Prices in the Electric Power Sector. 2017. Available online: https://media.rff.org/documents/RFF-Rpt-Kruger-Internal20Carbon20Pricing.pdf (accessed on 17 November 2021).
21. Barradale, M.J. Investment under uncertain climate policy: A practitioners' perspective on carbon risk. *Energy Policy* **2014**, *69*, 520–535. [CrossRef]
22. Fuss, S.; Johansson, D.J.; Szolgayova, J.; Obersteiner, M. Impact of climate policy uncertainty on the adoption of electricity generating technologies. *Energy Policy* **2009**, *37*, 733–743. [CrossRef]
23. Yang, M.; Blyth, W.; Bradley, R.; Bunn, D.; Clarke, C.; Wilson, T. Evaluating the power investment options with uncertainty in climate policy. *Energy Econ.* **2008**, *30*, 1933–1950. [CrossRef]
24. Brealey, R.A.; Myers, S.C.; Allen, F. *Principles of Corporate Finance*, 11th ed.; McGraw-Hill Education: New York, NY, USA, 2014.

25. Konstantin, P.; Konstantin, M. Investment Appraisal Methods. In *Power and Energy Systems Engineering Economics: Best Practice Manual*; Springer International Publishing: Cham, Switzerland, 2018; pp. 39–64.
26. IEA. *World Energy Investment 2019*; IEA: Paris, France, 2019.
27. IRENA. *Renewable Power Generation Costs in 2017*; International Renewable Energy Agency: Abu Dhabi, United Arab Emirates, 2018.
28. EIA. *The Electricity Market Module of the National Energy Modeling System: Model Documentation 2018*; U.S. Energy Information Administration: Washington, DC, USA, 2018.
29. Noothout, P.; de Jager, D.; Tesnière, L.; van Rooijen, S.; Karypidis, N.; Brückmann, R.; Jirouš, F.; Breitschop, B.; Angelopoulos, D.; Doukas, H.; et al. *The Impact of Risks in Renewable Energy Investments and the Role of Smart Policies*; Fraunhofer ISI: Brussels, Belgium, 2016.
30. BP, P.L.C. *Statistical Review of World Energy*; BP: London, UK, 2021; Available online: https://www.bp.com/en/global/corporate/energy-economics/statistical-review-of-world-energy.html (accessed on 11 October 2021).
31. Capion, K. Guest Post: Why German Coal Power is Falling Fast in 2019. 2019. Available online: https://www.carbonbrief.org/guest-post-why-german-coal-power-is-falling-fast-in-2019 (accessed on 20 May 2021).
32. Joskow, P.L. Comparing the Costs of Intermittent and Dispatchable Electricity Generating Technologies. *Am. Econ. Rev.* **2011**, *101*, 238–241. [CrossRef]
33. Sepulveda, N.A.; Jenkins, J.D.; de Sisternes, F.J.; Lester, R.K. The Role of Firm Low-Carbon Electricity Resources in Deep Decarbonization of Power Generation. *Joule* **2018**, *2*, 2403–2420. [CrossRef]
34. Anwar, M.B.; Stephen, G.; Dalvi, S.; Frew, B.; Ericson, S.; Brown, M.; O'Malley, M. Modeling investment decisions from heterogeneous firms under imperfect information and risk in wholesale electricity markets. *Appl. Energy* **2022**, *306*, 117908. [CrossRef]
35. Royal Dutch Shell P.L.C. *CDP Climate Change 2021 Information Request*; Royal Dutch Shell, P.L.C.: The Hague, The Netherlands, 2021; Available online: https://www.shell.com/sustainability/transparency-and-sustainability-reporting/performance-data/greenhouse-gas-emissions.html#vanity-aHR0cHM6Ly93d3cuc2hlbGwuY29tL2doZy5odG1s (accessed on 8 November 2021).
36. Sensfuß, F.; Ragwitz, M.; Genoese, M. The merit-order effect: A detailed analysis of the price effect of renewable electricity generation on spot market prices in Germany. *Energy Policy* **2008**, *36*, 3086–3094. [CrossRef]
37. Hirth, L. The market value of variable renewables: The effect of solar wind power variability on their relative price. *Energy Econ.* **2013**, *38*, 218–236. [CrossRef]
38. López Prol, J.; Steininger, K.W.; Zilberman, D. The cannibalization effect of wind and solar in the California wholesale electricity market. *Energy Eco.* **2020**, *85*, 104552. [CrossRef]
39. Brown, T.; Reichenberg, L. Decreasing market value of variable renewables is a result of policy. *Energy Econ.* **2021**, *100*, 105354. [CrossRef]
40. Ballester, C.; Furió, D. Effects of renewables on the stylized facts of electricity prices. *Renew. Sustain. Energy Rev.* **2015**, *52*, 1596–1609. [CrossRef]
41. Ketterer, J.C. The impact of wind power generation on the electricity price in Germany. *Energy Econ.* **2014**, *44*, 270–280. [CrossRef]
42. Woo, C.K.; Moore, J.; Schneiderman, B.; Ho, T.; Olson, A.; Alagappan, L.; Chawla, K.; Toyama, N.; Zarnikau, J. Merit-order effects of renewable energy and price divergence in California's day-ahead and real-time electricity markets. *Energy Policy* **2016**, *92*, 299–312. [CrossRef]
43. Auer, H.; Haas, R. On integrating large shares of variable renewables into the electricity system. *Energy* **2016**, *115*, 1592–1601. [CrossRef]
44. Brown, S. *German State Awards EUR 317 Million To Loss-Making Coal Plants*; Ember: London, UK, 2021; Available online: https://ember-climate.org/commentary/2020/12/08/german-hard-coal/ (accessed on 11 October 2021).
45. Carbon Tracker Initiative. *Four in Five EU Coal Plants Unprofitable as Renewables and Gas Power Ahead*; Carbon Tracker Initiative: London, UK, 2019; Available online: https://carbontracker.org/four-in-five-eu-coal-plants-unprofitable-as-renewables-and-gas-power-ahead/ (accessed on 2 November 2021).
46. Gils, H.C. Economic potential for future demand response in Germany—Modeling approach and case study. *Appl. Energy* **2016**, *162*, 401–415. [CrossRef]
47. Göransson, L.; Johnsson, F. A comparison of variation management strategies for wind power integration in different electricity system contexts. *Wind. Energy* **2018**, *21*, 837–854. [CrossRef]
48. Schill, W.-P. Electricity Storage and the Renewable Energy Transition. *Joule* **2020**, *4*, 2059–2064. [CrossRef]

energies MDPI

Article

How to Reach the New Green Deal Targets: Analysing the Necessary Burden Sharing within the EU Using a Multi-Model Approach

Felix Kattelmann *, Jonathan Siegle, Roland Cunha Montenegro, Vera Sehn, Markus Blesl and Ulrich Fahl

Institute of Energy Economics and Rational Energy Use (IER), University of Stuttgart, 70565 Stuttgart, Germany; jonathan.siegle@ier.uni-stuttgart.de (J.S.); rolandcmontenegro@gmail.com (R.C.M.); vera.sehn@ier.uni-stuttgart.de (V.S.); Markus.Blesl@ier.uni-stuttgart.de (M.B.); ulrich.fahl@ier.uni-stuttgart.de (U.F.)

* Correspondence: felix.kattelmann@ier.uni-stuttgart.de

Abstract: The Green Deal of the European Union defines extremely ambitious climate targets for 2030 (−55% emissions compared to 1990) and 2050 (−100%), which go far beyond the current goals that the EU member states have agreed on thus far. The question of which sectors contribute how much has already been discussed, but is far from decided, while the question of which countries shoulder how much of the tightened reduction targets has hardly been discussed. We want to contribute significantly to answering these policy questions by analysing the necessary burden sharing within the EU from both an energy system and an overall macroeconomic perspective. For this purpose, we use the energy system model TIMES PanEU and the computational general equilibrium model NEWAGE. Our results show that excessively strong targets for the Emission Trading System (ETS) in 2030 are not system-optimal for achieving the 55% overall target, reductions should be made in such a way that an emissions budget ratio of 39 (ETS sector) to 61 (Non-ETS sector) results. Economically weaker regions would have to reduce their CO_2 emissions until 2030 by up to 33% on top of the currently decided targets in the Effort Sharing Regulation, which leads to higher energy system costs as well as losses in gross domestic product (GDP). Depending on the policy scenario applied, GDP losses in the range of −0.79% and −1.95% relative to baseline can be found for single EU regions. In the long-term, an equally strict mitigation regime for all countries in 2050 is not optimal from a system perspective; total system costs would be higher by 1.5%. Instead, some countries should generate negative net emissions to compensate for non-mitigable residual emissions from other countries.

Keywords: Green Deal; burden sharing; effort sharing regulation; emissions trading system; energy system analysis; TIMES PanEU; NEWAGE

Citation: Kattelmann, F.; Siegle, J.; Cunha Montenegro, R.; Sehn, V.; Blesl, M.; Fahl, U. How to Reach the New Green Deal Targets: Analysing the Necessary Burden Sharing within the EU Using a Multi-Model Approach. *Energies* **2021**, *14*, 7971. https://doi.org/10.3390/en14237971

Academic Editor: Chi-Ming Lai

Received: 31 October 2021
Accepted: 22 November 2021
Published: 29 November 2021

1. Introduction

The Green Deal [1] of the European Union (EU) specifies a very ambitious reduction in greenhouse gas (GHG) emissions by 2030 (−55% compared to 1990), while complete climate neutrality in the EU is to be achieved by 2050.

The discrepancy between current targets for 2030 established in the European Union's Effort Sharing Regulation (ESR) and the new 2030 target imposed by the Green Deal raises the question of how and, more importantly, by whom this gap is to be closed.

On one hand, it must be discussed how the additional reductions are to be distributed between the European Union Emissions Trading System (ETS) and the Non-ETS sector (also referred to as the ESR sector in the following). While the EU Commission's proposals envisage a relatively balanced ratio of the burdens between the two systems, other studies see the necessity of significantly stronger reduction contributions in the ETS sector.

On the other hand, the question emerges as to which countries will contribute these additional reductions. According to the effort-sharing that has been enacted thus far, the reductions in the Non-ETS sector are not distributed equally among the countries, but take into account "the different capacities of Member States to take action by differentiating targets according to gross domestic product (GDP) per capita across Member States" as well as the "cost-effectiveness for those Member States with an above average GDP per capita" [2].

This leads to the issue of how far this distribution between countries should be adapted, taking into account the new, significantly more ambitious targets under the Green Deal. The focus of this paper is on the tightening of targets for 2030, but the long-term mitigation burdens that will result from achieving climate neutrality in 2050 are also addressed.

As shown in Section 2, we could not find any modelling study concerning the Green Deal that addresses burden-sharing between EU regions, neither with energy system models, nor with general equilibrium models. For this reason, we want to obtain new scientific insights with our paper by quantifying the required burden sharing within the EU because of the Green Deal and analysing the resulting economic implications in the European regions with regard to energy system costs and GDP.

Many studies have considered stronger mitigation in the ETS sector to be favourable to achieving the 2030 targets. We aim to contribute to this with TIMES PanEU by shedding light on the optimal distribution of mitigation burden between ETS and Non-ETS sectors from a system perspective.

Furthermore, we use TIMES PanEU to examine the techno-economic requirements for the energy system emerging from the general, EU-wide climate target for 2030. The computable general equilibrium model NEWAGE provides an independent macroeconomic perspective on this target. In this way, we can reach a comprehensive overall view of the Green Deal goal.

With TIMES PanEU, the Green Deal can be studied in terms of its impact on the energy system with a high level of technological detail. Since all sectors are mapped, it is suitable for the investigation of a far-reaching goal such as climate neutrality for the EU. We distinguish ourselves from the existing literature by not only considering the EU as a single entity, but by also mapping and analysing the individual countries of the EU in detail. With this approach, we are able to identify opposing effects of the Green Deal on individual countries or regions in Europe. We want to go into regional depth from an energy system perspective in order to explore the reduction burden of the individual countries and regions more precisely within the EU.

With NEWAGE, we look at macroeconomic developments, which particularly concern burden sharing between countries. NEWAGE does not have the same level of detail and precision in its representation of the energy system, but it takes repercussions between developments in different economic sectors into account, and thereby provides an independent view on the overall economic effects on EU regions.

The three key findings of the paper can be summarized as follows:

- An excessively strong focus on mitigation in the ETS sector in 2030 is not cost-optimal. Reductions should be made in a way between the ESR and ETS sectors so that an emissions budget ratio of 61 to 39 results, similar to what the EU Commission also proposes. Nevertheless, the ETS sector always provides the major contribution to emissions reductions in all scenarios until 2030, with −60% emissions in the optimal scenario compared to 2005.
- From an energy system perspective, economically weaker countries should reduce their emissions significantly more by 2030 than previously envisaged in the ESR targets to achieve the EU-wide targets at optimal cost. Their respective additional reductions range from 27% to 33%, depending on the regions. However, the macroeconomic studies show the high economic burdens that result from distributing emission budgets according to a gross EU27 + UK optimum, which makes support via compensation measures absolutely necessary. Depending on the policy scenario applied, GDP losses

in the range of −0.79% and −1.95% relative to baseline can be found for single EU regions.

- To achieve climate neutrality in 2050 for the entire EU, an equally strict mitigation regime for all countries is not optimal from a system perspective, as total annualized system costs would be higher by 1.5% in this case. In particular, countries with large shares of agricultural emissions in their total emissions should not be given excessively strong targets. In contrast, countries with high biomass potentials should generate negative emissions to compensate for these residual emissions.

The rest of this work is structured as follows. In Section 2, we conduct an extensive literature review. Section 3 contains a brief description about the two models used and the general scenario framework. We then conduct our analyses of burden sharing in 2030 in Section 4, while we turn to the long-term analyses for the year 2050 in Section 5. Finally, we provide a short discussion of our results including an outlook for further research in Section 6 and the conclusions in Section 7.

2. Literature Review

In the literature before 2019, the energy scenario definitions mainly focussed on greenhouse gas reduction goals for Europe between 80 and 85% [3,4] or on analysing the renewable energies integration in the energy system [5–7]. There is a wide variety of modelling tools for energy scenario assessments [8], but we focussed on the ones calculating deep decarbonisation scenarios. The existing literature can be further divided into studies analysing the near future [9,10] or the long-term effects. Energy system research, which considers the actual European targets for 2050 or the global target to limit climate change to 1.5 °C, shows a widely varying demand for bioenergy plus carbon capture and storage (BECCS) to reach deep decarbonisation, reflecting the uncertainty of the remaining carbon dioxide (CO_2) budgets.

Pietzcker et al. [11] analysed the tightening of the European targets for 2030 and 2050 using the electricity market model LIMES EU. Their analysis showed a faster transformation of the ETS sector with an earlier phase-out of coal and accelerated expansion of renewable energy sources. The more ambitious targets led to the deployment of BECCS above a carbon price level of 100€/t CO_2 Furthermore, they found a rather limited effect of BECCS availability on carbon emissions as well as on carbon and electricity prices.

Luderer et al. [12] used seven integrated assessment models (IAM) to analyse residual CO_2 emissions from fossil fuels under the 1.5 °C goal. They assumed a global budget of 200 Gt CO_2 between 2016 and 2100 to keep global warming below 1.5 °C. According to their research, a significant amount of ca. 800 Gt of residual CO_2 emissions will need to be stored, even under strict mitigation policies. Delayed policy action further increases this amount substantially.

In a study with the energy system model PROMETHEUS, Fragkos evaluated higher CO_2 budgets to limit the global warming to 1.5 °C [13] to account for new budget estimates from the IPCC Special Report on the impacts of global warming of 1.5 °C [14]. The author suggests that, for a CO_2 budget of 860 Gt and depending on the respective scenario assumptions, the quantity of CO_2 captured by BECCS to reach net zero emissions varies between 0 and 205 Gt.

In a study with the model GENeSYS-MOD, the implications of the 2 °C and a 1.5 °C climate scenario on the European energy system were analysed and compared to a business-as-usual scenario [15]. Hainsch et al. concluded that the transformation to renewable energies is mostly market driven, while further decarbonisation requires policy action. BECCS technology is only applied in the 1.5 °C scenario, but the need for carbon capture technologies and their cost-effectiveness might be underestimated because the model does not include some sectors of the economy (e.g., agriculture and a few branches of the industry).

A similar study was performed with the PyPSA model, which represents the energy, heat, and transport sector [16]. An early and steady decarbonisation pathway of the

European energy system was compared with a late and rapid decarbonisation, while the carbon budget for minimizing climate change to 1.75 °C was applied to both scenarios. They identified the early decarbonisation scenario as the one with the lowest system costs and the energy mix being mainly based on photovoltaic and wind technologies while deployment of BECCS is not required.

In most of the studies, deep decarbonisation requires not only the integration of renewable energies, but also of several hydrogen and negative emission technologies. Sgobbi et al. investigated several hydrogen technologies within the JRC-EU-TIMES model and emphasised its importance to decarbonise the transport and industry sector, gaining even more relevance if negative emission technologies are to be lately deployed [17].

Klein et al. analysed the deployment of biofuelled integrated-gasifiation-combined-cycle (Bio-IGCC) with Carbon Capture & Storage (CCS) technology in the integrated assessment model REMIND and reviewed that the deployment was more sensitive to price changes of the biomass than to different techno-economic parameters of the Bio-IGCC process itself [18]. Davis et al. investigated the limits of BECCS technologies and argued that their potential was not only limited by the geological storage capacity of CO_2, but also by the required land area, nutrient demand, and water use. Regarding the economic aspects, BECCS seems to be a cost-effective mitigation option for the energy system because the deployment of the BECCS technology can be observed even in less ambitious climate scenarios, but it appears as a risky mitigation option to the authors in comparison to an immediate and strong GHG reduction [19].

In its Impact Assessment on the revised 2030 climate target, the European Commission provides an in-depth analysis of consequences in various aspects [20]. The European Commission employs three modelling tools to evaluate macro-economic effects of the new target under different assumptions. Results from all three modelling tools presented only moderate GDP effects at the EU level. GDP deviation from baseline was below 1% for all models and assumptions, for some slightly negative and for others slightly positive. Regarding sectoral output developments, the results of the JRC-GEM-E3 CGE model showed severe losses in the fossil fuel sectors. Developments in energy-intensive sectors with strong international competition depend heavily on the level of global climate action. Deviations from baseline are usually negative for fragmented action and positive for global action. Observed impacts on employment are generally very limited. However, employment impacts in the fossil fuels sectors are strong, especially in the coal sector.

The European Commission also evaluates several options of future ETS and ESR design. Continuation of the current ETS and ESR scope would require significant tightening of the reduction targets in one or both of the systems. Adjusting the ETS to the revised 2030 target could include an adjustment of the linear reduction factor or a one-off cap-reduction. The Impact Assessment also mentions the option of not strengthening the current ESR targets at all, at the expense of even further tightened ETS targets.

Furthermore, the Impact Assessment discusses in detail a scope extension of the ETS. This approach is subject to several publications (e.g., by Meyer-Ohlendorf and Barth [21]). According to the analysis of the European Commission, extension of the ETS sectoral coverage to buildings and road transport induces lower emission reductions in these sectors. In particular, the transport sectors' response to ETS carbon prices is considered weak due to the already high level of energy and national carbon taxation within Member States.

Current ESR targets were established with respect to GDP per capita within Member States to ensure fairness between higher and lower income countries [2]. The EU Commission also provided an Inception Impact Assessment specifically on the review of the ESR in light of the revised 2030 target and the goal of climate neutrality by 2050 [22]. According to this document, the review of the ESR has four objectives on a more specific level: (...) "incentives for the necessary additional action in the effort sharing sectors should be provided, cost-effective solutions should be promoted, Member States' efforts should be shared in a fair and consistent manner, and coherence with related legislation should be maintained". Recently, the European Commission's proposal for the buildings and transport sector seems to shape up as an inclusion of these sectors in the ETS, possibly in a separated system [23].

Babonneau et al. evaluated the 2016 effort sharing suggestions of the EU commission related to the 80% GHG mitigation target in 2050 related to 1990 [24]. According to their results, application of burden sharing to all sectors is more beneficial in terms of welfare for low income member states, while high income member states benefit from an application of the ETS to all sectors.

We could not find any modelling study concerning the Green Deal that addresses burden-sharing between EU regions, neither with energy system models, nor with general equilibrium models. This paper aims to deepen the understanding of the implications of achieving the 2030 target through a combined exploration of the Green Deal from both an energy system and an economics perspective.

Multiple studies consider stronger mitigation in the ETS sector to be favourable for achieving the 2030 targets. We aim to contribute to this with TIMES PanEU by shedding light on the optimal distribution of mitigation burden between ETS and Non-ETS sectors from a system perspective.

By comparing the literature, one can observe that the need for negative emissions relies on how many sectors of the energy system are covered within the model. While assessments with IAMs opt for the deployment of BECCS technologies, the studies with energy system models covering not all sectors review these technologies as unnecessary. As a full sectoral coverage is crucial for the energy scenario analysis, this work further investigates the significance of negative emissions for the implementation of the Green Deal.

3. Methodology

For this study, we employed an energy system model, TIMES-PanEU, and the computable general equilibrium (CGE) model NEWAGE. Captured emissions, determined by TIMES, were used as input for NEWAGE. This section further explains the two modelling tools.

3.1. TIMES PanEU

Energy system models are a widely used tool for analysing the techno-economic implications of policy imperatives such as emission targets for countries or regions. They have been successfully employed to analyse emission targets in a large number of publications (e.g., [13,15,18,25]) and the European Commission's Joint Research Centre (JRC) uses a different model [26], which was derived from the same model framework utilised in this study.

TIMES PanEU was applied in this study as an energy system model. This section provides a very brief description of the underlying mechanics, basic correlations within the energy system, and information about the assumptions on the costs and potentials that we established for this study. For a more detailed description of the model, please refer to [25,27,28].

The fundamental framework of the model is the reference energy system (RES). As can be seen in Figure 1, the RES maps all energy carriers, technologies, materials, emission flows, and service demands, which are necessary to thoroughly depict the energy system. It covers the complete energy system, beginning with the supply of resources and energy carriers and ending with the fulfilment of the defined demands. Primary energy can be converted into secondary energy before being used as final energy, multiple times, taking into account the respective costs and efficiencies of the conversion steps. The consumption of fossil energy sources causes emissions, whereby the greenhouse gases CO_2, methane (CH_4), and nitrous oxide (N_2O) were considered in the scope of this study.

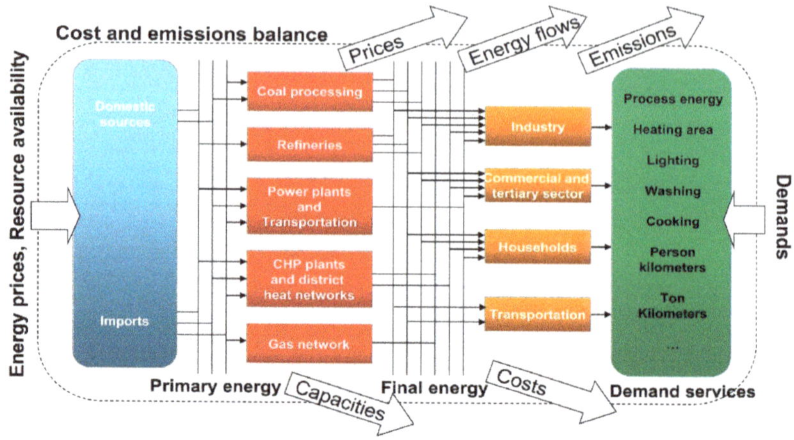

Figure 1. TIMES PanEU reference energy system [28].

Technologies are mapped with investment costs, fixed and variable costs, lifetimes and efficiencies, energy carriers can be mined within a region, or imported at specified costs. Ultimately, all costs are included in the system cost function, which is to be minimised.

TIMES PanEU is a linear optimiser that aims to minimise the total discounted system cost in a given timeframe to meet exogenously given service demands [29]. It has perfect foresight over the whole modelled time horizon.

Our model covers all of the EU plus the United Kingdom (UK), Norway (NO), and Switzerland (CH), with each country represented as an individual region with its unique energy system. Energy service demands must be met for every region, and interactions between the regions are mapped via trade in electricity, bioenergy as well as emissions.

The model horizon spans from 2010 to 2050, split into periods with a length of five years each. Periods are represented by a single year, with each year being divided into 12 time slices: one day for every season divided into a slice for the day, one for the night, and one peak hour, which covers the time of the day where maximal load occurs.

Additional to the described model structure in [25,27,30], the TIMES PanEU model was further developed to consider the latest technology updates and policy-relevant developments.

Domestic hydrogen production was implemented. In addition to the hydrolysis, biomass and gas gasification options were integrated into the model to produce hydrogen. The techno-economic characteristics of the production technologies were taken from [26]. Hydrogen-fuelled technologies are also defined in each sector. Ammonia production is implemented into the model based on the Haber–Bosch process in the hydrogen and nitrogen [31]. Synfuels are employed to provide additional decarbonisation options, especially in the transport and industry sectors. For these energy carriers, import processes are defined for synthetic gas, synthetic kerosene, synthetic diesel, synthetic fuel oil, and synthetic gasoline. They are implemented as zero emission energy carriers.

Coal phase out commitments of the different Member States were integrated, based on [32]. By considering these discussions, coal and lignite CCS technologies were not defined as investment options for the respective countries.

The technological option to generate negative emissions was given to the model via the combination of electricity generation from biomass with downstream CCS, further referred to as BECCS. Biomass potential for every country was taken from [33]. Across the analysis, high biomass potential curves were integrated to the existing model structure. Renewable energy potentials were based on ENSRPESO [34] and livestock demand was reduced by 50% until 2050, following the AT-Kearney Study [35]. However, this leaves residual emissions (especially CH_4) that can be reduced by (costly and complex) technical measures, but not completely. A base amount of about 25% of agricultural emissions cannot be reduced.

The GHG abatement options for the process emissions in the agriculture sector were derived from [36]. Import prices for fossil fuels were taken from the "Sustainable Development Scenario" in [37].

3.2. NEWAGE

For the analysis of the macroeconomic effects of different scenarios, we employed the CGE model NEWAGE (Model website as of 25.11.2021: https://www.ier.uni-stuttgart.de/en/research/models/NEWAGE/, accessed on 21 Novemeber 2021). CGE models are a well-established class of macroeconomic models already used previously in the evaluation of current EU effort sharing [24]. One of the models used in the European Commission's Impact Assessment is also a CGE model (JRC-GEM-E3) [38].

NEWAGE's representation of the energy sector is not as precise as TIMES Pan-EU's. However, NEWAGE takes income and demand effects into account. Economy is modelled in a "closed loop" with its interconnections between consumers and industry sectors. Moreover, NEWAGE covers not only the EU, but the whole world. Therefore, it facilitates the analysis of repercussions of energy-related policy decisions in the worldwide economy. In the following, a brief overview on the basic features of the model is given.

NEWAGE is applied in a recursive-dynamic manner and does not have foresight. The base year is 2011, followed by 2015, and further five-year time steps until 2050. Production of goods and services is split into 23 sectors. Underlying trade data are taken from the GTAP 9 [39] and EXIOBASE 3 [40] databases. Production is modelled with Constant Elasticity of Substitution (CES) functions. A special feature of NEWAGE is the representation of the electricity sector with 18 different electricity generation technologies. In its current version, NEWAGE includes not the full spectrum of greenhouse gas emissions, but only energetic CO_2 emissions. For more details on the structure of production and electricity generation in NEWAGE, see Appendix B.

In NEWAGE, the world is represented by 18 regions. Some large countries make up a region by themselves, but most countries are aggregated into regions. For example, the Scandinavian and Baltic countries are combined with Ireland in the Northern EU region. Austria, Czech Republic, Hungary, Slovakia, Slovenia, Croatia, Romania, Bulgaria, Greece, Cyprus, and Malta make up the south-eastern EU region (see Figure 2 below).

Region	Short Form	Colour
United Kingdom	UKI	
Germany	DEU	
France	FRA	
Italy	ITA	
Poland	POL	
Spain + Portugal	ESP	
Benelux	BNL	
Northern EU	EUN	
South-Eastern EU	EUS	

Figure 2. NEWAGE regions in EU 27 + the United Kingdom.

NEWAGE calculations consider trade flows among world regions. In all scenarios calculated for this article, prices of fossil fuels in the regions with the largest fossil fuel resources were fixed at values given by the sustainable development scenario of IEA [37].

3.3. Scenario Framework

For this study, we examined three main scenarios. All three scenarios were based on the goals of the EU Commission's Green Deal (i.e., 100% greenhouse gas neutrality for the entire EU is achieved by 2050).

- Optimal (OPT):
  Reaching the primary climate targets without additional restrictions regarding the distribution between ETS and ESR or between the countries exceeding the already agreed distributions.
- ETS first:
  Reaching the climate targets, but with a major contribution from the ETS sector that has to be completely GHG-neutral by 2050. However, there are no restrictions regarding the burden sharing between the countries in the ETS sector. Reductions already agreed upon in the ESR sector will be extrapolated until 2050.
- ESR more:
  Meeting the climate targets, but with a major contribution from the ESR sectors, these must reduce GHG emissions by 95% until 2050 (compared to 2005). However, each country must reduce its emissions by at least 80%. The ETS sector must achieve slightly higher reductions than the optimal scenario.

The precise reduction targets we have specified for each scenario can be found in Table 1. Please note that the targets refer to the EU plus the UK. We included the UK alongside the EU in the targets of the scenarios to ensure comparability with the current Effort Sharing Regulation, which still includes the UK. In NEWAGE, CO_2 reduction goals for regions outside EU + the United Kingdom were derived from the sustainable development scenario of IEA [37].

Table 1. GHG emission targets in the three scenarios. Emissions from waste and LULUCF are not included.

Year	Scenario	ETS Sector [41]	ESR Sector [1] [2]	Overall GHG [42,43]
2005	Statistics	2360 Mt	2677 Mt	5037 Mt
2030	OPT	−43%	−30%	−55%
	ETS first	−77%	−30%	
	ESR more	−58%	−47%	
2050	OPT	−43%	−30%	−100%
	ETS first	−100%	−30%	
	ESR more	−58%	−95%	

[1] The country-specific ESR targets can be found in Table A1 in Appendix A.

4. Optimal Burden Sharing in 2030 to Reach the Goal of 55% Reduction

The objective of the following first part of the results analysis is to provide answers to these two questions:

- What mitigation in the ETS and ESR sectors is optimal from a system perspective in 2030?
- Which countries or regions should shoulder which burden in 2030?

We examined both issues always under the condition that the 55% reduction target in 2030 for the EU as a whole is achieved.

We begin the analyses of burden sharing with the allocation of the mitigation efforts between the sectors covered by the ETS and the sectors not covered by the ETS, the

ESR sectors. We then move on to the burden sharing between the countries of the EU, first from an energy system point of view before supplementing the assessment with a macroeconomic perspective.

4.1. Burden Sharing between ETS and ESR Sectors

We conducted the analysis of the optimal allocation of abatements between ETS and ESR by examining the distribution of the 2030 emissions budget between the ETS and ESR sectors. The corresponding budgets for the three scenarios can be found in Figure 3.

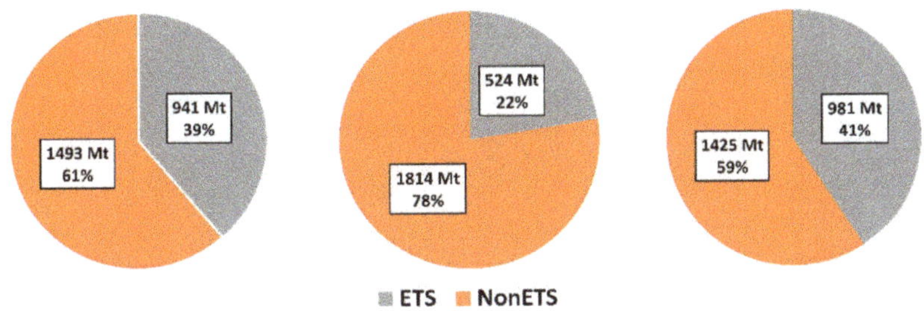

Figure 3. Emission budget for the ETS and Non-ETS in 2030 (Optimal, ETS first, ESR more).

Our model calculations with TIMES PanEU indicated a share for the ESR sector of the overall GHG emissions of about 61% in the optimal scenario for 2030. Currently, this share is at 57%, and the proposals of the EU Commission would also yield to a similar share of 61% in the medium-term. Hence, this is already the first interesting finding, as other studies [44] have derived significantly higher shares of about 80% for the budget of the ESR sector. Unfortunately, it is not clear from the study cited how the applied model is structured, so we cannot conclusively determine whether these differences are due to different implementations of ETS or ESR in the model.

The comparison of the three scenarios showed particularly large deviations in ETS first compared to the other two. In Figure 4, it is evident that the ETS sector mitigated significantly more in relation to 2005, and thus also departed substantially from the ratios in the Optimal and ESR more. However, all three scenarios shared in common that the ETS sector always contributed the greater part of the reductions, ranging from −60% (Optimal) to −78% (ETS first). Even in ESR more, with −58%, ETS sector reductions were greater than the −55% specified in the scenario framework.

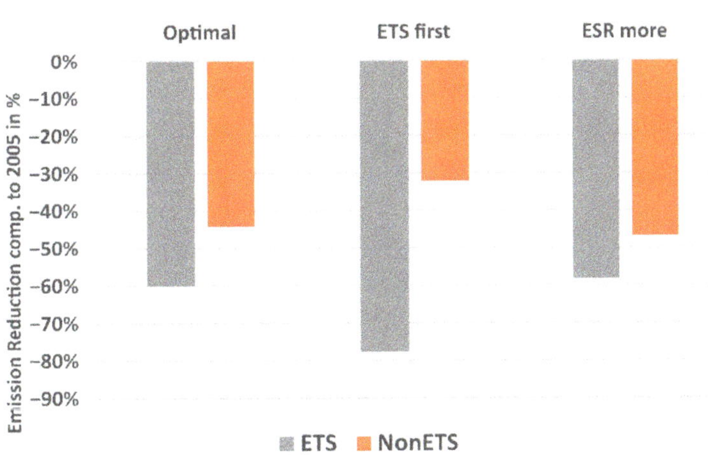

Figure 4. Greenhous gas emissions reduction in 2030.

Nevertheless, with a reduction of 44%, the ESR sector made a substantial contribution to the overall emission mitigation in our cost-optimal case. We conclude that a strong focus on the ETS sectors, as the ETS first scenario stipulates, is disadvantageous from a system perspective, which can also be deduced from the annuated system costs in 2030 where ETS first had 2.5% higher costs than the reference scenario compared to 1.3% for the Optimal scenario or 1.5% in ESR more.

We have identified two main reasons for this effect, which we will elaborate in the following:

1. In the optimal case, the building sector can cheaply reduce emissions to a certain amount through district heating; with tough ETS targets, this is limited in the medium-term.
2. By burdening the power sector with the ETS targets, we obtain a higher electricity price, which ultimately leads to a decreased use of electricity-based technologies in the ESR sectors, particularly in the transport sector.

The expansion of district heating is a central component of the transformation of the energy system. The share of district heating in the final energy consumption of buildings rose between 2020 and 2030 from approximately 7% to roughly 12% in the Optimal scenario. For the building sector, this technology option represents a good option for reducing emissions in the medium- and long-term, especially as it is largely provided by efficient gas combined heat and power (CHP) plants in the medium-term. In the long-term, the district heating supply is then defossilised by shifting production to large-scale heat pumps, biomass-fired CHP plants, or geothermal energy.

Given a strong focus on the ETS sector, a large part of the gas-fired power plants cannot be operated in 2030 in order to achieve the reduction targets, as the ETS first scenario demonstrates. As can be seen in Figure 5a, this led to a considerably low share of heat in final energy consumption in the building sector, which dropped from around 12% (Optimal) to around 8% (ETS first).

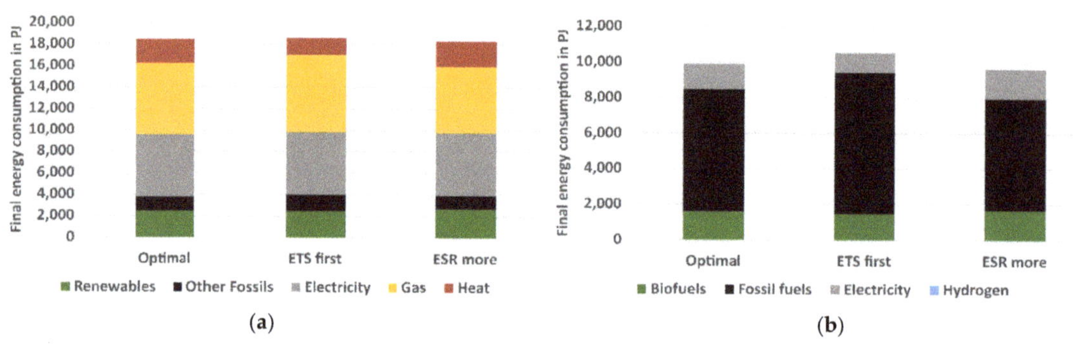

Figure 5. Final energy consumption in 2030 for (**a**) residential and commercial buildings and (**b**) the transport sector.

It emerges that in the case of ETS first, district heating was predominantly replaced by the direct combustion of gas. As a result, emissions in the ESR sector increased, while in the ETS sectors, emissions decreased by approx. 140 Mt CO_2. This "shifting" of emissions ultimately accomplished the ETS goals, but makes little sense from an energy system perspective.

However, the building sector saw only a negligible decline in electricity consumption, which can be explained by the concurrent decrease in district heating. On one hand, the requirements imposed by ETS first increased the marginal costs for the supply of electricity, which in fact made electricity less attractive for the building sector. On the other hand, this effect could be found to a much greater extent in district heating. We therefore observed two effects running in opposite directions: District heating becomes less attractive for the building sector, but this does not trigger a stronger electrification due to the simultaneously rising electricity prices; the gap is filled by fossil energy sources.

As a second in-depth investigation, the impact of the ETS first on the transport sector will be elaborated by looking again at the final energy consumption in Figure 5b. We can see that in the case of the optimum in 2030, a significant share of the transport sector was already electrified (approx. 15% of final energy). In TIMES PanEU, we assumed cost parity between electric and combustion engines until 2025. In the Optimal scenario, this led to an early electrification of the transport sector, resulting in 33 million fully electrically powered cars in the EU in 2030.

In ETS first, the share of electricity in the final energy consumption dropped to only 11% of the final energy consumption and the 4% difference to the Optimal scenario was completely replaced by fossil fuels. The higher costs of electricity generation due to the high reduction pressure in ETS first led to a decreasing economic appeal of electrical alternatives in transport. Here, just as in the building sector, a non-cost-optimal shift of emissions to the ESR sector takes place.

Overall, ETS first led to a lower electricity consumption of about 100 TWh compared to Optimal and also ESR more, which corresponded to a relative deviation of about 3%. However, this difference occurred almost exclusively in the transport sector; other sectors were affected to a much lesser extent, as is described above for the building sector.

The conclusions of this chapter can thus be summarised as follows:

- Although a cost-optimal reduction to achieve the EU targets in 2030 leads to a relatively stronger reduction in the ETS sector, the ESR sector should also make its significant contribution, leading to a ratio of emissions of 39% to 61% (ETS/ESR).
- A too heavy focus on reductions in the ETS sector in 2030 leads to two negative effects: first, district heating, which optimally contributes to decarbonisation, is deployed less in the building sector (there was 8% of final energy consumption in ETS first compared to 12% in the Optimal scenario). Second, tightened targets for the power sector lead to higher electricity prices, meaning that electric options are less deployed in the transport sector, which results in higher emissions in that sector.

4.2. Burden Sharing between the European Regions in 2030 from an Energy System Perspective

This section analyses the burden sharing between the countries in 2030, which has become necessary through the tightened targets of the Green Deal.

As a first step, a comparison of the needed emission reductions in the EU Member States in the Optimal scenario with the targets for these countries that have been agreed in the ESR thus far was carried out for Non-ETS sectors. On one hand, the aim was to review the targets with regard to their suitability to achieve the 55% target for Europe in 2030. On the other hand, it is to be examined which countries should reduce their Non-ETS sector by how much from a system perspective. The comparison, shown in Figure 6, of the current targets with the Optimal shows that all countries reduced more than they are currently required to do by the Effort Sharing Regulation. The majority of countries even needed to reduce significantly more, so the ESR targets adopted thus far are nowhere near sufficient to achieve the 55% target in 2030.

It is striking that Poland, Romania, and Bulgaria contributed a significantly higher reduction in the Optimal scenario. From a system perspective, countries that are "spared" in the ESR should actually reduce significantly more in the ESR sector. From a system perspective, economically weaker countries should contribute more to mitigation at an early stage than previously envisaged.

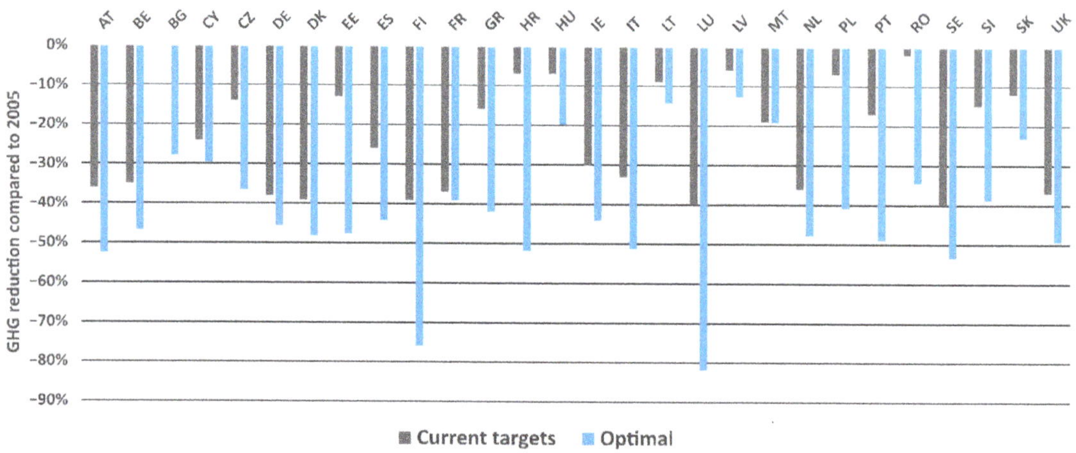

Figure 6. Country specific greenhouse gas reduction in the ESR sector against 2005 in the Optimal scenario in 2030 compared to the current targets, agreed in the Effort Sharing Regulation.

To examine the burden sharing between the countries and regions of Europe more deeply, we continued by determining the total CO_2 reductions that the countries will have to provide in 2030 compared to a reference case to achieve the climate targets in this year. In this context, we only evaluated the CO_2 emissions to facilitate comparisons with the NEWAGE results. However, the two remaining greenhouse gases were always part of the reduction requirement, regardless of this particular evaluation. To achieve better comparability between the models, we implemented a reference scenario to which we could compare the other scenarios. The scenario was defined as a business-as-usual scenario in which no reductions beyond the already adopted ETS and ESR targets are defined. NEWAGE does not assume any CO_2 reduction targets outside the EU + United Kingdom in this scenario. The results of this analysis are shown in Figure 7. The countries of the EU were aggregated here according to the NEWAGE standard (see Section 3.2. NEWAGE) in order to be able to subsequently better compare the effects with NEWAGE.

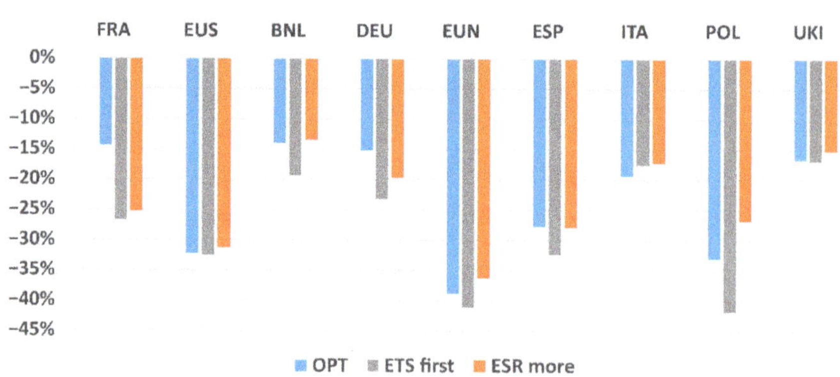

Figure 7. Reduction of CO_2 in 2030 compared to the reference scenario.

Looking at the additionally required reductions in the Optimal scenario compared to the reference scenario, it is apparent that in particular, economically weaker regions such as Poland, Spain, and Portugal, south-eastern EU as well as the rather strong northern EU (which also includes the Baltic countries) had to reduce more in relation to the reference scenario, which is in line with the findings from the comparison between the Optimal and current targets in Figure 6. The economically weaker regions reduced their emissions in the Optimal scenario by −27% up to −33%, compared to the previous targets of the reference scenario, which means that they had to contribute significantly larger reductions than regions such as Germany (−15%), the Benelux countries (−14%), France (−14%), or the United Kingdom (−17%).

By taking the annualised system costs (system costs include all fixed, variable, or investment costs of the technologies as well as the costs for import and distribution of energy carriers) (Figure 8) as a measure for the burden placed on the countries by the reductions in the three scenarios, we can see the reason for these unequally distributed burdens. Economically stronger countries such as Germany or France coped much better with the additional reductions than, for example, Poland, south-eastern EU, or northern EU.

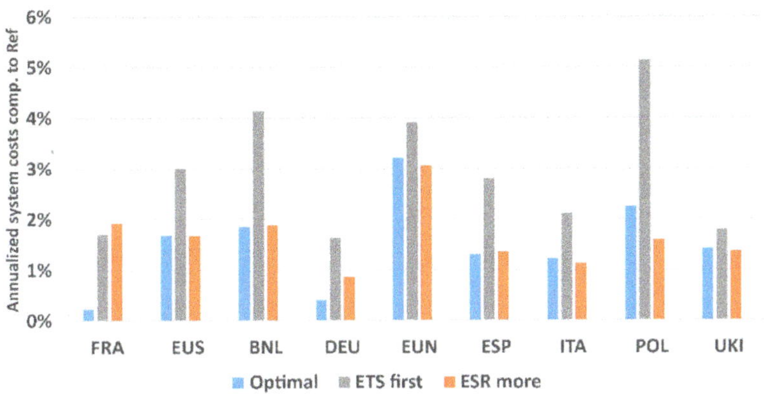

Figure 8. Deviation of the annualised system costs in 2030 compared to the reference scenario.

In the Optimal scenario, it is these regions (and Benelux) that had to relatively shoulder the largest increases in system costs. From a system perspective, these regions should mitigate more in order to achieve the set climate targets, but from a political perspective, it is also clear that these regions should be supported in this endeavour. If the system targets are to be achieved, there must either be European compensation mechanisms so that all countries are able to make their reductions, or (as in the ESR more), an uneven distribution of the reduction burden is imposed for the Non-ETS sector in order to relieve the weaker countries of some of their system costs. In this case, however, it may be accepted that the cost-optimal path is not followed.

However, this also shows the limits of energy system analysis with regard to the evaluation of political decision-making processes. With the help of TIMES PanEU, it is possible to very precisely analyse which sectors and regions have to bear which reduction burden from a system perspective in order to achieve the overarching reduction target.

TIMES PanEU can only partially analyse the interaction of the economic system with changes in the energy system; impacts on the economic structure from the demand effects are not taken into account. This is where the market-economic analysis with NEWAGE comes into play. NEWAGE has a less detailed depiction of the energy system but includes the economy as a whole in the analysis and can depict feedback effects between economic sectors. The analysis of the 2030 burden sharing within the EU is therefore continued in Section 4.3.

The conclusions of this section can thus be summarised as follows:

- For the −55% target in 2030, all countries must contribute significantly more in the ESR sector than agreed under the ESR targets. The economically weaker regions needed to additionally reduce their emissions in the Optimal scenario by −27% up to −33% compared to the current targets, which means that they had to contribute significantly larger relative reductions than regions such as Germany (−15%), the Benelux countries (−14%), France (−14%), or the United Kingdom (−17%).

- However, this leads to disproportionately high increases in the system costs of these countries, respectively, regions. If the system targets are to be met, there must either be European compensation mechanisms so that all countries are able to achieve their reductions. Alternatively, an uneven distribution of the reduction burdens can be prescribed for the Non-ETS sector in order to relieve the weaker countries of some of their system costs.

4.3. Burden Sharing between the Regions in 2030 from a Macroeconomic Point of View

To complement the analyses carried out with TIMES PanEU by adding an independent macroeconomic perspective, the same scenarios were calculated for 2030. In the following, effects are discussed in comparison to the reference scenario. Therefore, the mentioned effects occur on top of existing effects originating from the already adopted ETS and ESR targets.

From an EU-wide perspective, all of the three scenarios led to comparable losses in gross domestic product (GDP) in 2030 relative to the reference scenario. The Optimal scenario harmed EU-wide GDP development least (−1.20%). However, the losses in ETS first (−1.41%) exceed those of ESR more (−1.26%) in 2030.

Gross value added (GVA) of the Non-ETS sector showed the highest losses in ETS first and the lowest in ESR more. GVA of the ETS sector was increased and almost on the same level for all of the three scenarios. The main driver behind this positive ETS sector development was electricity production.

EU prices of almost all goods from ETS sectors increased in the three scenarios relative to the reference scenario, with the strongest rise in ETS first. In contrast, EU prices of almost all goods from the Non-ETS sector decreased. ETS industries could successfully impose higher prices, but other industries further down the value chain could not do this and had to carry the burden.

Among the rising prices from the ETS sector in ETS first, electricity prices stand out with very strong increases between 22% and 107%. For all EU regions, electricity prices clearly rose the most in ETS first and the least in ESR more.

Among the Green Deal scenarios, fossil fuel consumption was highest in ETS first and lowest in ESR more. The opposite was true for electricity consumption: it was highest for ESR more and lowest for ETS first. This points to the influence of electricity prices on electrification. If strong mitigation targets are not accompanied by moderate electricity prices, electrification could be impeded.

The regionally disaggregated GDP (Figure 9) view revealed losses between −0.79% and −1.95%.

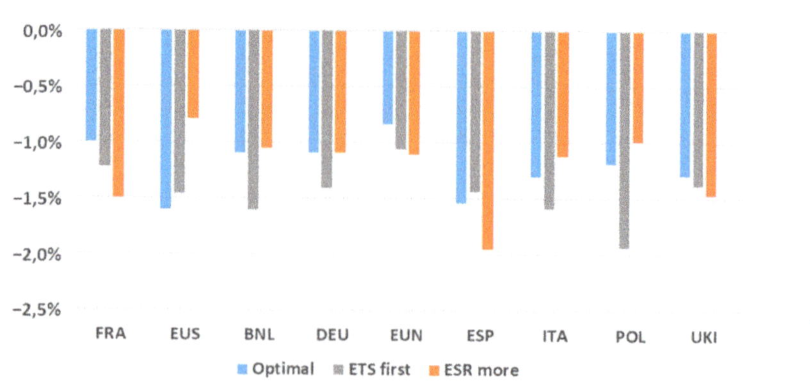

Figure 9. Deviation of EU regional GDP in 2030 compared to the reference scenario.

In the Optimal scenario, south-eastern EU faced the strongest relative GDP loss (−1.60%) compared to the other EU regions. Northern EU was affected the least (−0.84%). For no region except the south-eastern EU, Optimal was the worst scenario among the three, but was the best one for the northern EU, France, and UKI.

The Optimal scenario led to more levelled absolute CO_2 prices among the regions than the ETS first and ESR more scenarios. CO_2 abatement relative to the reference scenario was comparably high for the south-eastern EU in all three scenarios. However, compared with the other regions, absolute weighted CO_2 prices in ETS first and ESR were still rather low for the south-eastern EU (see Appendix C). Absolute Non-ETS prices were lowest among EU regions for the south-eastern EU in ETS first and ESR more. The south-eastern EU was "treated with care" in ETS first and ESR more by low Non-ETS prices, but not to the same extent in the Optimal scenario.

Northern EU reached comparably moderate relative abatement under the Optimal scenario prices. The more levelled CO_2 prices helped northern EU. In the Optimal scenario, absolute weighted and Non-ETS CO_2 prices did not exceed those of most other regions as much as in ETS first and in ESR more. Export from northern EU remained strong in the Optimal scenario, and losses of export value relative to the reference scenario were second lowest among the EU regions in this scenario.

Macroeconomic impacts in the ETS first scenario were most harmful for Poland (−1.94%) and least for the northern EU (−1.06%) relative to the reference scenario. From a macroeconomic perspective, it was the worst among the three scenarios for Poland, Benelux, Italy, and Germany, and the best one only for Spain and Portugal.

Poland was the EU region with the highest relative abatement among EU regions in ETS first. ETS abatement translates to high national abatement for Poland as it is the region with the highest GVA share of ETS sectors. However, the ETS sector even experienced above EU27 + UK average relative GVA increase. Poland faced particularly strong relative GVA losses in Non-ETS sub-sectors such as the buildings and the service sector. Relative Non-ETS GVA losses were higher than in all other EU regions. While Poland is the EU region with the highest input share of electricity among EU regions, electricity prices climbed the highest in Poland for all scenarios, but in ETS first, electricity prices rose the most - by 107% relative to the reference scenario in Poland. Germany was the region with the second largest electricity price rise in all three scenarios.

Northern EU's relative CO_2 abatement in ETS first was not as high as Poland's, but still higher than the relative abatement of most other EU regions. The same applied for the GVA share of the ETS sector. Northern EU's ETS sector as a whole showed the best GVA development compared to the reference scenario in ETS first among the EU regions. In ESR more and Optimal, relative GVA increase in the ETS sector also exceeded that of all (ESR more) or most (Optimal) of the other regions, but not as much as in ETS first. Relative losses of the Non-ETS sector were rather strong, but not as strong as in Poland. Again, northern EU's exports were hardly concerned with northern EU's trade balance at its highest values relative to Reference.

In the ESR more scenario, the Spain and Portugal region experienced the highest relative GDP loss (−1.95%) among the EU regions and south-eastern EU the lowest (−0.79%). ESR more was the worst scenario from a macroeconomic view for Spain and Portugal, France, northern EU, and UK, and the best scenario for Poland, Italy, Benelux, and Germany.

The Spain and Portugal region has the second largest service sector among EU regions. In general, GVA share of Non-ETS industries in Spain and Portugal is slightly higher than in most other EU regions. At the same time, the Spain and Portugal regions provided the largest Non-ETS abatement relative to the reference scenario among EU regions in ESR more. After all, overall GVA of the Non-ETS sector faced the highest relative losses in Spain and Portugal compared to the EU regions in ESR more. For Spain and Portugal, France, northern EU, UKI, and Italy, service sector GVA declined most in ESR more.

South-eastern EU was least affected among EU regions in ESR more despite facing the highest relative overall and ETS-abatement compared to the other EU regions. Weighted and Non-ETS CO_2 prices showed the strongest relative rise among EU regions. However, absolute weighted CO_2 prices for south-eastern EU were still rather low, absolute Non-ETS CO_2 prices were even the lowest among EU regions. In the end, the ETS sector in south-eastern EU increased its GVA more than EU27 + UK average percentage-wise, and the Non-ETS sector experienced better relative GVA development than that of all other EU regions in ESR more.

The conclusions of this section can thus be summarised as follows:

- A strong reduction requirement on the ETS side can lead to an increase in electricity prices that could impede electrification;
- On the EU level, strong abatement requirement on the ETS side not only affects the energy-intensive ETS industries themselves. Depending on how well they are able to pass through increased costs, high abatement in the ETS sector can even harm the Non-ETS sector in particular;
- Economically weaker regions of the EU tend to have a greater additional economic burden in the Optimal and ETS first scenario than the others (see Table 2 below). Compensation mechanisms should be created to offset these burdens if the focus of additional abatement is not on the ESR side; and
- In the short-term (2030 perspective), a focus on increased ESR reduction instead of ETS reduction might be economically more favourable from an EU-wide perspective. From the perspective of the single EU regions, this cannot be said in all cases. For economically weak regions, economic losses tend to be limited in the short-term if stronger Non-ETS reduction with continued differentiated effort sharing is applied.

Table 2. 2030 GDP effects relative to the reference scenario. EU regions grouped according to NEWAGE base year GDP per capita.

GDP Per Capita	Included Regions	Optimal	ETS First	ESR More
Low	Poland, South-Eastern EU, Spain and Portugal	−1.51%	−1.53%	−1.30%
Medium	Italy, United Kingdom, France	−1.18%	−1.38%	−1.39%
High	Germany, Benelux, EU North	−1.03%	−1.36%	−1.08%

5. Burden Sharing in 2050

Although the focus of this paper was on the implications of the tightened targets of the Green Deal in 2030, we consider it imperative to also examine the long-term effects of the tightened targets (i.e., climate neutrality in 2050). For these considerations, only TIMES PanEU will be used in the following.

We begin with the total GHG emissions for the individual countries shown in Figure 10, which arise under the three different scenarios:

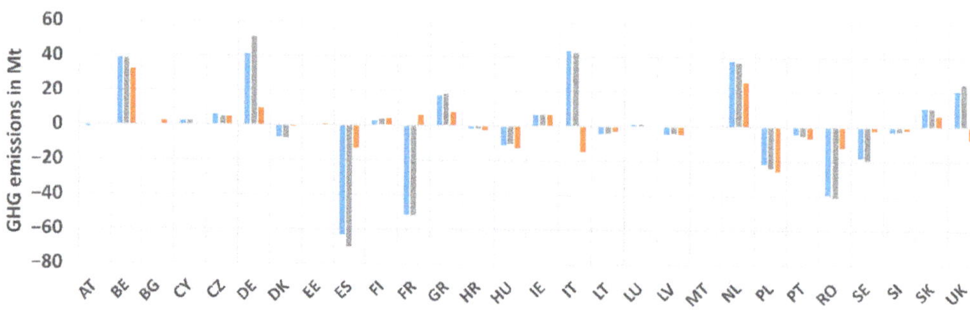

Figure 10. Total greenhouse gas emissions by country in 2050.

It can be observed that, from a cost-optimal system perspective, some of the countries had to bring their total GHG emissions into negative range by utilising electricity from biomass plus CCS in order to compensate for the remaining residual emissions of other countries, with Spain accounting for the largest total negative emissions of up to −65 Mt per year, depending on the scenario.

It should be noted that negative emissions from BECCS were used in all scenarios for the reasons mentioned in Section 3.1. For this reason, Optimal and ETS first barely differed in 2050, as in both scenarios, more than 100% mitigation took place in the ETS sector. It is noticeable that fewer negative emissions were needed in ESR more. This is mainly due to mitigation in the agricultural sector, where technological mitigation options are extremely costly. However, when they were necessary due to national requirements and the general ESR reduction of 95%, this also reduced the required BECCS use to compensate for agricultural emissions compared to the Optimal scenario.

The countries that dropped into the negative emissions range were, due to the use of BECCS, countries with high potential to grow biomass on arable land [36]: France, Spain, Romania, and Poland as well as partly Sweden or Hungary. In these countries, there was therefore automatically BECCS potential due to their biomass potential. It can also be seen here that it is predominantly these countries that benefited from ESR more, as it was no longer necessary to compensate for the residual emissions of other countries (except for Poland, which suffered from the 80% reduction in the ESR sector).

In accordance with the analysis in Section 4.2, we wanted to use not only the emissions, but also the impacts of the reductions in the system costs to evaluate the burden-sharing between the regions; these are shown for this reason in Figure 11.

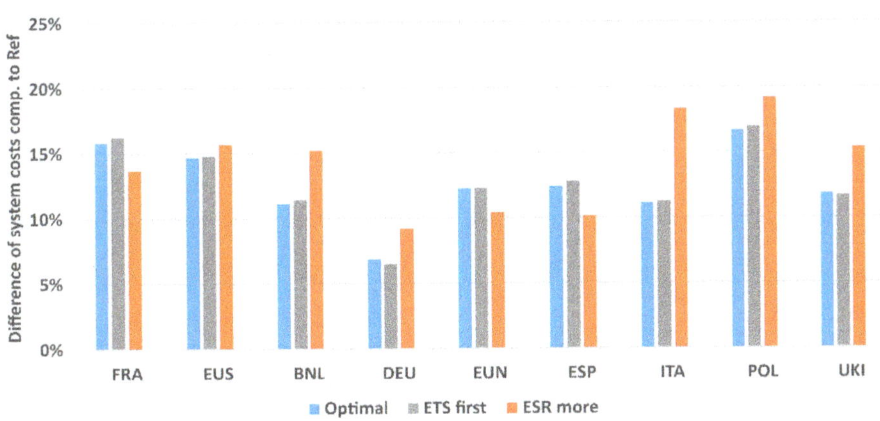

Figure 11. Deviation of the annualised system costs in 2050 compared to the reference scenario.

Due to the negative emissions, ETS first and Optimal hardly differed in system costs. There were only slight differences between the regions. In the case of the ESR more, however, there were very distinct differences in 2050. Due to the condition that all countries had to reduce at least 80% in the ESR sector, some regions were burdened significantly more than in the Optimal case. Italy, south-eastern EU, the Benelux countries, and Poland are particularly worth mentioning here.

Countries such as Bulgaria, Cyprus, Lithuania, Latvia, Malta, or Slovakia had to reduce significantly further in the strict regime of the ESR more than would have been optimal from a system perspective. The main reason for these different emission levels is the role of agriculture, which represents the most expensive reduction options in the system. A lot of technical effort has to be made here, which costs money. If livestock farming accounts for a relatively larger share of emissions, it becomes tremendously expensive for these countries to further mitigate these emissions. Therefore, these emissions were often not reduced to the maximum possible in the Optimal, but instead, were compensated by negative emissions.

However, due to the stipulation of a 95% reduction in the entire ESR regime, all countries were forced to make greater reductions, which is also reflected in 1.5% higher EU-wide system costs in ESR more compared to the Optimal scenario. Only countries with high negative emissions in the other scenarios benefited from this. Since these costs were not credited in any form according to the current status, but only burdened their own energy system, they benefited from a scenario in which residual emissions from the ESR were low (e.g., France, Spain, Romania).

The contrast to the Energy System in 2030 becomes obvious: while weaker countries should achieve more in the ESR in the medium-term from a system perspective, they should not be required to reduce too much here in the long-term. Especially in countries where agriculture accounts for a large share of the energy system and emissions, such targets would only be achievable with great effort and at high cost. In this case, the burden distribution between countries would be disadvantageous.

From a system perspective, the use of BECCS is unavoidable and makes sense in many countries. From an economic, political, and financial point of view, however, it is not yet reasonable for these countries. In order to realise the negative emissions necessary from a system perspective, regulatory or financial incentives must be established. It is conceivable, for instance, that the ETS could provide a payment for negative emissions that would make the use of the technology profitable. Regardless of the concrete design of these incentives, the generation of negative emissions must become economically attractive, otherwise the achievement of complete climate neutrality is unrealistic.

The conclusions of this section can thus be summarised as follows:

- A strict reduction regime that imposes strong mitigation targets on all countries is not optimal from a system perspective, which is reflected in 1.5% higher EU-wide system costs; in particular, countries with large shares of agricultural emissions should not be given overly ambitious targets; and
- To compensate for these residual emissions, countries with high biomass potentials should produce negative emissions. Countries that provide these additional mitigation burdens beyond climate neutrality must be financially compensated.

6. Discussion

In order to better correlate the TIMES and NEWAGE findings, central results of the models were compared and interpreted in the following.

In Section 4.3, NEWAGE determined a higher electricity price for the ETS first scenario, which may indicate a hindrance for electrification. TIMES identified an identical problem, since the transport sector is electrified to a much lesser extent in this scenario; in ETS first, electricity accounts for only 11% of the final energy consumption compared to 15% in the Optimal scenario (see Figure 5b). However, this effect was not observable in other sectors, as described in Section 4.1. Although higher electricity prices also resulted in TIMES, the prices for district heating, for example, rose much more sharply in the building sector, which is why the higher electricity price is not reflected here.

The GDP losses arising from burden sharing within the EU as calculated by NEWAGE (see Section 4.3) showed a slightly higher burden on the EU as a whole for ETS first than for the Optimal scenario for 2030 (−1.41% compared to −1.20%). Although TIMES PanEU in Section 4.2. saw a larger discrepancy in total system costs between these two scenarios of +2.6% for ETS first compared to +1.3% in the Optimal scenario or +1.5% in ESR more, the direction was the same. A too strong focus on mitigation in the ETS sector is not cost-optimal from both a macroeconomic and an energy system perspective to achieve the goals of the Green Deal.

We considered negative emissions due to agriculture to be indispensable, but there are certainly studies that have arrived at other results. The absolute level is, of course, directly dependent on the assumption of a decline in livestock farming. Should the 50%

prove to be unrealistic, the necessary compensation through negative emissions would be significantly higher, as the absolute values of negative emissions, distributed over the countries, is dependent on our assumed potentials for biomass.

As economically weak EU regions were impacted most negatively in the Optimal scenario from an energy system cost and macroeconomic perspective, a central finding of our work is the necessity of compensation mechanisms. Further research should be conducted in this field, especially with regard to the concrete design of these measures. An expansion of the Just Transition Fund would be conceivable, but more in-depth studies should be conducted on this.

Furthermore, we considered coal phaseouts as national climate measures in the context of this article. We are aware that there are a large number of national climate targets that are planned or have already been implemented. We have decided not to include these measures in order to limit the horizon of the study, but these national programmes should be examined in terms of their interactions with the Green Deal.

7. Conclusions

- A too heavy focus on reductions in the ETS sector in 2030 leads to two negative effects. First, district heating, which optimally contributes to decarbonisation, is deployed less in the building sector (8% of final energy consumption in ETS first compared to 12% in the Optimal scenario). Second, tightened targets for the power sector lead to higher electricity prices, meaning that electric options are less deployed in the transport sector, which results in higher emissions in that sector. Reductions should be made in a way between the ESR and ETS sectors so that an emissions budget ratio of 61 to 39 results, similar to what the EU Commission proposes.
- From an energy system perspective, economically weaker countries should reduce their emissions significantly more by 2030 than previously envisaged in the ESR targets in order to achieve the EU-wide −55% targets at optimal cost. The economically weaker regions need to additionally reduce their emissions in the Optimal scenario by up to −33% compared to the current targets.
- However, the macroeconomic studies show the high economic burdens that result from distributing emission budgets according to a gross EU27 + UK optimum, which makes support via compensation measures absolutely necessary. Depending on the policy scenario applied, GDP losses in the range of −0.79% and −1.95% relative to baseline are found for single EU regions.
- An equally strict mitigation regime for all countries in 2050 is not optimal from a system perspective, which is reflected in 1.5% higher EU-wide system costs. In particular, countries with large shares of agricultural emissions should not be given excessively strong targets.
- In contrast, countries with high biomass potentials should generate negative emissions to compensate for these residual emissions. Countries that shoulder these additional mitigation burdens beyond climate neutrality must be financially compensated.

Author Contributions: Conceptualization, F.K., R.C.M., V.S., M.B. and U.F.; Formal analysis, F.K. and J.S.; Investigation, F.K. and J.S.; Methodology, F.K. and J.S.; Project administration, M.B. and U.F.; Software, F.K. and J.S.; Supervision, M.B.; Validation, F.K., J.S., R.C.M., V.S., M.B. and U.F.; Visualization, F.K. and J.S.; Writing—original draft, F.K. and J.S.; Writing—review & editing, F.K., J.S., M.B. and U.F. All authors have read and agreed to the published version of the manuscript.

Funding: This research was funded by the German Federal Ministry of Education and Research in the ARIADNE Project, grant number 03SFK5H0.

Institutional Review Board Statement: Not applicable.

Informed Consent Statement: Not applicable.

Data Availability Statement: The data presented in this study are available on request from the corresponding author.

Acknowledgments: The authors gratefully acknowledge the support of the Open Access Publication Fund of the University of Stuttgart.

Conflicts of Interest: The authors declare no conflict of interest.

Appendix A. Scenario Framework by Country

Table A1. GHG emission targets in the ESR sector for the EU27 + UK countries (extension of Table 1).

	2005	2030		2050	
	Statistics [43]	OPT & ETS First [2]	ESR More [10]	OPT & ETS First	ESR More [10]
Austria	53.51 Mt	−36%	−51%	−36%	−80%
Belgium	76.59 Mt	−35%	−51%	−35%	
Bulgaria	20.79 Mt	±0%	−25%	±0%	
Cyprus	3.8 Mt	−24%	−45%	−24%	
Czech Rep.	57.48 Mt	−14%	−39%	−14%	
Germany	454.74 Mt	−38%	−51%	−38%	
Denmark	39.6 Mt	−39%	−53%	−39%	
Estonia	5.62 Mt	−13%	−36%	−13%	
Spain	225.89 Mt	−26%	−48%	−26%	
Finland	31.09 Mt	−39%	−51%	−39%	
France	378.2 Mt	−37%	−51%	−37%	
Greece	57.48 Mt	−16%	−39%	−16%	
Hungary	41.63 Mt	−7%	−31%	−7%	
Ireland	45.55 Mt	−30%	−55%	−30%	
Croatia	15.94 Mt	−7%	−31%	−7%	
Italy	314.26 Mt	−33%	−50%	−33%	
Lithuania	9.78 Mt	−9%	−33%	−9%	
Luxembourg	9.98 Mt	−40%	−51%	−40%	
Latvia	7.9 Mt	−6%	−31%	−6%	
Malta	0.82 Mt	−19%	−43%	−19%	
Netherlands	116.26 Mt	−36%	−52%	−36%	
Poland	166 Mt	−7%	−31%	−7%	
Portugal	40.27 Mt	−17%	−39%	−17%	
Romania	73.71 Mt	−2%	−28%	−2%	
Sweden	39.79 Mt	−40%	−52%	−40%	
Slovenia	10.96 Mt	−15%	−41%	−15%	
Slovakia	20.77 Mt	−12%	−36%	−12%	
UK	368.14 Mt	−37%	−50%	−37%	

Appendix B. NEWAGE Structure

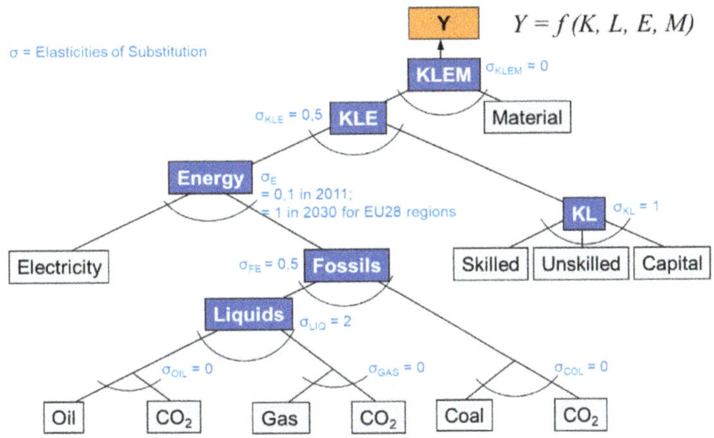

Figure A1. Nesting of NEWAGE CES production functions (given values for elasticities of substitution apply for 19 out of the 23 production sectors).

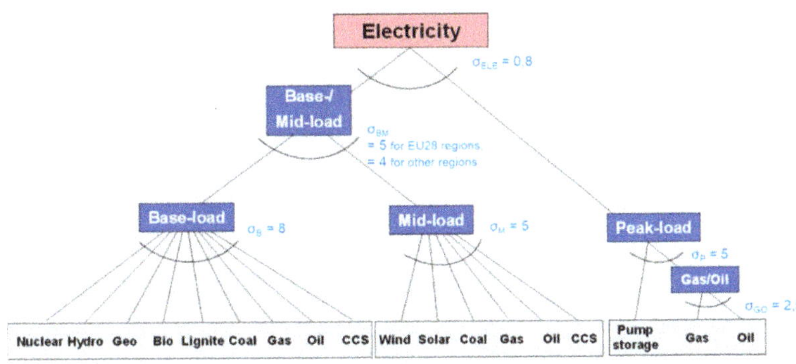

Figure A2. NEWAGE disaggregation of electricity generation.

Appendix C. Extended NEWAGE Scenario Results

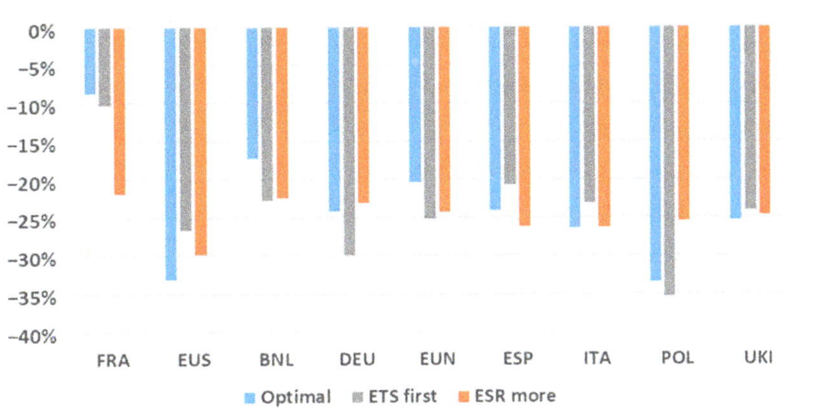

Figure A3. Deviation of EU27 + UK regional CO_2 emissions in 2030 relative to the reference scenario.

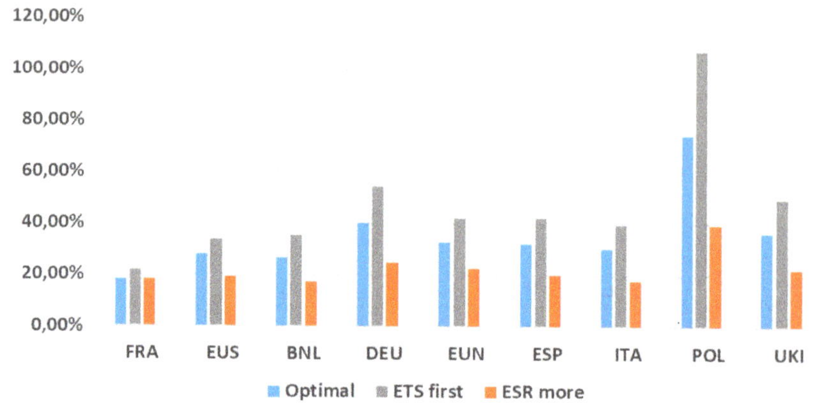

Figure A4. Deviation of EU27 + UK regional electricity prices in 2030 relative to the reference scenario.

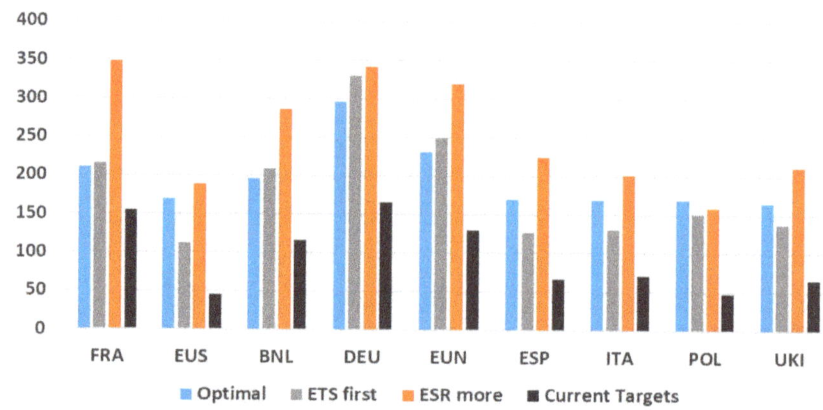

Figure A5. EU27 + UK regional absolute weighted CO_2 prices in 2030 [€2020].

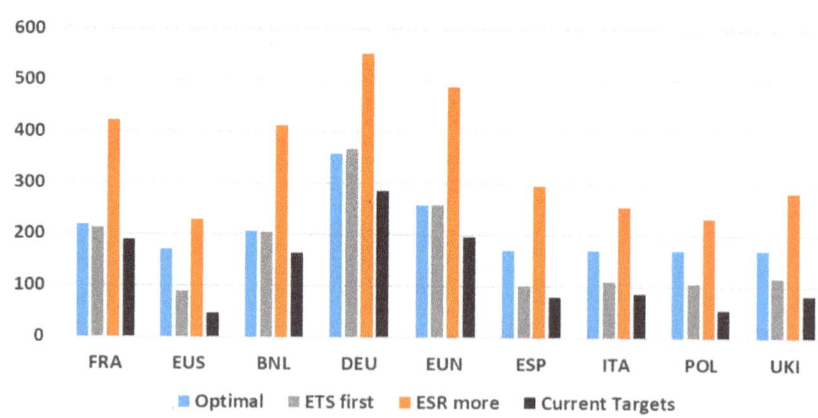

Figure A6. EU27 + UK regional absolute Non-ETS CO_2 prices in 2030 [€2020].

References

1. European Comission. *Communication from the Comission to the European Parliament, the European Council, the Council, the European Economic and Social Committee and the Committee of the Regions*; The European Green Deal: Brussels, Belgium, 2019.
2. European Comission Effort Sharing 2021–2030: Targets and Flexibilities. Available online: https://ec.europa.eu/clima/eu-action/effort-sharing-member-states-emission-targets/effort-sharing-2021-2030-targets-and_en (accessed on 29 October 2021).
3. Simoes, S.; Nijs, W.; Ruiz, P.; Sgobbi, A.; Thiel, C. Comparing Policy Routes for Low-Carbon Power Technology Deployment in EU—An Energy System Analysis. *Energy Policy* **2017**, *101*, 353–365. [CrossRef]
4. Knopf, B.; Chen, Y.-H.H.; De Cian, E.; Förster, H.; Kanudia, A.; Karkatsouli, I.; Keppo, I.; Koljonen, T.; Schumacher, K.; Van Vuuren, D.P. Beyond 2020—Strategies and Costs for Transforming the European Energy System. *Clim. Chang. Econ.* **2013**, *04*, 1340001. [CrossRef]
5. Gils, H.C.; Scholz, Y.; Pregger, T.; Luca de Tena, D.; Heide, D. Integrated Modelling of Variable Renewable Energy-Based Power Supply in Europe. *Energy* **2017**, *123*, 173–188. [CrossRef]
6. Ringkjøb, H.-K.; Haugan, P.M.; Seljom, P.; Lind, A.; Wagner, F.; Mesfun, S. Short-Term Solar and Wind Variability in Long-Term Energy System Models—A European Case Study. *Energy* **2020**, *209*, 118377. [CrossRef]
7. Oei, P.-Y.; Burandt, T.; Hainsch, K.; Löffler, K.; Kemfert, C. Lessons from Modeling 100% Renewable Scenarios Using GENeSYS-MOD. *EEEP* **2020**, *9*, 103–120. [CrossRef]
8. Connolly, D.; Lund, H.; Mathiesen, B.V.; Leahy, M. A Review of Computer Tools for Analysing the Integration of Renewable Energy into Various Energy Systems. *Appl. Energy* **2010**, *87*, 1059–1082. [CrossRef]
9. Jäger-Waldau, A.; Kougias, I.; Taylor, N.; Thiel, C. How Photovoltaics Can Contribute to GHG Emission Reductions of 55% in the EU by 2030. *Renew. Sustain. Energy Rev.* **2020**, *126*, 109836. [CrossRef]
10. Graichen, J.; Matthes, F.; Gores, S.; Fallasch, F. *How to Raise Europe's Climate Ambitions for 2030: Implementing a −55% Target in EU Policy Architecture*; Agora Energiewende: Berlin, Germany, 2020.
11. Pietzcker, R.C.; Osorio, S.; Rodrigues, R. Tightening EU ETS Targets in Line with the European Green Deal: Impacts on the Decarbonization of the EU Power Sector. *Appl. Energy* **2021**, *293*, 29. [CrossRef]
12. Luderer, G.; Vrontisi, Z.; Bertram, C.; Edelenbosch, O.Y.; Pietzcker, R.C.; Rogelj, J.; De Boer, H.S.; Drouet, L.; Emmerling, J.; Fricko, O.; et al. Residual Fossil CO_2 Emissions in 1.5–2 °C Pathways. *Nat. Clim. Chang.* **2018**, *8*, 626–633. [CrossRef]
13. Fragkos, P. Global Energy System Transformations to 1.5 °C: The Impact of Revised Intergovernmental Panel on Climate Change Carbon Budgets. *Energy Technol.* **2000**, *8*, 2000395. [CrossRef]
14. IPCC. *Global Warming of 1.5 °C.An IPCC Special Report on the Impacts of Global Warming of 1.5 °C above Pre-Industrial Levels and Related Global Greenhouse Gas Emission Pathways, in the Context of Strengthening the Global Response to the Threat of Climate Change, Sustainable Development, and Efforts to Eradicate Poverty*; Intergovernmental Panel on Climate Change (IPCC): Geneva, Switzerland, 2018.
15. Hainsch, K.; Burandt, T.; Löffler, K.; Kemfert, C.; Oei, P.-Y.; von Hirschhausen, C. Emission Pathways Towards a Low-Carbon Energy System for Europe: A Model-Based Analysis of Decarbonization Scenarios. *Energy J.* **2021**, *42*. [CrossRef]
16. Victoria, M.; Zhu, K.; Brown, T.; Andresen, G.B.; Greiner, M. Early Decarbonisation of the European Energy System Pays Off. *Nat. Commun.* **2020**, *11*, 6223. [CrossRef] [PubMed]
17. Sgobbi, A.; Nijs, W.; De Miglio, R.; Chiodi, A.; Gargiulo, M.; Thiel, C. How Far Away Is Hydrogen? Its Role in the Medium and Long-Term Decarbonisation of the European Energy System. *Int. J. Hydrogen Energy* **2016**, *41*, 19–35. [CrossRef]
18. Klein, D.; Bauer, N.; Bodirsky, B.; Dietrich, J.P.; Popp, A. Bio-IGCC with CCS as a Long-Term Mitigation Option in a Coupled Energy-System and Land-Use Model. *Energy Procedia* **2011**, *4*, 2933–2940. [CrossRef]
19. Smith, P.; Davis, S.J.; Creutzig, F.; Fuss, S.; Minx, J.; Gabrielle, B.; Kato, E.; Jackson, R.B.; Cowie, A.; Kriegler, E.; et al. Biophysical and Economic Limits to Negative CO_2 Emissions. *Nat. Clim. Chang.* **2016**, *6*, 42–50. [CrossRef]
20. *Impact Assessment Accompanying the Document Communication from the European Commission to the European Parliament, the Council, the European Economic and Social Committee and the Committee of the Regions Stepping up Europe's 2030 Climate Ambition Investing in a Climate-Neutral Future for the Benefit of Our People (Part1/2)*; European Commission: Brussels, Belgium, 2020.
21. Meyer-Ohlendorf, N.; Bart, I. *Climate Action Regulation 2.0—EU Framework for Making Non-ETS Sectors Climate Neutral*; Ecologic Institut: Berlin, Germany, 2020; p. 25.
22. European Comission. *Proposal for a Regulation of the European Parliament and of the Council Amending Regulation (EU) 2018/842 on Binding Annual Greenhouse Gas Emission Reductions by Member States from 2021 to 2030 Contributing to Climate Action to Meet Commitments under the Paris Agreement*; European Commission: Brusssels, Belgium, 2021.
23. EU Carbon Market Will Be Extended to Buildings and Transport, von Der Leyen Confirms. Available online: https://www.euractiv.com/section/energy/news/eu-carbon-market-will-be-extended-to-buildings-and-transport-von-der-leyen-confirms/ (accessed on 25 November 2021).
24. Babonneau, F.; Haurie, A.; Vielle, M. Welfare Implications of EU Effort Sharing Decision and Possible Impact of a Hard Brexit. *Energy Econ.* **2018**, *74*, 470–489. [CrossRef]
25. Korkmaz, P.; Gardumi, F.; Avgerinopoulos, G.; Blesl, M.; Fahl, U. A Comparison of Three Transformation Pathways towards a Sustainable European Society—An Integrated Analysis from an Energy System Perspective. *Energy Strategy Rev.* **2020**, *28*, 100461. [CrossRef]
26. Nijs, W.; Ruiz, P. 01_JRC-EU-TIMES Full Model. European Commission, Joint Research Centre (JRC) [Dataset]. Available online: http://data.europa.eu/89h/8141a398-41a8-42fa-81a4-5b825a51761b (accessed on 26 November 2021).

27. Korkmaz, P.; Cunha Montenegro, R.; Schmid, D.; Blesl, M.; Fahl, U. On the Way to a Sustainable European Energy System: Setting Up an Integrated Assessment Toolbox with TIMES PanEU as the Key Component. *Energies* **2020**, *13*, 707. [CrossRef]
28. Remme, U. *Overview of TIMES: Parameters, Primal Variables & Equations*; Presented at the ETSAP Workshop: Brasilia, Brazil, 2007.
29. Loulou, R.; Lehtilä, A.; Kanudia, A.; Remme, U.; Goldstein, G. *Documentation for the TIMES Model PART II*; Energy Technology Systems Analysis Programme (ETSAP): Paris, France, 2016; p. 38.
30. Schmid, D.; Korkmaz, P.; Blesl, M.; Fahl, U.; Friedrich, R. Analyzing Transformation Pathways to a Sustainable European Energy System—Internalization of Health Damage Costs Caused by Air Pollution. *Energy Strategy Rev.* **2019**, *26*, 100417. [CrossRef]
31. Bicer, Y.; Dincer, I.; Vezina, G. Comprehensive Evaluation of NH_3 Production and Utilization Options for Clean Energy Applications. 2017. Available online: https://zeropollution2050.com/wp-content/uploads/2021/01/MITACS-Canada-Ammonia-Report.pdf (accessed on 21 November 2021).
32. Europe Beyond Coal Overview of National Phase-out Announcements March 2021. Available online: https://beyond-coal.eu/2021/03/03/overview-of-national-phase-out-announcements-march-2021/ (accessed on 10 March 2021).
33. Ruiz, P.; Sgobbi, A.; Nijs, W.N.; Thiel, C.; Dalla Longa, F.; Kober, T.; Elbersen, B.; Hengeveld, G.; European Commission; Joint Research Centre; et al. *The JRC-EU-TIMES Model: Bioenergy Potentials for EU and Neighbouring Countries*; Publications Office: Luxembourg, 2015; ISBN 978-92-79-53879-7.
34. Ruiz, P.; Nijs, W.; Tarvydas, D.; Sgobbi, A.; Zucker, A.; Pilli, R.; Jonsson, R.; Camia, A.; Thiel, C.; Hoyer-Klick, C.; et al. ENSPRESO—An Open, EU-28 Wide, Transparent and Coherent Database of Wind, Solar and Biomass Energy Potentials. *Energy Strategy Rev.* **2019**, *26*, 100379. [CrossRef]
35. Gerhardt, C.; Suhlmann, G.; Ziemßen, F.; Donnan, D.; Warschun, M.; Kühnle, H.J. How Will Cultured Meat and Meat Alternatives Disrupt the Agricultural and Food Industry? *Ind. Biotechnol.* **2020**, *16*, 262–270. [CrossRef]
36. Lucas, P.L.; van Vuuren, D.P.; Olivier, J.G.J.; den Elzen, M.G.J. Long-Term Reduction Potential of Non-CO_2 Greenhouse Gases. *Environ. Sci. Policy* **2007**, *10*, 85–103. [CrossRef]
37. *IEA World Energy Outlook 2020*; IEA: Paris, France, 2020.
38. *Impact Assessment Accompanying the Document Communication from the European Commission to the European Parliament, the Council, the European Economic and Social Committee and the Committee of the Regions Stepping up Europe's 2030 Climate Ambition Investing in a Climate-Neutral Future for the Benefit of Our People (Part 2/2)*; European Commission: Brussels, Belgium, 2012.
39. Aguiar, A.; Narayanan, B.; McDougall, R. An Overview of the GTAP 9 Data Base. *J. Glob. Econ. Anal.* **2016**, *1*, 181–208. [CrossRef]
40. Stadler, K.; Wood, R.; Bulavskaya, T.; Södersten, C.-J.; Simas, M.; Schmidt, S.; Usubiaga, A.; Acosta-Fernández, J.; Kuenen, J.; Bruckner, M.; et al. EXIOBASE 3: Developing a Time Series of Detailed Environmentally Extended Multi-Regional Input-Output Tables. *J. Ind. Ecol.* **2018**, *22*, 502–515. [CrossRef]
41. Revision for Phase 4 (2021–2030). Available online: https://ec.europa.eu/clima/eu-action/eu-emissions-trading-system-eu-ets/revision-phase-4-2021-2030_en (accessed on 10 November 2021).
42. Stepping Up Europe's 2030 Climate Ambition Investing in a Climate-Neutral Future for the Benefit of Our People. Available online: https://eur-lex.europa.eu/legal-content/EN/TXT/?uri=CELEX%3A52020DC0562 (accessed on 18 November 2021).
43. National Inventory Submissions 2020 | UNFCCC. Available online: https://unfccc.int/ghg-inventories-annex-i-parties/2020 (accessed on 18 November 2021).
44. Abrell, J.; Rausch, S. A Smart Design of New EU Emissions Trading Could Save 61 Per Cent of Mitigation Costs. ZEW Policy Brief. 2021. Available online: https://ftp.zew.de/pub/zew-docs/policybrief/en/pb05-21.pdf (accessed on 29 October 2021).

energies

Article

The Role of BECCS in Achieving Climate Neutrality in the European Union

Igor Tatarewicz [1,*], Michał Lewarski [1], Sławomir Skwierz [1], Vitaliy Krupin [1,*], Robert Jeszke [1], Maciej Pyrka [1], Krystian Szczepański [2] and Monika Sekuła [1]

1 National Centre for Emissions Management (KOBiZE), Chmielna 132/134, 00-805 Warsaw, Poland; michal.lewarski@kobize.pl (M.L.); slawomir.skwierz@kobize.pl (S.S.); robert.jeszke@kobize.pl (R.J.); maciej.pyrka@kobize.pl (M.P.); monika.sekula@kobize.pl (M.S.)

2 Institute of Environmental Protection—National Research Institute (IEP-NRI), Krucza 5/11D, 00-548 Warsaw, Poland; krystian.szczepanski@ios.gov.pl

* Correspondence: igor.tatarewicz@kobize.pl (I.T.); vitaliy.krupin@kobize.pl (V.K.)

Abstract: The achievement of climate neutrality in the European Union by 2050 will not be possible solely through a reduction in fossil fuels and the development of energy generation from renewable sources. Large-scale implementation of various technologies is necessary, including bioenergy with carbon capture and storage (BECCS), carbon capture and storage (CCS), and carbon capture and utilisation (CCU), as well as industrial electrification, the use of hydrogen, the expansion of electromobility, low-emission agricultural practices, and afforestation. This research is devoted to an analysis of BECCS as a negative emissions technology (NET) and the assessment of its implementation impact upon the possibility of achieving climate neutrality in the EU. The modelling approach utilises tools developed within the LIFE Climate CAKE PL project and includes the MEESA energy model and the d-PLACE CGE economic model. This article identifies the scope of the required investment in generation capacity and the amount of electricity production from BECCS necessary to meet the greenhouse gas (GHG) emission reduction targets in the EU, examining the technology's impact on the overall system costs and marginal abatement costs (MACs). The modelling results confirm the key role of BECCS technology in achieving EU climate goals by 2050.

Keywords: BECCS; CCS; biomass; climate neutrality; greenhouse gas; emission; abatement cost; EU climate/energy policy; Fit for 55; European Union

Citation: Tatarewicz, I.; Lewarski, M.; Skwierz, S.; Krupin, V.; Jeszke, R.; Pyrka, M.; Szczepański, K.; Sekuła, M. The Role of BECCS in Achieving Climate Neutrality in the European Union. *Energies* **2021**, *14*, 7842. https://doi.org/10.3390/en14237842

Academic Editors: Panagiotis Fragkos and Pelopidas Siskos

Received: 8 October 2021
Accepted: 19 November 2021
Published: 23 November 2021

1. Introduction

The new 55% net greenhouse gas emission reduction target [1] and the Fit for 55 package [2] are key elements of the European Green Deal strategy [3]. This set of policy initiatives presents a milestone towards the vision of making Europe a climate-neutral continent by 2050, in line with the Paris Agreement (aiming to limit global warming to below 2 °C and preferably keep it at 1.5 °C), while at the same time increasing the competitiveness of European industry and aiming to ensure a just transition [4] for the affected regions.

Achieving this goal will require substantial efforts [5–8], including the application of new reduction technologies [9–11] that are currently in the early stages of technological development [12,13]. One such technology is bioenergy with carbon capture and storage (BECCS) [14–16], which is classified as a negative emission technology (NET) [17,18]. NETs are geo-engineering methods of removing greenhouse gases from the atmosphere and mitigating the impact of the energy system on global warming [19].

One of the main advantages of BECCS is the ability to generate negative greenhouse gas emissions [20,21] due to CO_2 removal and injection into geological formations or utilisation in industrial processes. The fuel in this process is the biomass, which absorbs CO_2 from the atmosphere in the process of photosynthesis. The CO_2 emitted during combustion is then captured. Negative emissions from this technology can compensate for emissions in areas where total reduction is impossible, such as agriculture and industry. The cost optimisation results obtained by [22] indicate the high competitiveness of BECCS technology in view of the high prices of CO_2 emission allowances.

BECCS appears in many climate stabilisation scenarios and is envisaged as a feature of the energy sector [23]. The IPCC Special Report [24] presents BECCS in three of four illustrative pathways as essential for the achievement of mitigation targets. The need to use BECCS is emerging in numerous research centres across the globe dealing with energy modelling [25–31]. Climate scenarios that keep global warming within the Paris Agreement limits rely on large-scale application of technologies that can remove CO_2 from the atmosphere in large volumes. This is necessary to compensate for the insufficiency of currently planned mitigation measures, which could lead to cumulative emissions of GHG overshooting the levels set by the EU climate legislation [32].

As part of the "Clean Planet for All" initiative, the European Commission (EC) published an in-depth analysis to support the long-term strategic vision of the economy [33]. This analysis shows reduction potentials of ca. 275 Mt CO_2 bioenergy with carbon capture, utilisation, and storage (BECCUS) and 178 Mt CO_2 for BECCS in the 1.5TECH scenario in the EU-28 by 2050. Scenarios ENTSO-E and ENTSO-G in the TYNDP 2020 report [34] calculate cumulative absorption by BECCS by 2050 within the EU-28 at the level of 808 Mt CO_2 in the Global Ambition scenario. BECCS in the energy sector is also an important part of all of the Climate Change Committee (CCC)'s Net-Zero scenarios, contributing to annual negative emissions in the range of 16–39 Mt CO_2 for the United Kingdom by 2050 [35] or even 51 Mt CO_2 of removal in the Further Ambition scenario by the same year [36]. Thus, the understanding of the feasibility of large-scale BECCS technology deployment in the EU, alongside deeper research of its possibilities, required conditions, and potential limitations, is important.

This article elaborates on selected research results obtained within the LIFE Climate CAKE PL project [37], focusing on BECCS technology's implementation and its role in the achievement of climate neutrality in the EU. The study examines the technical, economical, and environmental feasibility of BECCS and focuses on the application of this technology for the generation of electricity and district heat by power plants and combined heat and power (CHP) installations. It identifies barriers and indicates technological solutions in individual sectors of the economy that are influenced by the EU's energy and climate policy.

The article is divided into seven sections. Section 2 is devoted to a brief description of the analysed technology and its competitiveness; Section 3 focusses on the methods and materials used within the research; Section 4 provides in-depth modelling assumptions; Section 5 presents the research results; Section 6 discusses the results obtained, based on a comparison with other research, and aims to understand its advantages and drawbacks, as well as to define the peculiarities of the analysed technology's potential implementation on a large scale. Finally, Section 7 presents the research conclusions and limitations.

2. BECCS: The Technology and Factors of Its Competitiveness

BECCS is a technology that uses biomass (mainly wood and agricultural biomass, including energy crops) as fuel, equipped with a CCS installation that captures the carbon dioxide generated during the conversion to energy. Sequestrated CO_2 is then stored in geological formations or used to produce certain by products through CCU processes. The concept of BECCS became known in 1996 through the work of Robert H. Williams [38]. He assumed that biomass, combined with CCS, would remove carbon dioxide from the atmosphere, and, based on this finding, the concept of negative emissions was born [39]. Since then, BECCS has captured the attention of decision makers [40].

This technology operates within the following algorithm: At the first stage of the BECCS chain, CO_2 is absorbed from the atmosphere within the photosynthesis process, occurring during plant growth. It is then combusted in power plants equipped with technologies that capture the CO_2, thus preventing gas from being released into the atmosphere. The captured CO_2 is then injected into deep geological formations. The CO_2 is thus transferred from the atmosphere, if the emissions linked with supplying the biomass and capturing the CO_2 do not exceed the amount removed from the air via photosynthesis. Therefore, by delivering net-negative emissions, BECCS compensates for any increases in GHG emissions caused by delays in the implementation of climate policy, especially in sectors where GHG reduction is difficult because of technical limitations [41]. The full schematic of the process is shown in Figure 1.

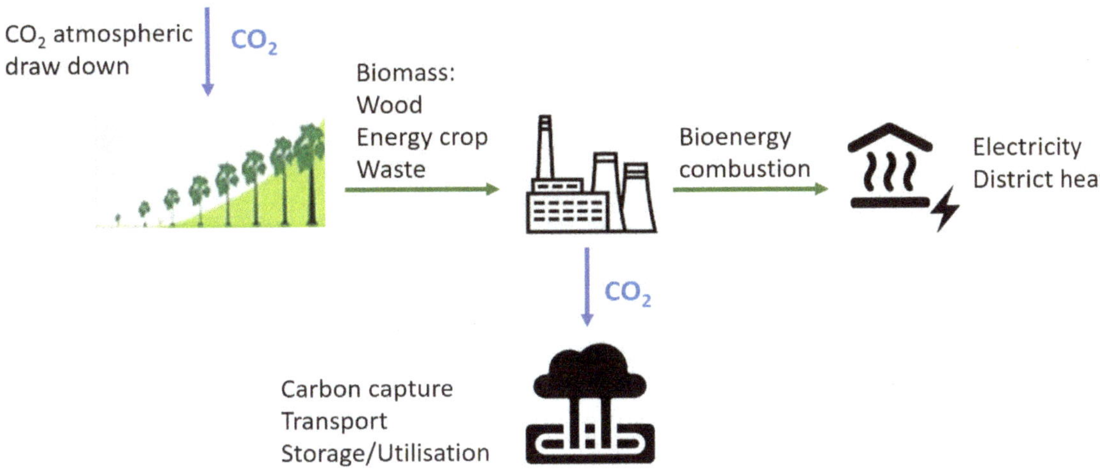

Figure 1. Bioenergy with carbon capture and storage (BECCS) technology. Source: own compilation based on [42].

The conversion of land for bioenergy production purposes may cause greenhouse gas emissions, which should be added to the supply chain emissions. The emissions can be direct, which refer to the change in total carbon stock present in the vegetation and soil of converted land, and indirect, which occur when the previous activity on land that is converted to bioenergy purposes moves to a different location and causes land use change and, therefore, emissions elsewhere [41].

Technology seems to be promising in light of the ambitious reduction targets. Next to afforestation and direct air capture with carbon storage (DACCS), it could be the main way of enabling negative CO_2 emissions, which are highly desirable when full carbon neutrality cannot be achieved in other sectors. Region-specific challenges may impact the technical possibility and efficiency of potential large-scale BECCS installations, including water availability and fertiliser needs.

The aforementioned factors are driving the growing interest from decision makers and investors. This raises the question of the technical potential of this technology and the existing limitations. The high uncertainty is primarily related to the limitations of the CO_2 storage potential, resulting from both technical and geological conditions and possible local social opposition, as well as others, such as biomass availability, conflict with food security and biodiversity goals, costs and financing options, competition for land, fertilisers, and water.

The economic effectiveness of BECCS depends on assumptions related to the type of biomass, the cost of transport, the technology solutions applied, and the fossil fuel emissions offset in the energy system. There are many types of biomass, and they can vary significantly in price. An important factor is the distance from which the biomass is transported and the means of transport, which influences the cost of the biomass. Probably the most significant challenge to the implementation of BECCS at bioenergy plants is the large investment and operational costs of CCS [43]. Thus, this paper presents estimates of the investment and operational costs of BECCS based on the assumptions of acknowledged research centres, data from which were the key input for the optimisation analyses of electricity and district heat generation structure in the EU.

Another important consideration is that the deployment of BECCS as a major climate mitigation solution will require planting bioenergy crops on large land areas. This aspect becomes particularly important in the context of the requirements for biomass in the EU's second Renewable Energy Directive (RED II) [44]. This directive, among others, sets renewable energy targets in the final gross consumption in the EU for 2021–2030. It is one of the main documents in the EU defining the scope, direction, and mechanisms of promotion of renewable energy sources (RESs).

According to the provisions of the aforementioned directive, biomass should meet the sustainability criteria, which means that obtaining it may turn out to be more difficult than it has been to date. Some part of biomass will not be qualified as renewable fuel, and biomass imports will also be limited. The construction of CCS installations can be economically justified in the case of large production units, for which ensuring an adequate level of biomass supply can require importing long distances, which will additionally generate a specific carbon footprint. According to the guidelines of RED II, this footprint must also be taken into account in the CO_2 balance. This is a significant factor hindering the large-scale application of BECCS technology.

From the standpoint of the cost efficiency of BECCS projects, there is currently no market for negative emissions. Currently, the negative emissions are not incentivised, and no remuneration or support exists for such actions; therefore, there is no adequate return for the investment needed to achieve them. Because different technologies have variable benefits and spill overs, at the early stage of implementing carbon payments, the full technology neutrality between greenhouse gas removal technologies is not desirable.

The following options can be considered to support BECCS technology in the short term [45]:

- Power Contract for Difference (CfD): the investor obtains the strike price for the generated electricity in the whole contract period. The strike price allows the investor to cover all costs related to electricity production (capital and operational). The difference between the market price and the strike price is usually covered by the government.
- Carbon payment: the investor obtains the strike price for the generated negative emissions in the whole contract period. The investor obtains predetermined remuneration only for negative emissions. This is a strike price for negative emission units.
- Carbon payment and power CfD: the investor obtains the strike price for the generated electricity and also for the negative emissions in the whole contract period. This is a combination of the two options mentioned above. Carbon payment provides remuneration for negative emissions, while power CfD covers electricity production costs.

Two potential NET-supporting systems can be analysed in the medium term, wherein all greenhouse gas removal technologies will compete, but payments for other complementary products should be implemented to avoid over-support for some technologies [45]:

- Carbon payment (and CfD for other complementary products, e.g., electricity or hydrogen): the investor obtains a fixed payment per tonne of generated negative emissions in the whole contract period.
- Negative CO_2 obligation scheme (and CfD for other complementary products, e.g., electricity or hydrogen): this option requires emitters to cover part of their emissions with negative emission certificates. NET investors earn these certificates and can sell them on the market.

The long-term goal for NET-supporting systems is to create a comprehensive CO_2 market [44] on which greenhouse gas removal technologies will compete with abatement technologies to reach the net-zero target.

From a regulatory point of view, there could be two main financing options for the long-term future of NETs in the system of emission management [46]:

- Separate NETs and EU ETS systems with different prices and goals—less cost effective, but BECCS technology could start its development using the opportunity caused by varying price levels in different systems.
- Inclusion of NET into the EU ETS—the most cost-effective way, but it could delay the start of the development of BECCS technology if the CO_2 price is too low in the first stage of BECCS development. Based on the existence of a link between emissions and CCS, and because the European Commission targets preserve the environmental integrity of the EU ETS (the most important policy instrument in the EU for reducing CO_2 emissions), this option seems to be the most beneficial for the future.

3. Materials and Methods

BECCS technology may be implemented in a variety of industries utilising various biomass feedstock and energy conversion methods [42]. In this paper, the research area is narrowed down to electricity and electricity and heat generation technologies; therefore, the term BECCS is used assuming only electricity and heat generation. One of the main advantages of BECCS is that it can be applied to a wide range of technologies with a varying coefficient of CO_2 emissions, e.g., combined heat and power plants (CHPs), dedicated or co-firing power plants (PPs), and other industry and biofuel production technologies, which are outside the scope of this article.

The role of BECCS in the electricity system of the EU was evaluated by applying an energy model utilising linear programming. The principle of the model is based on finding the lowest-cost feasible solution for investing in and operating a system under the given constraints for a year (the objective function is the lowest overall cost of the system over the entire analysed period). The model used, entitled Model for European Energy System Analysis (MEESA), is built on the basis of the OSeMOSYS [47], and was developed within the LIFE Climate CAKE project [37], its detailed description is available in [48]; thus, within this article, only key concepts and assumptions are listed.

MEESA is a model of the energy system of the EU Member States (EU-27), also including the United Kingdom, Switzerland, and Norway (thereafter called the EU+), designed for long-term planning. MEESA evaluates alternative energy supply strategies under user-defined constraints. Such restrictions are usually the following:

- Limits on new investment;
- Fuel availability and trade;
- Environmental regulations;
- Market regulations;
- Cross-border energy flow;
- Required levels of emission reduction;
- Required share of RESs in a given period, etc.

The model covers the most important dynamics and relations that reflect the functioning of the electricity and district heat sectors. The MEESA model is part of a complex toolkit designed for energy and economic systems analysis developed within the LIFE Climate CAKE project. In addition to the MEESA model, the toolkit also includes a global general equilibrium (CGE) d-PLACE model [49] and cooperative sectoral models: EPICA [50] for agriculture and TR^{3}E [51] for transport (Figure 2). The use of an integrated set of models allows for the analysis of aspects related to energy and climate policy in all sectors of the economy. Changes in one sector do not remain unaffected by other sectors, which, using the proposed set of tools, makes it possible to capture these changes. It uses computational loops with a specific sequence of actions, and the results obtained are the consequence of iterative processes.

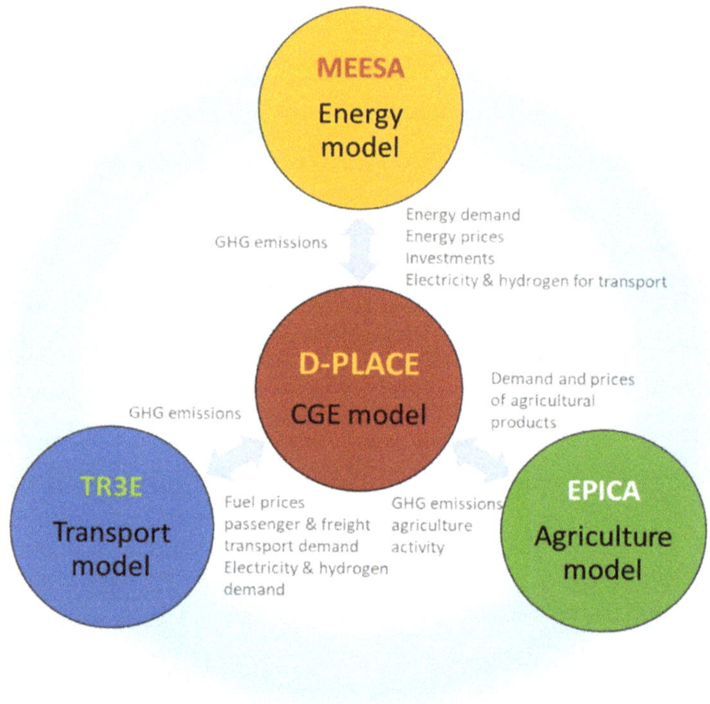

Figure 2. LIFE Climate CAKE PL project: general scheme of models and interactions. Source: own elaboration [37].

The MEESA model optimises the energy supply technology options meeting the given demand (electricity, district heat, and hydrogen) under a set of defined constraints. The main criterion of optimisation is the total discounted system cost in a given period of time (usually long term). The results are divided into four categories: activity, capacity, emissions, and costs. Activity means the production of various energy carriers, which, in the model, currently focusses on electricity, heat, and hydrogen.

The MEESA model is capable of working year by year, yet it was decided to implement a five-year resolution. Energy demand data are exogenous to the model (they come from the d-PLACE model in the five-year resolution). The MEESA allows for the modelling of all steps in the energy flows, starting from energy demand through transmission, distribution, and conversion to energy resources. Figure 3 shows the energy chain schematic representation defined in the MEESA model.

Approximately 50 different technology types are defined in the MEESA model, including existing and new conventional thermal units, RESs, energy storage, electrolysers, and demand side response (DSR) services. The hydrogen produced by electrolysers can be used in the model to produce electricity in gas turbines or directed to sectors where there is a demand for this energy carrier. Each technology defined in the model was assigned an appropriate CO_2 emissions factor related to its generating unit, which allows us to predict the total emissions from the energy sector and to include in the optimisation the costs related to the necessity of purchasing allowances on the market.

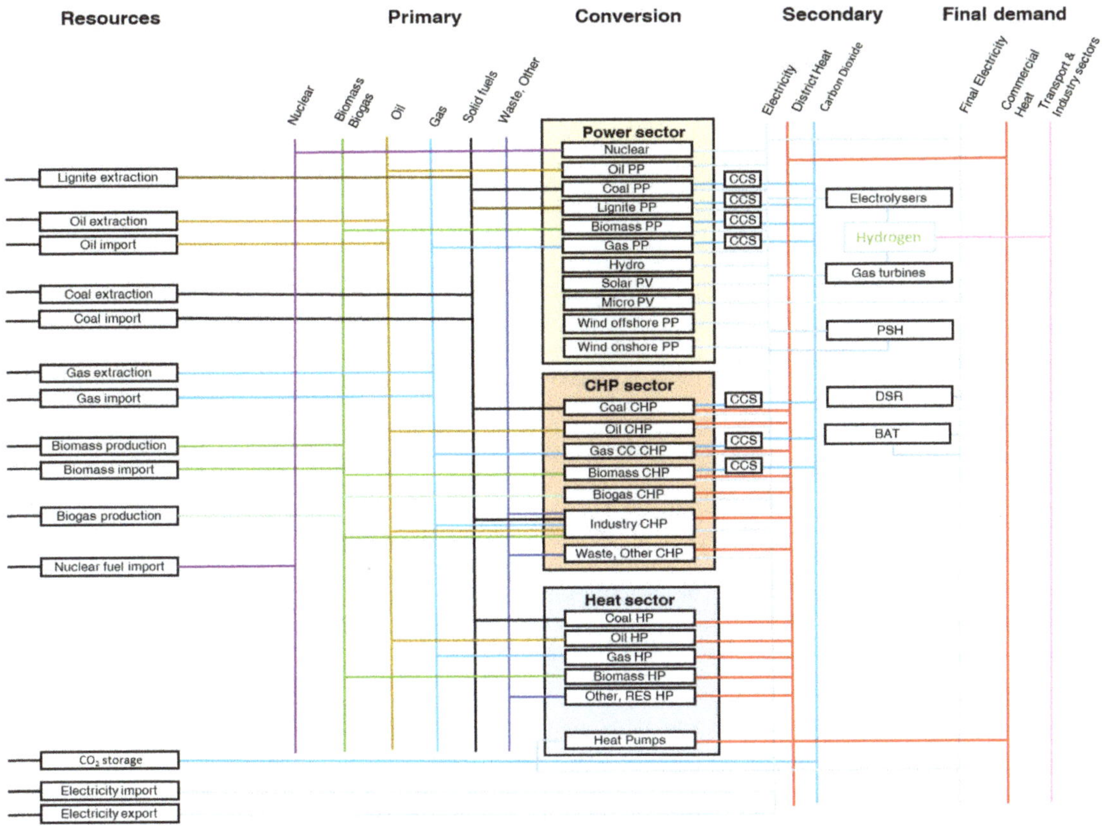

Figure 3. Schematic of energy chain implemented in the MEESA model. Source: own elaboration [48].

The model disaggregates demand in the optimised year for electricity and heat into 18 time slices based on historical data of the demand profile for each country according to seasons (winter and summer), types of days (low-, medium-, and high-demand or different RES productivity), and time of day (day, night, and peak demand period). This allows one to take into account both the average and extreme states of operation of PV and wind sources. This provides a basis for determining the mode of operation of individual units in the system. This solution also enables an analysis of the level and direction of intersystem electricity exchange, with each region being one node. The model ensures this by providing a 15% power margin. Each technology has a defined availability, e.g., photovoltaic 0%, wind onshore 10%, wind offshore 15%, BECCS CHP 70%, BECCS PP 80%, nuclear PP 90%, GAS PP 95%, and pumped-storage PP 100%. The MEESA model is integrated with an economic model that determines the economic impact of energy and climate policy changes and with sectoral models (TR^{3}E and EPICA). The d-PLACE model is a recursive dynamic global and multisector general equilibrium (CGE) model. The d-PLACE model is based on a static CGE model. The input data used to calibrate the base year (2015) come from the Global Trade Analysis Project (GTAP-10) database. The baseline scenario (up to 2050) is consistent with the external projections of changes in GDP, emission limits for the EU, and emission limits for the rest of the world. The effects of the individual regulations analysed in the climate policy scenarios are presented in this model as deviations from the baseline scenario. A characteristic feature of this model is the highly detailed elaboration of GHG emissions.

The model distinguishes carbon dioxide (CO_2), methane (CH_4), nitrous oxide (N_2O), and hydrofluorocarbons (HFCs). Emissions from the different gases are expressed as the CO_2 equivalent [52]. Emissions from fuel combustion and process emissions are modelled separately. Emissions are divided into two main categories:

- Emissions related to fuel combustion—emissions are proportional to energy/ fuel consumed;
- Process emissions (e.g., CO_2 emissions from cement production)—related to the level of activity and proportional to production.

Crucially, the d-PLACE model includes the full balance of emissions at the EU level and divides them into EU ETS and non-ETS, taking into account the targets set in the energy and climate policy of the EU. The model also includes emission reduction targets for regions outside the EU that are signatories of the Paris Agreement.

Energy consumption is modelled in detail in the d-PLACE model. The sectors defined in the model (industry, services, and households) respond to changes in the relative prices of various fuels (including the cost of emissions) and electricity and, in this way, adjust their energy demand patterns. Additionally, manufacturers can replace energy by capital and labour. The production process is modelled by a nested constant elasticity of substitution (CES) function and a Leontief production function [53]. To examine the impact of energy and climate policies, the model distinguishes between energy-intensive and carbon-intensive industries. The model also enables the analysis of the impact of climate and energy policies on aggregate household welfare, including the calculation of compensation mechanisms to offset increased product costs for consumers [49].

The d-PLACE model allows for an analysis of the relative emission reduction potentials across sectors and countries because it includes sector- and country-specific production technologies and consumption patterns. This allows us to look at environmental and climate policy goals from a cost-minimisation perspective, as well as to compare burdens across countries [22]. However, CGE models have significant limitations in terms of modelling specific aspects of the power system operation. By linking d-PLACE with MEESA, it is possible to overcome these limitations. Therefore, it is worth mentioning exactly what kind of data are being exchanged between the two models in an iterative process.

When the MEESA model is used in the connection mode, it uses marginal abatement cost in the EU ETS, demand for electricity and district heat, and demand for hydrogen obtained from the d-PLACE model. Electricity and hydrogen include the demand from the transport sector provided by the TR^3E model, uploaded to MEESA by d-PLACE, which in this case also serves as a hub for information exchange with other models. After the MEESA model results are obtained, the following data are transferred to the d-PLACE model: average prices of electricity, district heat and hydrogen, use of fuels in the electricity and district heat generation, investments in the energy sector, and CO_2 emissions related to fuel consumption (including negative emissions associated with the use of BECCS technology). This iterative process is conducted until a convergence of results is achieved.

Due to the use of several cooperating models (i.e., macroeconomic d-PLACE and sectoral ones: energy MEESA, transport TR^3E, and agriculture EPICA), it was possible to show interactions between various sectors and take into account how changes in one sector affect the possible development of other branches of the economy, as well as household consumption and GDP value.

It is important to note that this article presents the results of BECCS technology modelling based on the last of the NETs' financing schemes mentioned in Section 2, assuming that revenues for negative emissions come from the EU ETS system and depend on the CO_2 price determined within this system.

The applied approach, based on modelling the electricity and district heat system in the EU, with connection to other sectors through the EU ETS system, allows us to consider the role of BECCS technologies not only in the energy system but in the entire economy. The necessary scope of investments, the related costs, and the expected GHG reductions were indicated, taking into account the entire GHG emission chain (resulting from land use change and removals during growth, acquisition, transport, and use of biomass in energy boilers). Despite the identified barriers to the development of this technology, in the conditions of high prices of allowances and striving for a reduction in carbon dioxide in the atmosphere, this technology is competitive and plays an important role in the EU energy mix.

4. Modelling Assumptions

4.1. Scenarios

Within the conducted analysis, two scenarios were developed, a comparison of which enables the assessment of the impact of BECCS technology on energy generation structure, CO_2 emission reduction level, and costs, respectively:

1. The EU Climate Neutrality Scenario (NEU) is a baseline scenario assuming ca. 90% emission reductions in 2050 vs. 1990 and net-zero emissions (including removals) throughout the EU economy. In this scenario, no restrictions are placed on the development of any available technologies. The only limitations are imposed on the projected technical and investment potential. This scenario assumes achievement of the targets set in the Fit for 55 package for a given timeframe and strives towards realisation of the climate neutrality target by 2050.
2. Scenario with no BECCS technology (NO BECCS)—the assumptions for this scenario were exactly as above, except that a complete limitation on BECCS technology was implemented. This scenario is necessary for comparison purposes and to determine the impact of BECCS technology on electricity generation and overall system costs.

The assumptions that have the greatest impact on the modelling results obtained include electricity and district heating demand, technical and economic parameters of BECCS (including capital expenditures, operating costs, energy conversion efficiency, and efficiency of CO_2 capture from flue gases), biomass prices with transport, CO_2 emissions at different stages of biomass utilisation (according to the scheme presented in Figure 1, which illustrates the complete CO_2 cycle for BECCS technology), and the method of accounting for negative emissions in the EU ETS system. These assumptions are described below.

4.2. Electricity, District Heat, and Hydrogen Demand

Electricity, district heat, and hydrogen demand are the input to the MEESA from the d-PLACE macroeconomic model. In general, the electricity demand for the MEESA model is exogenous, but MEESA also calculates additional demand for heat pumps, energy storages, and hydrogen production; therefore, the final electricity demand is a sum of the demand provided by the d-PLACE model and the demand generated internally within the MEESA model. The situation is similar with hydrogen—generally, the demand for industry and transport sectors is provided by the d-PLACE model, but hydrogen generated in the MEESA model can also be used internally for electricity and heat generation, as a form of energy storage. Hydrogen is produced in the process of electrolysis, preferably during the periods of the day in which the electricity price is at the lowest level, to cover the hydrogen demand. The prices of electricity, heat, and hydrogen generated in the MEESA model for specific demand levels and for a particular year are sent back to d-PLACE and affect the estimated new level of demand; this iterative process is repeated until equilibrium between models is achieved.

4.3. Techno-Economic Parameters

The technical and economic parameters and assumptions for electricity and heat generation technologies defined in the MEESA model were based on the final assumptions adopted in the new PRIMES Reference Scenario 2020 [53]. Table 1 presents the set of techno-economic parameters of technologies adopted for model calculations.

Table 1. Techno-economic parameters of key technologies covered by the model for selected years. Source: Primes [54] and own assumptions.

Technology	Overnight Investment Cost *, EUR/kW			Fixed Operation and Maintenance Cost, EUR/kW_{yr}			Variable Cost, EUR/MWh			Electrical Efficiency (Net) in Optimal Load Operation, Ratio			Technical Lifetime, Years
	2030	2040	2050	2030	2040	2050	2030	2040	2050	2030	2040	2050	
BECCS_PP	3700	3300	3200	69	63	61	5.9	5.8	5.8	0.31	0.32	0.32	40
BECCS_CHP	5000	4500	4300	93	85	82	8.0	7.8	7.8	0.25	0.26	0.26	40
Biomass_PP	1800	1700	1700	40	39	38	3.6	3.6	3.6	0.39	0.40	0.40	40
Biomass_CHP	2450	2300	2300	54	53	52	4.9	4.9	4.9	0.30	0.30	0.30	30
Gas_PP	580	575	570	21	20	19	1.9	1.8	1.7	0.61	0.62	0.63	30
GAS_CHP	780	775	770	28	27	26	2.6	2.4	2.3	0.48	0.48	0.48	30
GAS_PP_CCS	1625	1500	1500	38	35	34	3.0	2.9	2.8	0.50	0.50	0.50	30
GAS_CHP_CCS	2200	2025	2025	52	47	46	4.1	3.9	3.8	0.32	0.32	0.32	30
Lignite_PP_CCS	3340	3250	3150	65	62	61	5.1	3.6	3.4	0.33	0.34	0.35	40
Coal_PP_CCS	3150	2890	2850	65	56	54	5.0	4.8	4.8	0.37	0.38	0.38	40
Biogas	465	458	450	24	24	23	2.6	2.6	2.6	0.38	0.39	0.39	25
Nuclear	5100	4900	4700	115	108	105	7.4	7.6	7.8	0.38	0.38	0.38	60
Wind—onshore	1175	1150	1100	13	12	12	0.2	0.2	0.2	1.00	1.00	1.00	30
Wind—offshore	1650	1577	1503	27	26	26	0.4	0.4	0.4	1.00	1.00	1.00	30
Solar PV	551	529	507	15	11	9	0.0	0.0	0.0	1.00	1.00	1.00	30
Solar PV small	543	522	500	15	11	9	0.0	0.0	0.0	1.00	1.00	1.00	30
Hydro	1670	1660	1650	8	8	8	0.0	0.0	0.0	1.00	1.00	1.00	50

* Note: excluding financial costs during construction.

There are also technologies in a different data layout, e.g., batteries have a 115 thousand EUR/MWh investment cost and 6.5 EUR/kW_{yr} in operation and maintenance costs. In storage technologies, as well as in hydrogen generating technology, the price of electricity for a particular time slice is crucial. Whenever particular necessary data were missing from the PRIMES Reference Scenario 2020, the MEESA model used data from recognised research institutions, dealing with energy modelling and investment processes, such as the National Technical University of Athens (NTUA), Tractebel, Ecofys, International Energy Agency (IEA), Joint Research Centre (JRC), and Frontier Economics.

A limited decrease in future investment costs was assumed due to the potential of the economic scale effect for CCS installations that have not yet been fully commercialised. Since the techno-economic data for BECCS provided by PRIMES Reference Scenario 2020 show the investment costs for power plants, the costs for CHPs were assumed to be 35% higher. The energy conversion efficiency of BECCS was lower by 8% compared to the technology without CCS due to the high coefficient of CO_2 capture process needs. The overall efficiency of CO_2 capture from flue gases was assumed to be 87%, which is a rather conservative assumption given the early stage of development of this technology [55].

4.4. Fuel Prices

Primary fuel prices, excluding biomass, were assumed for the calculations based mainly on projections coming from the World Energy Outlook [56]. The prices for biomass were defined on different sources and are basically the result of an expert assessment based on historical data and the analysis of forecasts published by recognised research centres. Biomass prices depend on a number of factors, including the type of biomass (e.g., wood, agricultural products, and solid waste), availability, direct and indirect production support system, demand, and import possibilities. As a result, biomass prices tend to vary from country to country. The potential of biomass was determined for the purpose of this paper based on the following analysis [57,58]. The first publication is a part of the "Biomass role in achieving the Climate Change & Renewables EU policy targets" (BIOMASS FUTURES) project. This publication provides an overview of the potential of different types of biomass feedstock throughout the European Union. This information was then combined with extraction cost estimates to make supply–demand curves. The resulting prices were then averaged to create a trajectory, the same for all EU countries, to enter into the model. This is a simplified approach but acceptable given the purpose of the analysis. The trajectory adopted in the calculations assumes a moderate increase in biomass prices from 5.3 EUR'2015/GJ in 2015 to 12.5 EUR'2015/GJ in 2050. The potential assumed from the first publication has been revised based on more recent sources. The final assumed potential for modelling purposes is in line with the average potential from the 2015 JRC study [58]. It should be noted that the potential in this study may be overestimated, as it does not take into account the need for sustainability criteria, included in Annex IX of RED II [44]. It was assumed in the MEESA model that a maximum of 60% of the biomass production potential in the EU can be used for electricity and heat; the rest of the potential remains available for industry and households.

4.5. EU ETS Allowance Prices

CO_2 allowance prices are the result of an iteration process between the MEESA, d-PLACE, and other sectoral models. Shifts in this parameter cause changes in the energy mix and affect the balance of allowances in the EU ETS, which has a certain impact on prices. Therefore, the prices of CO_2 emission allowances in the proposed methodology are not exogenous data, but the result of model calculations that take into account emission reduction targets along with changes in the fuel mix and process emissions in the EU ETS sectors. This method allows us to establish the marginal cost of CO_2 emissions in a given year.

4.6. Net Emissions Accounting

A key aspect with respect to the viability of investing in BECCS is accounting for negative emissions. As previously mentioned, currently, negative emissions are not incentivised, and no remuneration or support exists for such actions. It is likely (in the authors' opinion) that in the long term, NETs would be implemented into the EU ETS system. Thus, model calculations were based on this assumption. The accounting method adopted allows for a full reflection of GHG emissions from the entire cycle of biomass generation, processing, and transport, as well as revenues from the sale of negative emissions in the market. Emissions data for each part of the emission chain were based on the information contained in Annex VI of RED II and also on the IPCC report [59]. The IPCC calculated the median lifecycle emissions for biomass slightly above 20 g CO_2eq/MJ. The biomass emission range in the RED II is between 3 and 54 g CO_2eq/MJ. In the analysis, it was assumed that only 80% of the absorbed emissions can be considered as avoided; i.e., the indirect emissions of the entire biomass life cycle will be in the range of 22 g CO_2eq/MJ (the EU+). Taking into account the indirect emissions related to biomass acquisition and the efficiency of the CO_2 carbon capture unit, the overall amount of assumed negative emissions is about 70% of the direct CO_2 emissions of a biomass power plant without CCS (per fuel input; this share calculated per energy output would be even smaller due to the energy consumption by the CCS unit).

5. Results

The following section presents the modelling results regarding the electricity demand and the structure of electricity generation in the EU+ within two scenarios (NEU and NO BECCS). The aim of the comparison was to capture significant differences in the two scenarios, enabling the assessment of the impact of BECCS technology on the power generation mix, CO_2 marginal abatement costs, and the overall system costs.

In the NEU scenario, the range of necessary investments in BECCS capacity in the EU+ is 39 GW and 52 GW in 2040 and 2050, respectively (Figure 4). Before 2035, this technology plays a marginal role due to its high costs and too-low CO_2 prices to ensure the competitiveness of these installations. After 2035, BECCS starts to develop rapidly in the EU+. In this period, the decline in unit construction investment costs and the increase in MAC are considerable. At the same time, a further tightening of reduction targets in all sectors of the economy makes it necessary to look for negative emissions that can in some way offset emissions from sectors where reduction is difficult or sometimes even impossible for technical reasons. BECCS technologies, along with gas and nuclear, are becoming a major source of stable and predictable electricity generation (which is important in view of the high share of RESs, such as wind and solar), while at the same time providing the possibility of accounting for negative emissions that have been captured and stored in geological formations. Although they do not constitute a substantial group in quantitative terms, they play an important role in the EU energy system.

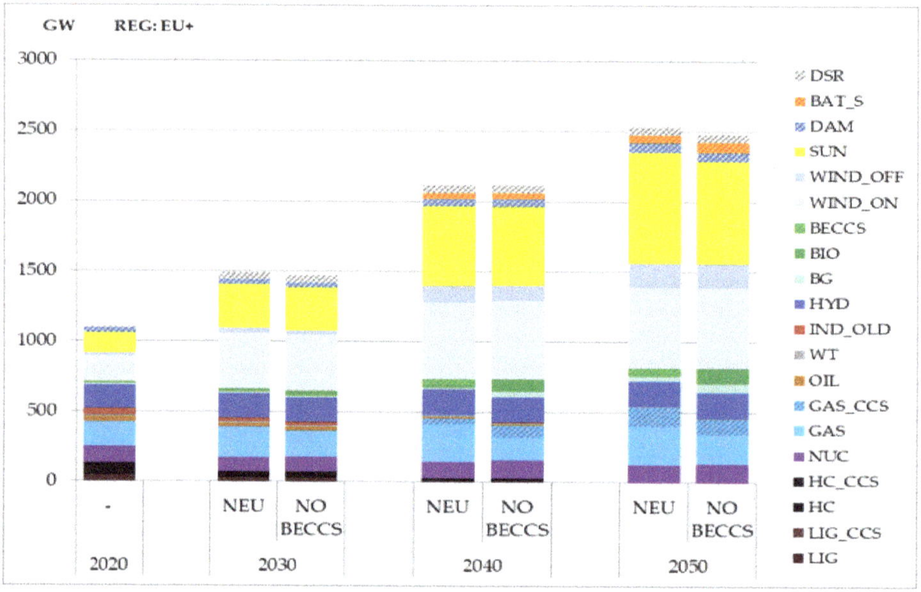

Legend:

DSR	Demand Side Response potential
BAT_S	Batteries
DAM	Hydro-Pumped Storage PP
SUN	PV_Large and PV_Small
WIND_OFF	Wind Offshore
WIND_ON	Wind Onshore
BECCS	Biomass PP_CHP with CCS
BIO	Biomass PP_CHP
BG	Biogas CHP
HYD	Hydro PP
IND_OLD	Industrial_PP_CHP (existing)
WT	Waste CHP
OIL	Oil PP_CHP
GAS_CCS	Gas PP_CHP with CCS
GAS	Gas PP_CHP
NUC	Nuclear PP
HC_CCS	Hard Coal PP_CHP with CCS
HC	Hard Coal PP_CHP
LIG_CCS	Lignite PP with CCS
LIG	Lignite PP

Figure 4. Installed electricity generation capacity by fuels and technologies in the EU+ (in GW). Source: own calculations based on MEESA model results.

According to the results presented in Figure 5, the electricity production of BECCS plants will be 257 TWh and 286 TWh in 2040 and 2050, respectively. BECCS technologies act both as a baseload power source and as a component of a wider mitigation strategy. It can be said that the role of BECCS in providing negative emissions is almost as important as energy production. The scenario with BECCS enables higher use of fossil fuels (because of high CO_2 prices, mostly natural gas) in the energy sector. Without BECCS, biomass PPs play an important role in power generation during the winter, as they are not complemented by the GAS PPs (due to very high CO_2 prices), thus requiring more capacity than is available. In the summer, overall demand is lower, and photovoltaics (combined with batteries) cover a large part of the energy demand; therefore, biomass PPs are mainly needed in the winter. Fossil fuel power without CCS is used very rarely in 2040–2050; CCS is very important for reserve power balance. The lack of BECCS technology also increases the use of hydrogen in the power sector.

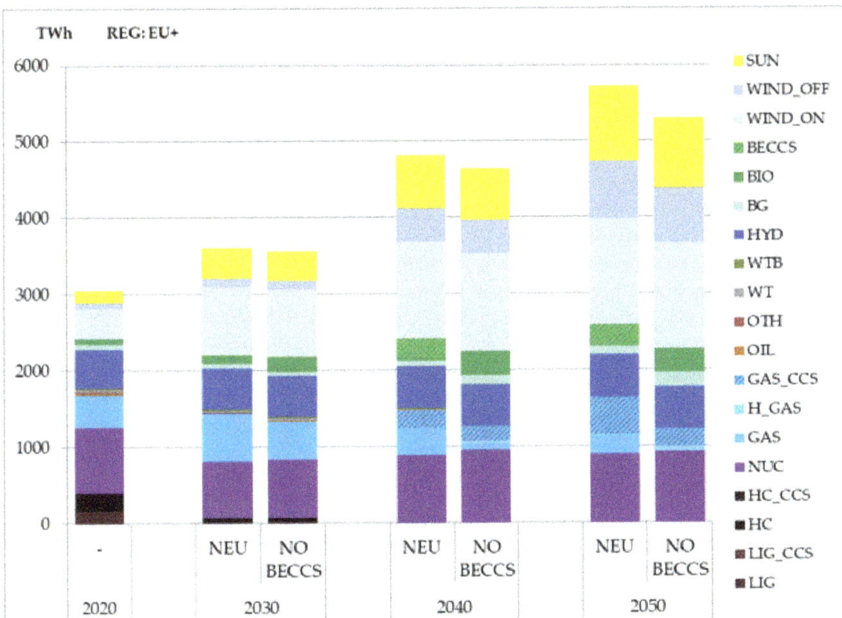

Legend:

SUN	PV_Large and PV_Small
WIND_OFF	Wind Offshore
WIND_ON	Wind Onshore
BECCS	Biomass PP_CHP with CCS
BIO	Biomass PP_CHP
BG	Biogas CHP
HYD	Hydro PP
WTB	Renewable Waste CHP
WT	Waste CHP
OTH	Other_Fuels CHP
OIL	Oil PP_CHP
GAS_CCS	Gas PP_CHP with CCS
H_GAS	Hydrogen+gas PP_CHP
GAS	Gas PP_CHP
NUC	Nuclear PP
HC_CCS	Hard Coal PP_CHP with CCS
HC	Hard Coal PP_CHP
LIG_CCS	Lignite PP with CCS
LIG	Lignite PP

Figure 5. Electricity generation by fuels and sources in the EU+ (in MWh). Source: own calculations based on MEESA model results.

Figure 6 depicts the use of biomass in power plants (PPs), combined heat and power plants (CHPs), and heat plants (HPs). In both scenarios, the utilisation of this fuel increases significantly from the 2050 perspective, and it is more or less at the same level. This means almost complete utilisation of the assumed potential of biomass [58]. However, the distribution of biomass use between PPs, CHPs, and HPs differs greatly. The NEU scenario is dominated by PPs and CHPs, which means that HPs (which are not equipped with CCS) are less cost effective. This scenario, which assumes full CCS availability, shows that under high CO_2 allowance prices, CCS installation becomes economically viable. In 2050, more than half of biomass use comes from power plants. In the NO BECCS scenario, the share of CHPs and HPs in biomass consumption is greater than in NEU. This is probably due to the fact that in NO BECCS, biomass becomes more attractive than other options in heat production. With high CO_2 prices, gas-fired CHPs and HPs in NO BECCS practically disappear, and also, heat pumps are less competitive due to higher electricity prices. In this situation, most of the available biomass is used in district heating, as there are more options for producing low-emission energy in electricity.

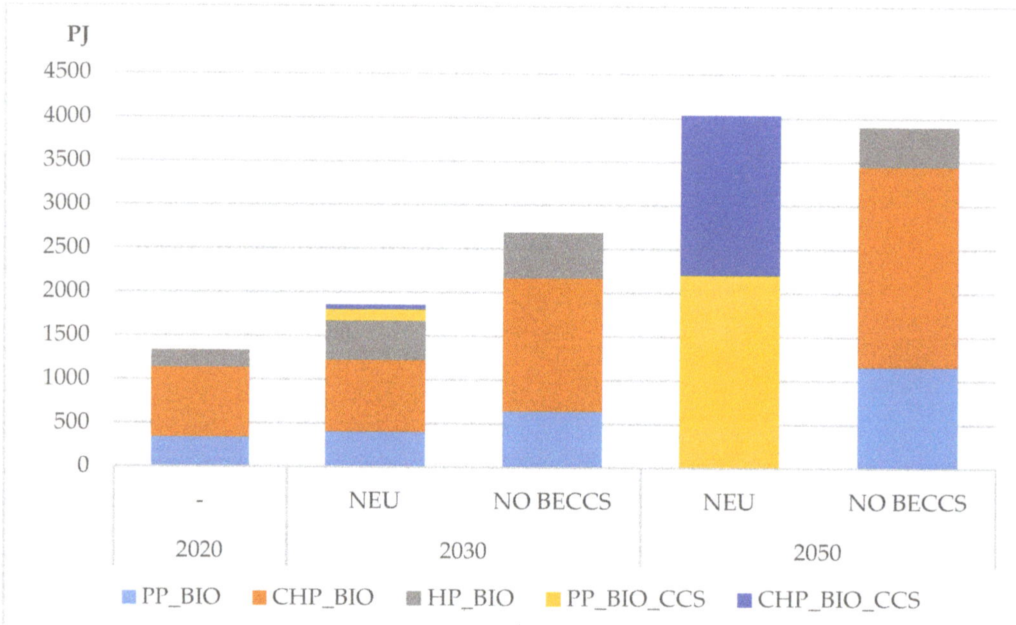

Figure 6. Use of biomass in the considered scenarios (in PJ). Source: own calculations based on MEESA model results.

It is worth mentioning that the method used, based on the combination of sectoral models and the core CGE model, allows one to estimate the impact of increased energy prices on energy demand. In the NO BECCS scenario, the electricity demand is nearly 8% lower than in the NEU scenario (Figure 7). Changes in electricity demand, resulting from CGE model simulations, can be interpreted as a compound effect of economic growth, energy efficiency improvement, substitution of other energy sources for electricity, and changes in the structure of production. A distinctively high increase in electricity use is observed in the household and transport service sectors, largely attributed to the expansion of electric vehicles. On the other hand, demand from energy-intensive industries grows much more slowly, partly because the share of those industries (particularly the extractives) in total output decreases. Under the no BECCS scenario, the demand increase is weaker than in the NEU scenario, due to both higher electricity prices and more moderate economic growth.

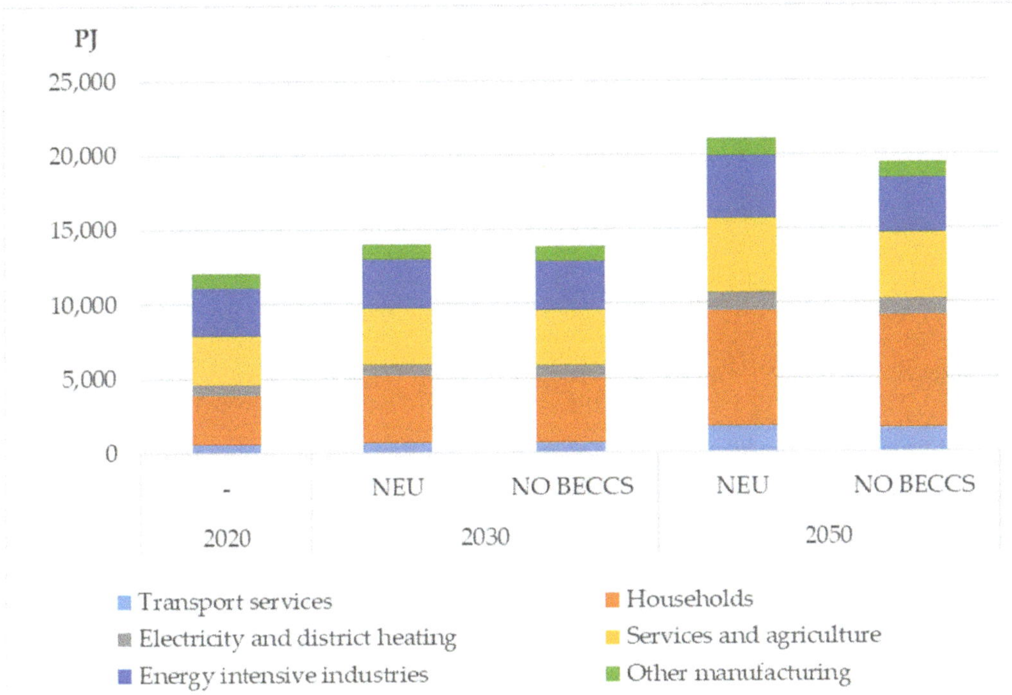

Figure 7. Electricity demand in the considered scenarios (in PJ). Source: own calculations based on d-PLACE model results.

The average electricity generation costs in the EU+ countries (Figure 8) differ between scenarios, and in NO BECCS, they are ca. 10% higher within the 2040–2050 period compared to the NEU scenario. The increase in energy costs may intensify energy efficiency improvement but could also have a negative impact on industrial production.

However, for industry, the increased cost of CO_2 allowances is much more important. Without BECCS, this cost becomes significant. Marginal abatement costs of CO_2 emissions arise from the iteration process between the d-PLACE and sectoral models (Figure 9). The obtained results for the CO_2 reduction costs are ca. 100 EUR/tCO_2 in 2030 for both scenarios, 200 EUR/tCO_2 for NEU and almost 600 EUR/tCO_2 for NO BECCS in 2040, and 400 EUR/tCO_2 in NEU and 2000 EUR/tCO_2 in NO BECCS in 2050. Such high price levels should be treated with caution due to the methodological limitations of the modelling toolkit used. Nevertheless, this shows that BECCS technology can have a significant impact on the marginal costs of abatement. It must be noted that DACCS technology has not been included in the model. Perhaps adding this technology could mitigate the increase in CO_2 abatement cost. This technology, due to its early stage of development and uncertainty, was not considered. However, it could be a direction of the future development of the MEESA model.

The CO_2 emissions differ for the energy sector depending on the scenario (Figure 10). Crucially, as soon as 2040, negative emissions are viable in the power sector, which makes it possible to offset emissions from other sectors. The overall negative emissions in the EU+ achieved in the scenario associated with BECCS technology were almost 300 Mt CO_2 (total negative emissions in the EU+ take into account positive emissions in the energy sector in the NEU scenario at ca. 180 Mt CO_2). This shows that BECCS technology is a promising option to support the EU's climate neutrality goal.

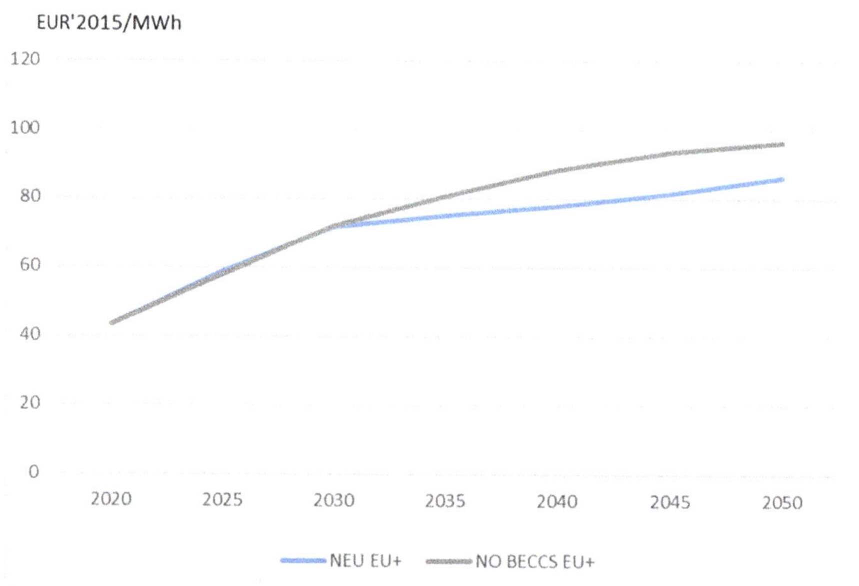

Figure 8. Average electricity generation cost in the EU+ countries (in EUR'2015/MWh). Source: own calculations based on MEESA model results.

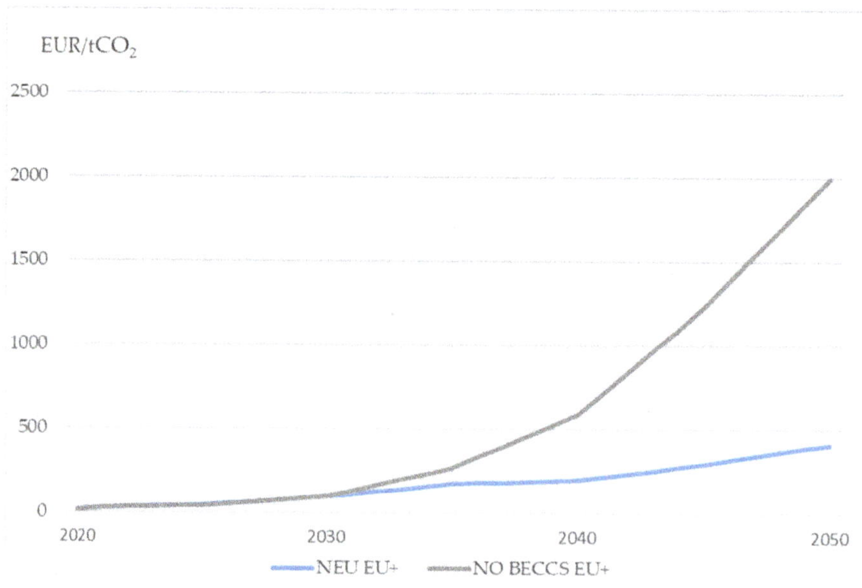

Figure 9. CO_2 price in the EU+ (in EUR/tCO_2). Source: own calculations based on MEESA and d-PLACE models.

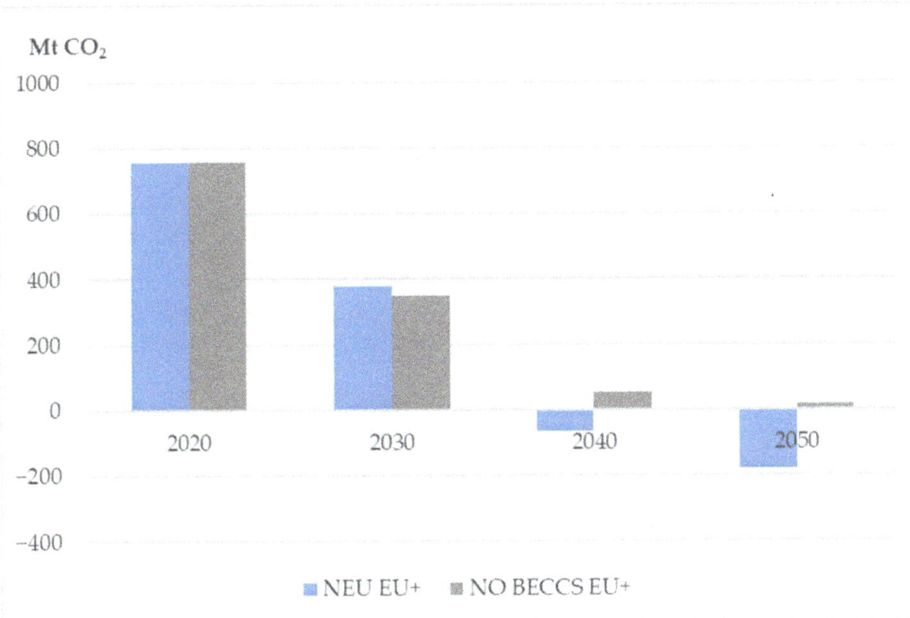

Figure 10. Total CO_2 emissions in the EU+ energy sector (in Mt CO_2). Source: own calculations based on MEESA model.

6. Discussion

The results of the analysis conducted indicate the important role that BECCS could play in achieving the EU's climate goals by 2050. The biggest advantage, assuming large-scale implementation of these technologies, is the generation of negative emissions, which are necessary to compensate for CO_2 emissions in hard-to-decarbonise sectors. Without this technology, negative emissions in the energy sector would be difficult to achieve. Furthermore, the impact of this technology in reducing marginal abatement costs across the economy is evident from the results obtained. A comparison of results from the NEU and NO BECCS scenarios (Figure 7) shows the significant impact of this technology. In the NO BECCS scenario, achieving climate neutrality requires relatively expensive solutions that increase the overall abatement cost to the economy. An analysis prepared using a set of integrated energy and economic models (MEESA and d-PLACE) indicates the need for the deployment of BECCS in the EU on a large scale. Production needs have been set at 250-300 TWh annually after 2040 in order to fulfil climate neutrality goals and secure energy supply safety. Early large-scale deployment of BECCS technology is unlikely due to the high cost and long investment process involved in building the CCS facilities.

The results obtained and presented in this article are, to an extent, in line with the conclusions of other research in this field referred to in the Introduction [33–36]. BECCS is the most scalable negative emissions technology available to remove CO_2 from the atmosphere in the near future. Bioenergy is already widely used, and biomass energy conversion technologies are mature. However, it is important to keep in mind that the component technology for BECCS, carbon capture and sequestration, although relatively well understood, for economic reasons, has limited commercial utilisation [60]. The problem is the high installation costs, both capital and operational. The use of CCS installations reduces the efficiency of fuel conversion into energy by 8–12%, which also affects the economic efficiency of such projects. Nevertheless, given ambitious GHG emission reduction targets and high carbon prices, BECCS could gradually become more economically competitive.

Despite the fact that, from this perspective, BECCS technology looks promising, one cannot forget the barriers and drawbacks that exist. In addition to cost considerations, these elements will undoubtedly have a significant impact on the feasibility of large-scale application. One barrier may be the lack of public acceptance of CO_2 capture and storage [61–63]. A "Not in My Backyard (NIMBY) effect" is found both for pipelines and storage in respondent surveys [64]. Transporting and injecting CO_2 into geological reservoirs raises concerns about carbon leakage, seismic activity, and water pollution. A solution to this problem could be placing the generating units near storage sites.

Social resistance can be overcome by creating mutual trust between stakeholders and commitment to each other and to the project. This can be accomplished by including all stakeholders in the project process at an early stage and communicating about the project and its process to the community [65,66]. State officials, local authorities, self-government representatives, and potential investors should convey knowledge in a gradual and elaborate manner, as well as involve local communities in the planning of BECCS investments and environmental impact assessment processes. Moreover, new incentives and enabling reforms to existing policy instruments are needed [67].

Besides the problem of the transport and storage of CO_2, there is controversy arising from the competition for arable land and fresh water [68], as well as additional risks for biodiversity [24]. Converting large areas of land to bioenergy crops could increase food prices. The U.S. National Academy of Sciences found that, in negative emissions scenarios using BECCS, every Gt of CO_2 stored per year requires approximately 30–40 million hectares of BECCS feedstock [69]. According to the Carbon Sequestration Leadership Forum (CSLF), this equates to ca. 300–700 million hectares of land dedicated to bioenergy crops [70]. Taking these figures as correct, in the NEU scenario, where the annual emissions removed by BECCS technology equal 250 Mt CO_2 in 2040 and 300 Mt in 2050, the total land requirement for energy crops would be 75–175 million hectares in 2040 and 90–210 million hectares in 2050 only to meet the needs of BECCS plants. These are large numbers, but they are much lower than those presented in other sources. Limited biomass potential also means that not all of it can be used for electricity generation; a significant amount will also be used for district heat.

Nevertheless, according to the results obtained, BECCS technology appears to be an important element of the energy generation system, without which it would be difficult to achieve negative emissions in the power industry and, thus, fail to meet the ambitious climate goals set for 2050. However, it certainly cannot be perceived as a "golden mean" solving most climate challenges. The results of the analysis take into account the development of a number of other technologies and solutions in the energy sector, such as energy efficiency improvement, the use of hydrogen, DSR services, and gas units equipped with CCS. There are also other technologies not yet covered by the model, such as direct air carbon capture and storage (DACCS); however, an initial analysis has shown that, at this point in the technology's development, the cost of large-scale installation is lower for other NET options [71]. Overall, this demonstrates the importance of complementary solutions in achieving climate goals rather than concentrating on a limited number of technologies.

The assumption in the MEESA model of the inclusion of NETs into the EU ETS allows us to specify the level of support needed for the development and implementation of BECCS technology. The modelling results show a swift increase in installed BECCS capacity after 2030 and prices in the EU ETS system over 100 EUR/tCO_2; the earlier development of this technology on lower CO_2 levels would obviously require some additional financing. This information obtained from the MEESA model could be used by decision makers to accelerate and direct the development of BECCS.

A problem worth mentioning, not yet analysed in the research, is connected to the amount of support to be transferred to BECCS installations in the long term. As the EU approaches the net-zero target in terms of emissions, the price of CO_2 would intensely increase and could be significantly higher than that necessary for installations such as

BECCS. The EU should start addressing this potential problem at the stage of designing the future mechanism of integrating negative emissions into the EU ETS.

7. Conclusions

The results of our study confirm that BECCS technology could play an important role in reducing GHG emissions and achieving climate neutrality by 2050. Its main advantage is the possibility of providing negative emissions. The results of the analysis of the EU's energy system indicate the necessity of developing this technology on a large scale in order to compensate for GHG emissions from other sectors. The electricity and district heat sectors are the only ones where negative emissions are possible on a large scale, except for afforestation and inventions aimed at CO_2 capture from the air (which are currently still in the experimental stage, i.e., DACCS).

Additionally, BECCS installations provide a stable energy supply, which is particularly important in energy systems with a high share of intermittent RESs. They represent a valid sustainable alternative to natural gas and nuclear power plants. Moreover, biomass can be used from local sources of supply, which has a positive impact on regional economic development and job creation.

However, it is important to keep in mind that, despite the many advantages, there are barriers associated with the development of this technology that may prove difficult to overcome. One of the most important things from the point of view of the net-zero emissions goal for BECCS technology is the origin of biomass and its lifecycle CO_2 emissions. The problem is noted in RED II and will play an important role in the coming years. In order to maintain the consistency of the EU ETS system, the emissions actually avoided should be carefully calculated. According to the provisions of the aforementioned directive, biomass must also meet sustainability criteria, which may limit its supply in the future and affect the increase in raw material prices.

Moreover, BECCS is a relatively expensive technology, especially when it comes to the capital expenditures to build CCS installations. While this may change over time with the technology's development, currently, it can significantly affect the economic efficiency of such projects. CCS also reduces the efficiency of the conversion process. The deployment of BECCS will require public policy interventions at different levels. At the early stage, there is a need for additional financing to de-risk and/or co-finance investments in large-scale demonstration units. In our opinion, there is an urgent need to implement policy mechanisms that take into account and reward negative emissions.

Author Contributions: Conceptualisation, I.T., M.L., S.S., V.K. and R.J.; methodology, I.T., M.L., S.S. and R.J.; writing—original draft preparation, I.T., M.L., S.S. and V.K.; writing—review and editing, I.T., M.L., S.S., V.K., R.J., M.P., K.S. and M.S.; visualisation, I.T., M.L. and S.S.; supervision, I.T., M.L. and R.J. All authors have read and agreed to the published version of the manuscript.

Funding: This research has received funding from the European Union's LIFE Programme and the National Fund for Environmental Protection and Water Management under grant LIFE16 GIC/PL/000031.s.

Institutional Review Board Statement: Not applicable.

Informed Consent Statement: Not applicable.

Data Availability Statement: Not applicable.

Acknowledgments: The article was prepared within the Centre for Climate and Policy Analysis (CAKE) set up by the National Centre for Emission Management (KOBiZE), which is part of the Institute of Environmental Protection–National Research Institute (IEP–NRI). It was prepared within the project "The system of providing and exchanging information in order to strategically support implementation of the climate and energy policy (LIFE Climate CAKE PL)" (LIFE16 GIC/PL/000031), which is co-financed by the EU LIFE programme and by funds from the National Fund for Environmental Protection and Water Management. More about the project can be found at http://climatecake.pl (accessed on 14 November 2021).

Conflicts of Interest: The authors declare no conflict of interest.

References

1. Stepping Up Europe's 2030 Climate Ambition. Investing in a Climate-Neutral Future for the Benefit of Our People. Communication from the Commission to the European Parliament, the Council, the European Economic and Social Committee and the Committee of the Regions of 17 September 2020, COM(2020) 562 Final. Available online: https://eur-lex.europa.eu/legal-content/EN/TXT/?uri=COM:2020:562:FIN (accessed on 10 September 2021).
2. "Fit for 55": Delivering the EU's 2030 Climate Target on the Way to Climate Neutrality. Communication from the Commission to the European Parliament, the Council, the European Economic and Social Committee and the Committee of the Regions of 14 July 2021, COM(2021) 550 Final. Available online: https://eur-lex.europa.eu/legal-content/EN/TXT/?uri=CELEX%3A52021DC0550 (accessed on 12 September 2021).
3. The European Green Deal. Communication from the Commission to the European Parliament, the European Council, the Council, the European Economic and Social Committee and the Committee of the Regions of 11 December 2019, COM(2019) 640 Final. Available online: https://eur-lex.europa.eu/legal-content/EN/TXT/?uri=CELEX%3A52019DC0640 (accessed on 10 September 2021).
4. European Commission. The Just Transition Mechanism: Making Sure No One Is Left Behind. 2021. Available online: https://ec.europa.eu/info/strategy/priorities-2019-2024/european-green-deal/finance-and-green-deal/just-transition-mechanism_en (accessed on 4 September 2021).
5. Guterres, A.; United Nations. Carbon Neutrality by 2050: The World's Most Urgent Mission. 2020. Available online: https://www.un.org/sg/en/content/sg/articles/2020-12-11/carbon-neutrality-2050-the-world%E2%80%99s-most-urgent-mission (accessed on 10 September 2021).
6. Huang, M.-T.; Zhai, P.-M. Achieving Paris Agreement temperature goals requires carbon neutrality by middle century with far-reaching transitions in the whole society. *Adv. Clim. Chang. Res.* **2021**, *12*, 281–286. [CrossRef]
7. Chappell, B. To Be Carbon-Neutral By 2050, No New Oil and Coal Projects, Report Says. NPR. 2021. Available online: https://www.npr.org/2021/05/18/997834721/no-new-oil-and-coal-projects-now-to-be-carbon-neutral-by-2050-report-says?t=1633534811152 (accessed on 18 September 2021).
8. Levin, K.; Fransen, T.; Schumer, S.; Davis, C.; World Resources Institute. What Does "Net-Zero Emissions" Mean? 8 Common Questions, Answered. 2021. Available online: https://www.wri.org/insights/net-zero-ghg-emissions-questions-answered (accessed on 8 September 2021).
9. Tso, K.; Krol, A.; Plata, D.; MIT Climate Portal. Do We Have the Technology to Go Carbon Neutral Today? Available online: https://climate.mit.edu/ask-mit/do-we-have-technology-go-carbon-neutral-today (accessed on 24 September 2021).
10. Wettengel, J.; Clean Energy Wire. Negative Emissions Technologies Must Be Deployed Early for Climate Neutrality. 2021. Available online: https://www.cleanenergywire.org/news/negative-emissions-technologies-must-be-deployed-early-climate-neutrality-report (accessed on 21 September 2021).
11. Wolf, S.; Teitge, J.; Mielke, J.; Schütze, F.; Jaeger, C. The European Green Deal—More Than Climate Neutrality. *Intereconomics* **2021**, *56*, 99–107. [CrossRef] [PubMed]
12. Budinis, S.; International Energy Agency. Going Carbon Negative: What Are the Technology Options? 2020. Available online: https://www.iea.org/commentaries/going-carbon-negative-what-are-the-technology-options (accessed on 15 September 2021).
13. Nicolaides, C. Mission Possible—The Mission on Climate Neutral and Smart Cities A new approach to sustainable urban transformation and urban transition to climate neutrality. *Renew. Energy Sustain. Dev.* **2021**, *7*, 41–42.
14. Babin, A.; Vaneeckhaute, C.; Iliuta, M.C. Potential and challenges of bioenergy with carbon capture and storage as a carbon-negative energy source: A review. *Biomass Bioenergy* **2021**, *146*, 105968. [CrossRef]
15. Li, M.; Lu, Y.; Huang, M. Evolution patterns of bioenergy with carbon capture and storage (BECCS) from a science mapping perspective. *Sci. Total Environ.* **2021**, *766*, 144318. [CrossRef] [PubMed]
16. Fajardy, M.; Morris, J.; Gurgel, A.; Herzog, H.; Mac Dowell, N.; Paltsev, S. The economics of bioenergy with carbon capture and storage (BECCS) deployment in a 1.5 °C or 2 °C world. *Glob. Environ. Chang.* **2021**, *68*, 102262. [CrossRef]
17. Talei, S.; Soleimani, Z. An Overview of Bioenergy with Carbon Capture and Storage Process as a Negative Emission Technology. *J. Environ. Agric. Biol. Sci.* **2021**, *3*, 1–11. [CrossRef]
18. Diyary, H.A. Bioenergy with Carbon Capture and Storage (BECCS) as an Approach to Achieve Negative Emissions in Europe. Master's Thesis. 2021. Available online: https://dspace.library.uu.nl/handle/1874/402526 (accessed on 21 September 2021).
19. Nicolle, W. Four Negative Emission Technologies (NETs) That Could Get Us to Net Zero. Policy Exchange. 2020. Available online: https://policyexchange.org.uk/four-negative-emission-technologies-nets-that-could-get-us-to-net-zero (accessed on 1 October 2021).
20. Tanzer, S.E.; Blok, K.; Ramírez, A. Decarbonising Industry via BECCS: Promising Sectors, Challenges, and Techno-economic Limits of Negative Emissions. *Curr. Sustain./Renew. Energy Rep.* **2021**. [CrossRef]
21. Rosa, L.; Sanchez, D.; Mazzotti, M. The Role of Beccs to Deliver Negative CO_2 Emissions in Europe. In Proceedings of the TCCS-11—Trondheim Conference on CO_2 Capture, Transport and Storage, Trondheim, Norway, 21–23 June 2021. Available online: https://hdl.handle.net/11250/2780214 (accessed on 9 September 2021).

22. Pyrka, M.; Jeszke, R.; Boratyński, J.; Tatarewicz, I.; Witajewski-Baltvilks, J.; Rabiega, W.; Wąs, A.; Kobus, P.; Lewarski, M.; Skwierz, S.; et al. Mapa Drogowa Osiągnięcia Wspólnotowych Celów Polityki Klimatycznej dla Polski do 2050 r. Instytut Ochrony Środowiska—Państwowy Instytut Badawczy/Krajowy Ośrodek Bilansowania i Zarządzania Emisjami (KOBiZE), Warszawa. 2021. Available online: https://climatecake.ios.edu.pl/wp-content/uploads/2021/07/CAKE_Mapa-drogowa-net-zero-dla-PL.pdf (accessed on 15 September 2021).
23. Creutzig, F.; Erb, K.-H.; Haberl, H.; Hof, C.; Hunsberger, C.; Roe, S. Considering sustainability thresholds for BECCS in IPCC and biodiversity assessments. *GCB Bioenergy* **2021**, *13*, 510–515. [CrossRef]
24. IPCC. 2018: Summary for Policymakers. In *Global Warming of 1.5 °C. An IPCC Special Report on the Impacts of Global Warming of 1.5 °C Above Pre-Industrial Levels and Related Global Greenhouse Gas Emission Pathways, in the Context of Strengthening the Global Response to the Threat of Climate Change, Sustainable Development, and Efforts to Eradicate Poverty*; Masson-Delmotte, V., Zhai, P., Pörtner, H.-O., Roberts, D., Skea, J., Shukla, P.R., Pirani, A., Moufouma-Okia, W., Péan, C., Pidcock, R., et al., Eds.; in press; Intergovernmental Panel on Climate Change: Geneva, Switzerland.
25. IEA. Carbon Removal through BECCS and DACS in the Sustainable Development Scenario and IPCC SR1.5 Scenarios, 2030–2100. Paris, France, 2020. Updated 23 September 2020. Available online: https://www.iea.org/data-and-statistics/charts/carbon-removal-through-beccs-and-dacs-in-the-sustainable-development-scenario-and-ipcc-sr1-5-scenarios-2030-2100 (accessed on 26 September 2021).
26. Global Energy Assessment Scenario Database. 2019. Available online: https://iiasa.ac.at/web/home/research/researchPrograms/Energy/Global-Energy-Assessment-Database.en.html (accessed on 30 September 2021).
27. Energy Modeling Forum. EMF 28: The Effects of Technology Choices on EU Climate Policy. Available online: https://emf.stanford.edu/projects/emf-28-effects-technology-choices-eu-climate-policy (accessed on 28 September 2021).
28. Science for Global Insight. Public GEA Scenario Database. Available online: https://tntcat.iiasa.ac.at/geadb/dsd?Action=htmlpage&page=about (accessed on 27 September 2021).
29. Science for Global Insight. AMPERE Project. Available online: https://tntcat.iiasa.ac.at/AMPEREDB/dsd?Action=htmlpage&page=about (accessed on 27 September 2021).
30. Science for Global Insight. CD-LINKS Scenario Database (Version 1.0). Available online: https://db1.ene.iiasa.ac.at/CDLINKSDB/dsd?Action=htmlpage&page=10 (accessed on 27 September 2021).
31. NGFS Climate Scenarios. Network for Greening the Financial System. Available online: https://www.ngfs.net/en/publications/ngfs-climate-scenarios (accessed on 28 September 2021).
32. Regulation (EU) 2021/1119 of the European Parliament and of the Council of 30 June 2021 Establishing the Framework for Achieving Climate Neutrality and Amending Regulations (EC) No 401/2009 and (EU) 2018/1999 (European Climate Law). Available online: https://eur-lex.europa.eu/legal-content/EN/TXT/?uri=CELEX%3A32021R1119 (accessed on 12 September 2021).
33. A Clean Planet for All, A European Strategic Long-Term Vision for a Prosperous, Modern, Competitive and Climate Neutral Economy. Communication from the Commission of 28 November 2018, COM(2018), 773 Final, Brussels, Belgium. Available online: https://eur-lex.europa.eu/legal-content/EN/TXT/?uri=CELEX%3A52018DC0773 (accessed on 3 September 2021).
34. Ten Year Europe's Network Development Plan to 2025, 2030 and 2040. ENTSO-E, 2020, Brussels, Belgium. Available online: https://tyndp.entsoe.eu (accessed on 30 September 2021).
35. Climate Change Committee. The Sixth Carbon Budget, Greenhouse Gas Removals. 2020. Available online: https://www.theccc.org.uk/wp-content/uploads/2020/12/Sector-summary-GHG-removals.pdf (accessed on 17 September 2021).
36. Climate Change Committee. Net Zero: The UK's Contribution to Stopping Global Warming. 2019. Available online: https://www.theccc.org.uk/publication/net-zero-the-uks-contribution-to-stopping-global-warming (accessed on 17 September 2021).
37. System of Providing and Disseminating Information in Order to Support the Strategic Implementation of Climate Policy (LIFE Climate CAKE PL) Project. Available online: http://climatecake.pl (accessed on 27 September 2021).
38. Williams, R.H. *Fuel Decarbonization for Fuel Cell Applications and Sequestration of the Separated CO_2*; PU/CEES Report No. 295; Center for Energy and Environmental Studies, Princeton University: Princetown, NJ, USA, 1996.
39. Carton, W.; Asiyanbi, A.; Beck, S.; Buck, H.J.; Lund, J.F. Negative emissions and the long history of carbon removal. *WIREs Clim. Chang.* **2020**, *11*, e671. [CrossRef]
40. Vandermel, J. *Preventing Climate Change with BECCS: Bioenergy Carbon Capture and Storage*; PSCI, Princeton University: Princetown, NJ, USA, 2020. Available online: https://psci.princeton.edu/tips/2020/11/15/preventing-climate-change-with-beccs-bioenergy-with-carbon-capture-and-storage (accessed on 23 September 2021).
41. Fajardy, M.; Kőberle, A.; Mac Dowell, N.; Fantuzzi, A. *BECCS Deployment: A Reality Check*; Briefing Paper No 28; Imperial College London, Graham Institute: London, UK, 2020.
42. Consoli, C. *Bioenergy and Carbon Capture and Storage: 2019 Perspective*; Global CCS Institute: Melbourne, Australia, 2019. Available online: https://www.globalccsinstitute.com/wp-content/uploads/2019/03/BECCS-Perspective_FINAL_PDF.pdf (accessed on 7 September 2021).
43. Budinis, S.; Krevor, S.; Dowell, N.M.; Brandon, N.; Hawkes, A. An assessment of CCS costs, barriers and potential. *Energy Strategy Rev.* **2018**, *22*, 61–81. [CrossRef]
44. Directive (EU) 2018/2001 of the European Parliament and of the Council of 11 December 2018 on the Promotion of the Use of Energy from Renewable Sources, 11 December 2018. Available online: https://eur-lex.europa.eu/legal-content/EN/TXT/?uri=uriserv:OJ.L_.2018.328.01.0082.01.ENG (accessed on 21 September 2021).

45. Supporting the Deployment of Bioenergy Carbon Capture and Storage (BECCS) in the UK: Business Model Options. Frontier Economics. A report for Drax, March 2021. Available online: https://www.drax.com/wp-content/uploads/2021/04/Frontier-Economics-Supporting-the-Deployment-of-BECCS.pdf (accessed on 26 September 2021).
46. Rickels, W.; Proelß, A.; Geden, O.; Burhenne, J.; Fridahl, M. *The Future of (Negative) Emissions Trading in the European Union*; Kiel Working Paper, No. 2164; Kiel Institute for the World Economy (IfW): Kiel, German, 2020. Available online: http://pure.iiasa.ac.at/id/eprint/16704 (accessed on 25 September 2021).
47. Howells, M.; Rogner, H.; Strachan, N.; Heaps, C.; Huntington, H.; Kypreos, S.; Hughes, A.; Silveira, S.; DeCarolis, J.; Bazilian, M.; et al. OSeMOSYS: The Open Source Energy Modeling System: An introduction to its ethos, structure and development. *Energy Policy* **2011**, *39*, 5850–5870. [CrossRef]
48. Tatarewicz, I.; Lewarski, M.; Skwierz, S. *The MEESA Model, Ver. 1.0*; Institute of Environmental Protection—National Research Institute/National Centre for Emissions Management (KOBiZE): Warsaw, Poland, 2020.
49. Gąska, J.; Pyrka, M.; Rabiega, W.; Jeszke, R. *The CGE Model d-PLACE, Ver.1.0*; Institute of Environmental Protection—National Research Institute/National Centre for Emissions Management (KOBiZE): Warsaw, Poland, 2020.
50. Wąs, A.; Witajewski-Baltvilks, J.; Krupin, V.; Kobus, P. *The EPICA Model, Ver. 1.0*; Institute of Environmental Protection—National Research Institute/National Centre for Emissions Management (KOBiZE): Warsaw, Poland, 2020.
51. Rabiega, W.; Sikora, P.; Gąska, J. *The TR^3E Model, Ver.1.0*; Institute of Environmental Protection—National Research Institute/National Centre for Emissions Management (KOBiZE): Warsaw, Poland, 2020.
52. IPCC. Climate Change 2007: Synthesis Report. In *Contribution of Working Groups I, II and III to the Fourth Assessment Report of the Intergovernmental Panel on Climate Change*; Core Writing Team, Pachauri, R.K., Reisinger, A., Eds.; IPCC: Geneva, Switzerland, 2007.
53. Dixon, P.B.; Jorgenson, D. (Eds.) *Handbook of Computable General Equilibrium Modelling*; Elsevier: Amsterdam, The Netherlands, 2012; 888p.
54. Primes Reference Scenario 2020. Final Assumptions, E3-Modelling; Brussels, Belgium. 2020. Available online: https://ec.europa.eu/energy/data-analysis/energy-modelling/eu-reference-scenario-2020_en (accessed on 20 September 2021).
55. Carbon Capture, Utilisation and Storage. Joint Research Centre, EU Science Hub. 2021. Available online: https://ec.europa.eu/jrc/en/research-topic/carbon-capture-utilisation-and-storage (accessed on 30 September 2021).
56. *World Energy Outlook 2017*; International Energy Agency: Paris, France, 2018. Available online: https://www.iea.org/reports/world-energy-outlook-2017 (accessed on 30 September 2021).
57. Elbersen, B.; Startisky, I.; Hengeveld, G.; Schelhaas, M.-J.; Naeff, H. *Atlas of EU Biomass Potentials. Deliverable 3.3: Spatially Detailed and Quantified Overview of EU Biomass Potential Taking into Account the Main Criteria Determining Biomass Availability from Different Sources*; Alterra/IIASA: Wageningen, The Netherlands, 2012; 139p.
58. Ruiz, P.; Sgobbi, A.; Nijs, W.; Thiel, C.; Longa, F.D.; Kober, T.; Elbersen, B.; Hengeveld, G. *The JRC-EU-TIMES model. Bioenergy Potentials for EU and Neighbouring Countries*; Publications Office of the European Union: Luxembourg, 2015.
59. Schlömer, S.; Bruckner, T.; Fulton, L.; Hertwich, E.; McKinnon, A.; Perczyk, D.; Roy, J.; Schaeffer, R.; Sims, R.; Smith, P.; et al. Annex III: Technology-specific cost and performance parameters. In *Climate Change 2014: Mitigation of Climate Change. Contribution of Working Group III to the Fifth Assessment Report of the Intergovernmental Panel on Climate Change*; Edenhofer, O., Pichs-Madruga, R., Sokona, Y., Farahani, E., Kadner, S., Seyboth, K., Adler, A., Baum, I., Brunner, S., Eickemeier, P., et al., Eds.; Cambridge University Press: Cambridge, UK; New York, NY, USA, 2014.
60. *What Is BECCS? Fact Sheet 2020*; American University: Washington, DC, USA, 2020.
61. Fridahl, M.; Lehtveer, M. Bioenergy with carbon capture and storage (BECCS): Global potential, investment preferences, and deployment barriers. *Energy Res. Soc. Sci.* **2018**, *42*, 155–165. [CrossRef]
62. Bellamy, R.; Lezaun, J.; Palmer, J. Perceptions of bioenergy with carbon capture and storage in different policy scenarios. *Nat. Commun.* **2019**, *10*, 743. [CrossRef] [PubMed]
63. Gough, C.; Vaughan, N.E. Synthesising Existing Knowledge on the Feasibility of BECCS. AVOID 2 Programme. 2015. Available online: http://avoid-net-uk.cc.ic.ac.uk/wp-content/uploads/delightful-downloads/2015/07/Synthesising-existing-knowledge-on-the-feasibility-of-BECCS-AVOID-2_WPD1a_v1.pdf (accessed on 2 September 2021).
64. Wallquist, L.; Seigo, S.L.; Visschers, V.; Siegrist, M. Public acceptance of CCS system elements: A conjoint measurement. *Int. J. Greenh. Gas Control* **2012**, *6*, 77–83. [CrossRef]
65. Parmiter, P.; Bell, R. Public perception of CCS: A Review of Public Engagement for CCS Projects. 2nd Report of the Thematic Working Group on: Policy, Regulation and Public Perception. CCUS Project Network. 2020. Available online: https://www.ccusnetwork.eu/sites/default/files/TG1_Briefing-Report-Public-Perception-of-CCS.pdf (accessed on 8 September 2021).
66. Jasiński, J.; Kozakiewicz, M.; Sołtysik, M. Determinants of Energy Cooperatives' Development in Rural Areas—Evidence from Poland. *Energies* **2021**, *14*, 319. [CrossRef]
67. Bellamy, R.; Fridahl, M.; Lezaun, J.; Palmer, J.; Rodriguez, E.; Lefvert, A.; Hansson, A.; Grönkvist, S.; Haikola, S. Incentivising bioenergy with carbon capture and storage (BECCS) responsibly: Comparing stakeholder policy preferences in the United Kingdom and Sweden. *Environ. Sci. Policy* **2021**, *116*, 47–55. [CrossRef]
68. Fajardy, M.; Mac Dowell, N. Can BECCS Deliver Sustainable and Resource Efficient Negative Emissions? *Energy Environ. Sci.* **2017**, *10*, 1389–1426. [CrossRef]
69. National Academy of Sciences (NAS). *Negative Emissions Technologies and Reliable Sequestration: A Research Agenda*; The National Academies Press: Washington, DC, USA, 2019. [CrossRef]

70. Carbon Sequestration Leadership Forum. Technical Summary of Bioenergy Carbon Capture and Storage (BECCS). 2018. Available online: https://www.cslforum.org/cslf/sites/default/files/documents/Publications/BECCS_Task_Force_Report_2018-04-04.pdf (accessed on 7 September 2021).
71. Realmonte, G.; Drouet, L.; Gambhir, A.; Glynn, J.; Hawkes, A.; Köberle, A.C.; Tavoni, M. An inter-model assessment of the role of direct air capture in deep mitigation pathways. *Nat. Commun.* **2019**, *10*, 3277. [CrossRef] [PubMed]

energies

MDPI

Article

Automated Scheduling Approach under Smart Contract for Remote Wind Farms with Power-to-Gas Systems in Multiple Energy Markets

Zhenya Ji, Zishan Guo *, Hao Li and Qi Wang

School of Electrical and Automation Engineering, Nanjing Normal University, No. 2 Xueyuan Road, Nanjing 210046, China; jizhenya@njnu.edu.cn (Z.J.); shadowix_lh@outlook.com (H.L.); wangqi@njnu.edu.cn (Q.W.)

* Correspondence: zishanbang@126.com

Citation: Ji, Z.; Guo, Z.; Li, H.; Wang, Q. Automated Scheduling Approach under Smart Contract for Remote Wind Farms with Power-to-Gas Systems in Multiple Energy Markets. *Energies* **2021**, *14*, 6781. https://doi.org/10.3390/en14206781

Academic Editors: Panagiotis Fragkos and Pelopidas Siskos

Received: 13 September 2021
Accepted: 14 October 2021
Published: 18 October 2021

Abstract: The promising power-to-gas (P2G) technology makes it possible for wind farms to absorb carbon and trade in multiple energy markets. Considering the remoteness of wind farms equipped with P2G systems and the isolation of different energy markets, the scheduling process may suffer from inefficient coordination and unstable information. An automated scheduling approach is thus proposed. Firstly, an automated scheduling framework enabled by smart contract is established for reliable coordination between wind farms and multiple energy markets. Considering the limited logic complexity and insufficient calculation of smart contracts, an off-chain procedure as a workaround is proposed to avoid complex on-chain solutions. Next, a non-linear model of the P2G system is developed to enhance the accuracy of scheduling results. The scheduling strategy takes into account not only the revenues from multiple energy trades, but also the penalties for violating contract items in smart contracts. Then, the implementation of smart contracts under a blockchain environment is presented with multiple participants, including voting in an agreed scheduling result as the plan. Finally, the case study is conducted in a typical two-stage scheduling process—i.e., day-ahead and real-time scheduling—and the results verify the efficiency of the proposed approach.

Keywords: integrated energy system; scheduling; energy trade; smart contract

1. Introduction

With the aggravation of global energy security and environmental pollution problem, various renewable energy types—especially the wind energy—have become the focus on large-scale development and utilization. Due to the limitation for local absorption of intermittent wind power in remote areas, the phenomenon of power curtailment exists in large quantities, which compromise the carbon reduction and economic benefits of wind farms [1]. The power-to-gas (P2G) technology, with its advantages of reducing renewable energy curtailment and consuming carbon, has become a necessary supplement for remote wind farms [2]. By taking P2G in the scheduling plans in multiple energy forms—which include electricity, gas, and carbon—it can further enhance the economic benefits, while improving the wind power accommodation and reducing carbon emissions [3].

The introduction of P2G systems makes the scheduling of power from wind farms not only a matter of the electricity market, but also encompasses gas market and carbon market. A typical scenario for wind farms with P2G embedded is that surplus power that is not sold in the electricity market due to bidding strategies, transmission constraints, etc., can be utilized through P2G [4], while also participating in the gas and the carbon markets. Many researchers have studied such kinds of optimal scheduling of wind farms with P2G embedded in multiple energy markets [5–7]. In [8], the market behaviors of power systems and natural gas systems which are coupled by P2G are described considering the influence of the market pricing mechanism on the coordinated optimal scheduling.

The concept of a combined P2G and gas-fired generator system is brought up in [9], and the optimal scheduling is studied considering renewable energy accommodation and the ability to reduce carbon emissions. The potential of P2G to absorb renewable energy is assessed in [10], and then the optimal scheduling results of the electricity and gas markets considering the impact of P2G on long/short term natural gas prices is analyzed. However, these studies described P2G with crude models. A P2G system has a high investment and operation cost [11], and the adoption of simplified P2G models would lead to inaccurate results. Improvements thus can be made to enhance the accuracy of scheduling results.

Moreover, most wind farms are geographically remote and spatially dispersed due to constraints in wind farm siting [12,13]. It leads to a decentralized form of information exchange, so cyber instabilities—such as delays, dropouts, and tampers, etc.—are hard to avoid and difficult to fix in time, even if they can be detected immediately. In this context, wind farms would face the issue of communication instability and insecurity. In addition, the electricity, gas, and carbon markets are generally managed by various organizations in different locations and are often with different temporal scales. When wind farms located in remote areas participate in multiple energy markets, the possibility of information dissonance increases. Any untimely or missing or falsified information with one market organizer can affect the scheduling results. Therefore, when the ideal stable and secure communication environment is not assumed, improvements need to be made regarding how to guarantee the effectiveness of scheduling plans in multiple energy markets [14].

To deal with the above-mentioned problems, an automated scheduling approach under blockchain-enabled smart contracts can provide an effective solution. Being a distributed database, the blockchain technology facilitates the prevention of information tampering, ensures the security of transactions, and provides the ability to automate the execution of transactions/settlements [15,16]. Meanwhile, smart contracts are able to execute pre-determined contracts automatically and securely. Smart contracts in the blockchain environment are thus able to automate the contract procedures and minimize interactions between market organizers [15,17].

Researchers have carried out exploratory studies and applications in related energy fields [14,18] and market transactions [19]. A pilot project for an energy system in Japan in [20] analyzes the multiple challenges to the expansion of blockchain in the energy sector from both technology and economy aspects. The Brooklyn microgrid project practices the application of blockchain for microgrid energy markets [21]. A blockchain-secured demand response scheme is proposed to promote individualized incentive pricing under a dual-incentive mechanism in [22]. Focusing on the resource-consuming drawbacks of blockchain itself, beneficial solutions to reduce frequent transactions on blockchain is proposed in [23].

However, the blockchain technology has its inherent weakness. In blockchain environment, smart contract can execute contract automatically and securely. For the sake of secure operation, some measures have been taken such as designing a so-called gasLimit variable in some blockchain environment, e.g., Ethereum. This variable restricts the number of computation steps, the logic of contract contents, and the complexity of contract logic. Hence, current smart contracts are able to support simple scripting language [24] but cannot support complex calculations. This limitation is very practical, considering that the consensus process also makes it difficult and unnecessary to include complex calculations. A consortium blockchain-enabled secure energy trading framework for electric vehicles is proposed in [25], and the contract optimization problem is solved by using the iterative convex–concave procedure algorithm. It demonstrates how to get contract items using off-chain computation, while not involving the implementation of smart contracts. A decentralized cooperative demand response framework is presented in [26] to manage the daily energy exchanges, and smart contract is utilized to enforce autonomous monitor and transaction. Currently, the respective on-chain and off-chain tasks and their cooperation in the blockchain environment are not widely discussed in the literature.

For remote wind farms with P2G systems which trade in multiple energy markets, an automated scheduling framework under blockchain-based smart contract is proposed in order to guarantee that transactions in multiple energy markets can still proceed under potential communication instability and insecurity at real-time schedule. Main contributions in this paper can be briefed as follows:

- A scheduling strategy considering the revenues of participating in multiple energy markets, the capability of reducing wind power curtailment, the penalizations of violating contract items, and the investment/operation cost of investing a wind farm equipping a P2G system is established, in which the non-linearity in the electrolysis of P2G system is considered with detailed models.
- An automated scheduling framework with both off-chain and on-chain procedures is proposed to ensure the applicability of smart contract in blockchain environment, especially in the case that the scheduling considers a non-linearity model of P2G system and trades in multiple energy markets.
- A modified smart contact protocol is adapted considering that more than one scheduling result from the wind farm can be submitted as potential contract items. Moreover, a two-stage scheduling processes and the off-chain/on-chain framework is simulated to compare the effectiveness of the proposed approach.

The rest of the paper is organized as follows. Section 2 describes the framework of smart-contract-enabled automatic scheduling for remote wind farms participating in multiple energy markets. Section 3 introduces the non-linear modeling and scheduling objectives for such a wind farm with the P2G system. Section 4 illustrates the implementation of the proposed framework with commonly used smart contracts. Section 5 provides results of simulations. Conclusions are discussed in Section 6.

2. Smart-Contract-Enabled Automated Scheduling Framework

For a wind farm, being equipped with a P2G can further help to enhance the scheduling economy, improve the wind power accommodation, and reduce carbon emissions. The amount of carbon P2G absorbs can be regarded as the permits of carbon emission P2G owns, which can be sold in carbon market. However, there are some concerns with this kind of scheduling involving multiple energy types. The first one is that energy markets are isolated from each other, which makes it difficult for wind farms to coordinate their scheduling plans. The second one is that, for remote wind farms, it is not easy to guarantee the stability and trustworthy of information considering the communication conditions.

Focusing on these concerns, a smart-contract-enabled automated scheduling framework for remote wind farms with P2G systems is established and the overall framework is shown in Figure 1. In blockchain environment, smart contract can only support simple scripting language considering the operation security. Due to the insufficient calculation capability of the smart contract, the scheduling objective with a non-linear P2G model are solved off-chain. The respective functionalities of on-chain and off-chain procedures are described as follows:

- The off-chain procedure is executed by the wind farm, and is able to find a set of potential scheduling results. Even without the framework proposed here, one wind farm is obliged to run a scheduling function and report its scheduling results in corresponding energy markets. In addition, since predictions on wind power output are often difficult to limit to one particular result, it is also very common to obtain a set of potential scheduling results based on multiple predicted wind power output curves. Although the objectives in [25] are electric vehicles, the process of obtaining results from off-chain procedure is similar to this paper. Details on obtaining contract items will be given in Section 3.
- The on-chain procedure is used to urge that one of these scheduling results can be recognized and executed between wind farms and multiple energy markets. Each participator in the blockchain—i.e., a wind farm owner and organizers of multiple energy markets—votes in one scheduling result from the set of potential scheduling

results, and automatically settles among participators based on the smart contract. Specifically, the Open Vote Network (OVN), i.e., a voting protocol as a smart contract in Ethereum [27], is adapted. Details on reclaiming this security and honesty through OVN will be explained in Section 4.

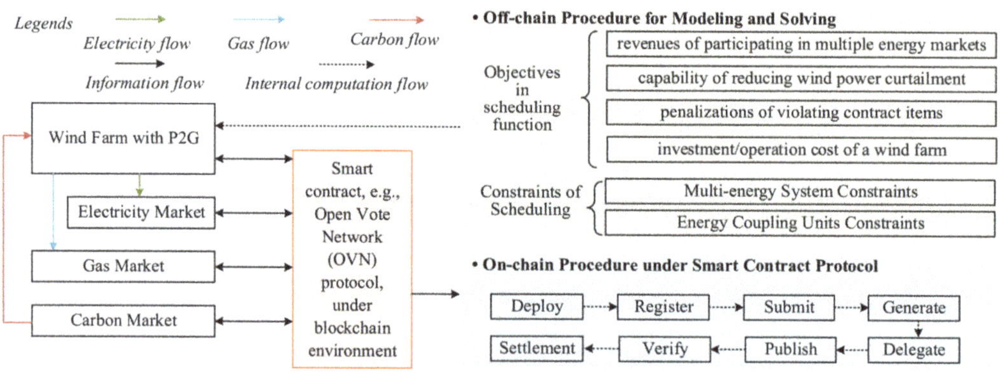

Figure 1. Framework of smart-contract-enabled automated scheduling for remote wind farms with P2G embedded considering multiple energy markets.

A typical two-stage scheduling process, i.e., day-ahead and real-time scheduling, is described for simplicity. In this framework, real-time scheduling can directly utilize the day-ahead scheduling result, eliminating the need for a new round of scheduling solving and confirmation with multiple energy markets. Moreover, in order to avoid that the retraction of deposits in the blockchain, the wind farm and multiple energy markets can trust each other to transfer a certain set of buying/selling volumes in real-time as agreed in the smart contract.

3. Off-Chain Modeling and Solving for Wind Farm with P2G System

In order to improve the accuracy of scheduling results, this section details the non-linear modeling of the P2G system. A scheduling objective function considering multiple energy markets is established for the whole wind farm with the P2G system.

3.1. Non-Linear Modeling of P2G System

Due to the uncertainty of wind power, bidding strategies, and congestion, a large amount of wind power is abandoned. In such a case, P2G system can sell an appropriate amount of carbon emission permits to obtain raw material of carbon, and can reduce wind power curtailment using electrolysis. The products of electrolysis can be combined with carbon to generate methane, which can be sold in the gas market. The P2G technology generally include two main steps of electrolysis ($2H_2O \rightarrow 2H_2 + O_2$) and methanation ($CO_2 + 4H_2 \rightarrow CH_4 + 2H_2O$). Typical energy conversion process of a P2G system is shown in Figure 2.

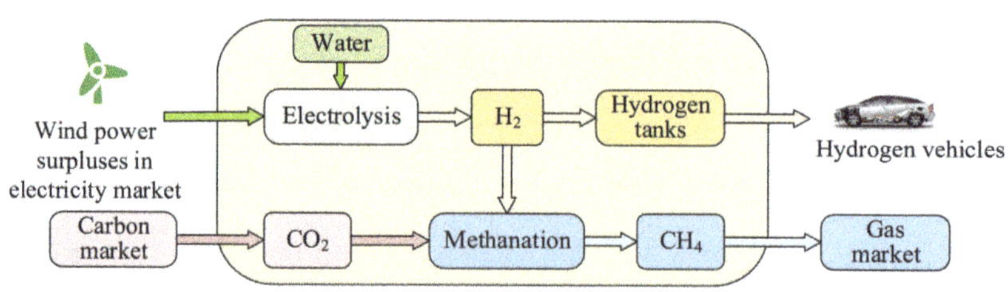

Figure 2. Diagram of energy conversion processes of a P2G system.

As shown in Figure 2, the first and crucial step is electrolysis, which consumes wind power to electrolyze water. The produced hydrogen can be pumped into hydrogen tanks for storage and to supply hydrogen loads—e.g., hydrogen powered vehicles. Meanwhile, an additional step of methanation is able to convert hydrogen and carbon into methane, which means this production process can absorb carbon. Therefore, the methanation process is considered in the P2G model in this paper, and all hydrogen produced by electrolysis is used to produce methane. All the rest units in the P2G system are summarized as the balance-of-plant (BoP) devices.

The proton exchange membrane electrolyze has the advantages of fast reaction, no pollution, and high efficiency, and thus is mostly used in recent engineering and research [28], which is also adopted for modeling in this paper. A detailed model considering the nonlinear nature of electrolysis is built as follows. It is composed of many electrolysis cells connected in series and in parallel. Through the continuous adjustment of cell currents, the electrolysis can promptly respond to the change of surplus wind powers, and accordingly adjust hydrogen production. For an electrolysis cell, its total voltage corresponds to the sum of open circuit voltage, activation overpotential, and ohmic overpotential, which is shown as

$$V_{\mathrm{cell},t} = V_{\mathrm{ocv},t} + \eta_{\mathrm{act},t} + \eta_{\mathrm{ohm},t} \tag{1}$$

where $V_{\mathrm{cell},t}$ is the total voltage of an electrolysis cell; $V_{\mathrm{OCV},t}$ is the open circuit voltage; $\eta_{\mathrm{act},\,t}$ is the activation overpotential; $\eta_{\mathrm{ohm},t}$ is the ohmic overpotential.

The open circuit voltage is calculated using Nernst equation [28,29], which is shown as

$$V_{\mathrm{ocv},t} = V_{\mathrm{eq}} + \frac{RT}{2F}\ln\left(\frac{p_{\mathrm{H_2}}p_{\mathrm{O_2}}^{1/2}}{p_{\mathrm{H_2O}}}\right) \tag{2}$$

$$V_{\mathrm{eq}} = 1.229 - 0.9 \times 10^{-3}(T - 298.15) \tag{3}$$

where V_{eq} is an equilibrium voltage related to cell temperature; R is the universal gas constant; T is the cell temperature; F is the Faraday constant; $p_{\mathrm{H_2}}$, $p_{\mathrm{O_2}}$ and $p_{\mathrm{H_2O}}$ are partial pressures of H_2, H_2O, and O_2, respectively.

The activation overpotential involves overcoming energy barriers at the reaction site. The relationship between the activation overpotential and current density is commonly described using Butler–Volmer equation [29], which is shown as

$$\eta_{\mathrm{act},t} = \frac{RT}{\alpha_{\mathrm{a}}F}\mathrm{arsinh}\left(\frac{i_{\mathrm{cell},t}}{2i_{\mathrm{a}}}\right) + \frac{RT}{\alpha_{\mathrm{c}}F}\mathrm{arsinh}\left(\frac{i_{\mathrm{cell},t}}{2i_{\mathrm{c}}}\right) \tag{4}$$

where α_{a} and α_{c} are charge transfer coefficients at the electrodes of anode and cathode; $i_{\mathrm{cell},t}$ is the cell current; i_{a} and i_{c} are exchange current densities at anode and cathode.

The ohmic overpotential occurs due to the electrical resistance of the electrolysis cell, which can be calculated by Ohm's law as

$$\eta_{\mathrm{ohm},t} = (R_{\mathrm{pem}} + R_{\mathrm{con}})i_{\mathrm{cell},t} = \left(\frac{\rho_{\mathrm{pem}}d}{A_{\mathrm{cell}}} + R_{\mathrm{con}}\right)i_{\mathrm{cell},t} \tag{5}$$

where R_{pem} and R_{con} are resistances of the proton exchange membrane and connections; ρ_{pem} is the resistivity of the proton exchange membrane; d is the thickness of the membrane; A_{cell} is the area of the membrane.

According to the connection mode of electrolysis cells and the relationship between the cell current and cell voltage, the power consumption and the amount of hydrogen production of the electrolysis are shown as

$$P_{\mathrm{E},t} = \mu_1 N_{\mathrm{stack}} N_{\mathrm{cell}} A_{\mathrm{cell}} V_{\mathrm{cell},t} i_{\mathrm{cell},t} \tag{6}$$

$$w_{\mathrm{E,H_2},t} = \mu_2 N_{\mathrm{stack}} N_{\mathrm{cell}} A_{\mathrm{cell}} \frac{\eta_{\mathrm{f}} i_{\mathrm{cell},t}}{2F} \tag{7}$$

where μ_1 and μ_2 are conversion factors; N_{stack} and N_{cell} are the number of parallel and series electrolysis cells, respectively; η_f is the Faraday efficiency.

The relationship between the efficiency of hydrogen production and the amount of hydrogen production is shown as

$$\eta_{E,H_2,t} = \frac{L_{HHV,H_2} w_{E,H_2,t} \Delta t}{\mu_3 P_{E,t} \Delta t} = \frac{L_{HHV,H_2} \mu_2 \eta_f}{2\mu_1 \mu_3 F V_{cell,t}} \tag{8}$$

where $L_{HHV,H2}$ is the higher heating value of hydrogen; μ_3 is a conversion factor.

The amount of methane production and carbon consumption of a methanation system are depicted as

$$w_{M,CH_4,t} = \frac{\mu_4 \eta_{M,CH_4} w_{E,H_2,t}}{\rho_{CH_4}} \tag{9}$$

$$w_{M,CO_2,t} = \frac{w_{E,H_2,t} M_{CO_2}}{\mu_5 \rho_{CO_2}} \tag{10}$$

where η_{M,CH_4} is the efficiency of methane production; ρ_{CH_4} is the density of methane at standard atmospheric pressure; M_{CO_2} is the molar mass of carbon; ρ_{CO_2} is the density of carbon at standard atmospheric pressure; μ_4 and μ_5 are molar conversion factors.

3.2. Objective Function

For a wind farm with P2G system embedded, the economic parameters include investment costs and operational costs for wind turbine generators (WTGs) and P2G equipment (i.e., electrolysis, methanation, and BoP devices). The cost of the system in a scheduling day is calculated as [30,31].

$$SIO = \sum_{i=1}^{I} \frac{CAP_i INVE_i \frac{\tau(1+\tau)^{y_i}}{(1+\tau)^{y_i}-1} + OPEX_i}{DAY} \tag{11}$$

where i is a component index (e.g., WTGs, electrolysis, methanation, and the BoP devices); CAP_i is the capacity of component i; $INVE_i$ is the investment cost of component i; τ is the interest rate; y_i is the lifetime of component i; $OPEX_i$ is the operation and maintenance expenditure of component i for a year; DAY is the number of available days in a year.

The P2G system has the ability to reduce emissions of carbon. From the carbon market perspective, the amount of carbon the P2G system absorbs can be regarded as the permits of carbon emission owned by the P2G system, which can be sold in carbon market to gain revenue. Therefore, the process of P2G participating in the carbon market can be seen as a process of selling carbon emission permits and gaining revenue.

The typical wind power curves can be generated by clustering. A simple example to select a typical wind power curve k_0 from some possible wind power outputs is according to the following judgement.

$$\sum_{t=1}^{T} (P_{tyw,k_0,t} - P_{pw,t})^2 \le \sum_{t=1}^{T} (P_{tyw,k,t} - P_{pw,t})^2, \forall k \in K \tag{12}$$

where $P_{tyw,k,t}$ is the kth possible wind power output at period interval t; $P_{pw,t}$ is the predicted wind power at period interval t; K is the set of typical wind power curves.

Then, the corresponding scheduling plans are executed as the optimal scheduling plans of the predicted wind power curves. Considering the day-ahead prediction deviations, there are penalizations for violating the contract items of executing scheduling plans of wind power curve k_0. The default volume of energy in corresponding market is shown as

$$p_{j,def,t} = \left| p_{j,t} - p_{j,k_0,t} \right| \tag{13}$$

where j is an energy market index (e.g., electricity, gas, and carbon market); $p_{j,t}$ is the volume of energy trading in energy market j at period interval t; $p_{j,k_0,t}$ is the volume of energy sold in energy market j according to the scheduling plans of wind power curve k_0.

A comprehensive scheduling objective for the remote wind farm considers the revenues of participating in multiple energy markets, the capability of reducing wind power curtailment, the penalizations of violating contract items, and the investment/operation cost of a wind farm equipped with a P2G system. The revenues of participating in multiple energy markets is characterized from energy trading. The capability of reducing wind power curtailment is characterized by the cost for wind power curtailment. The penalized deposit of violating contract items is characterized by the smart contract agreed between the wind farm and energy markets. Optimizing the scheduling objective aims to maximize the total net revenue in a scheduling day, which can be expressed as

$$\max Y = \sum_{t=1}^{T}\sum_{j=1}^{J}(c_j p_{j,t} - c_{j,\text{def}} p_{j,\text{def},t}) - \sum_{t=1}^{T} c_{\text{wp,curt}} p_{\text{wp,curt},t} - SIO \tag{14}$$

where c_j is the energy price in energy market j; $c_{j,\text{ def}}$ is the penalized deposit of violating contract items; $c_{\text{wp, curt}}$ is the cost for wind power curtailment; $p_{\text{wp, curt, }t}$ is the volume of wind power curtailment at period interval t.

3.3. Constraints

The constraints of P2G are depicted as

$$0 \le P_{\text{E},t} \le P_{\text{E,rated}} \tag{15}$$

$$-P_{\text{E,down}} \le P_{\text{E},t-1} - P_{\text{E},t-1} \le P_{\text{E,up}} \tag{16}$$

$$i_{\text{cell,min}} \le i_{\text{cell},t} \le i_{\text{cell,max}} \tag{17}$$

where $P_{\text{E,rated}}$ is the rated input power of the electrolysis; $P_{\text{E,down}}$ and $P_{\text{E,up}}$ are downward ramping rate and upward ramping rate, respectively; $i_{\text{cell,min}}$ and $i_{\text{cell,max}}$ are the minimum and maximum cell current.

The constraint of electricity network is shown as

$$P_{\text{e, min}} \le p_{\text{electricity, }t} \le P_{\text{e, max}} \tag{18}$$

where $P_{\text{e,min}}$ and $P_{\text{e,max}}$ are the minimum and maximum grid-connected power, respectively.

The congestion constraints of gas network are depicted as

$$0 \le w_{\text{M,CH4},t} \le Q_{\text{g,max}} \tag{19}$$

$$0 \le \sum_{t=1}^{T} w_{\text{M,CH}_4,t} \le C_{\text{g,max}} \tag{20}$$

where $Q_{\text{g,max}}$ is the maximum gas flow, $C_{\text{g,max}}$ is the upper limit of total flow on a trading day.

It should be noted that there are some simplifications in the objective function and the constraints:

1. Since methane can be pumped directly into existing natural gas pipelines for large-scale storage and long-distance transmission, the economic costs associated with constructing pipelines are not considered in Equation (11).
2. The wind farm participates in multiple energy markets as price takers, i.e., values of c_j in Equation (14) are parameters other than variables.
3. The wind farm is connected to the electricity/gas market by a single line/gas pipeline, as implied in Equations (18)–(20).

4. Implementation of Smart Contract under Blockchain Environment

4.1. Structure of Implemention

In general, a smart contract is a piece of codes running on blockchain environment whose logic defines it content. It can be used to receive and process information, store and transfer assets, etc. [32]. It is immutable once deployed and will remain dormant until some transactions submitted by a participator's account triggers it. A smart contract can contain a wide variety of contract items [25]. Therefore, for the implementation here, typical wind power curves and corresponding scheduling results in multiple energy markets can be defined as contract items.

Smart contracts are different from traditional programming language in that they are essentially a kind of agreement for the transfer of digital assets between untrusted participants. The simpler the code, the more secure and reliable it is. The smart contract is thus designed to support only simple scripting language in a particular environment—e.g., blockchain—and cannot support complex calculations. Considering the insufficient calculation capability of smart contracts, the complex modeling and solving process in Section 2 is implemented as an off-chain process, while only the process including submitting and confirming of energy transactions are put on-chain, as shown in Figure 3.

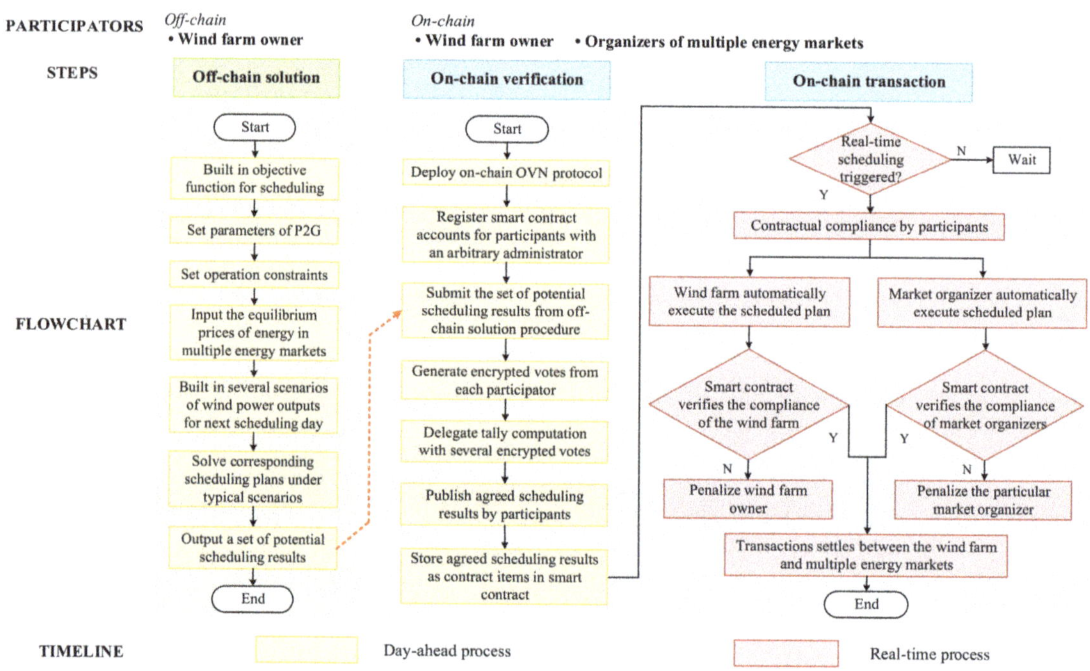

Figure 3. Processes in implementing the smart-contract-enabled automatic scheduling for a wind farm with P2G system in multiple energy markets.

Unlike the traditional scheduling approach, if a blockchain-backed smart contract is introduced, it will bring at least two benefits. On one hand, the traditional centralized multi-level structure among multiple markets to coordinate scheduling plans is changed. Under the blockchain environment, remote wind farms with P2G systems, as well as market organizers—i.e., electricity, gas, and carbon markets—can work as decentralized nodes. Wind farm owners, along with organizers of multiple energy markets, launch votes on a set of potential scheduling plans. With this structure, no one participant is the default dominance, guaranteeing fairness and reducing the need for complex collaboration. On the other hand, these transactions are arbitrated and recorded on-chain, and no matter what communication problems occur in real-time, the relevant transactions can be automatically executed in the same way as voted upon a day ahead, which can guarantee the timeliness and accuracy of real-time scheduling.

4.2. Off-Chain Procedure for Modeling and Solving

In the day-ahead process timeline, the wind farm collects a set of predicted wind power curves as well as equilibrium energy prices from multiple energy markets and operating parameters for power transmission lines and gas pipelines. As shown in Figure 3, the optimized scheduling objectives are modeled and solved off-chain by the wind farm alone.

4.3. On-Chain Procedure under Smart Contract Protocol

The on-chain processes in the automated scheduling for remote wind farms and multiple energy markets are divided into two timelines, which are given as follows.

4.3.1. Day-Ahead On-Chain Processes

The day-ahead on-chain processes focus on setting up the smart contract.

- Deployment of a smart contract protocol—Several smart contract protocols have been developed for different applications. The OVN protocol is able to provide a public bulletin board in a decentralized internet to support coordination among multiple participators [33]. All computations in OVN are written as a smart contract. The following processes are mainly developed under a standard OVN but with necessary modifications.
- Registration and deposition of participators—Like a permissioned blockchain, OVN-based smart contract only allows eligible participators. Although an administrator is required by the OVN protocol to authorize accounts, it is not necessarily a trusted authority. The following provision sets an arbitrary organizer from multiple energy markets as this administrator. The wind farm owner and other market organizers register as accounts participating in the smart contract.
- Submission of potential scheduling results as voting keys—The wind farm owner submits a set of scheduling results based on its off-chain solving. Through the restriction of a smart contract, dishonesty about how much energy the wind farm can provide will only result in penalties for the wind farm itself not being able to provide/absorb the corresponding physical energy, and the consideration of this kind of penalty is included in the objective function Equation (14).
- Generation of potential scheduling results as votes—After voting keys are submitted by the wind farm, all participators—i.e., market operators and wind farm itself— generate and broadcast their respective votes to the other nodes. If needed, an encryption can make the selections of participators anonymous and immutable along during broadcast among participators.
- Delegation and storage of selected scheduling result—The administrator delegates and publish all participators' votes, and all participators can examine as they wanted. The final voted scheduling result is casted as the contract items that stored in the smart contract.

Then, the on-chain process hangs until real-time scheduling triggers the next step.

4.3.2. Real-Time On-Chain Processes

The real-time on-chain processes focus on executing smart contract under the voted scheduling plan:

- Automated real-time schedule by participants—In the real-time schedule, any participant can automatically schedule based on contract items that have been agreed on-chain in day-ahead processes. However, defaults may happen. A typical default situation for a wind farm is that it fails to buy/sell the agreed energy volumes in a corresponding market. A typical default situation for a market is the inability to receive/supply the agreed energy volumes because of line constraints.

- Verification of compliance on energy trades by smart contract—The smart contract verifies whether the participants have strictly executed contract items. Unlike purely digital assets, energy volumes can be physical measured and difficult to tamper with. Moreover, in such a framework, even if information instability of remote wind farms occurs—e.g., delays—it only affects the settlement time of smart contract and not the timeliness of real-time scheduling.
- Settlement among participators—When all the scheduling hours of the real-time schedule finish, electricity, gas, and carbon markets settles with the wind farm respectively, including penalties for violating the agreed contract items.

5. Case Study

5.1. Parameter Settings

In this section, a 40MW wind farm with a 6MW P2G system is taken as an example to verify the effectiveness of the proposed scheduling approach. The day-ahead wind power output curve in the wind farm for real-time scheduling is shown in Figure A1 in Appendix A, which is adapted from [34] with some scaling according to their rated powers. Several typical day-ahead predicted curves of wind power output as the potential contract items are shown in Figure A2 in Appendix A, which are from [35] and some modifications have been made on this basis. Day-ahead scheduling results of the remote wind farm with a P2G system in multiple energy markets including electricity, gas, and carbon are solved off-chain by the wind farm owner. Four cases are presented to illustrate the effectiveness of the proposed scheduling model as follows.

Case 1: No P2G system is utilized for the wind farm, and thus trades only exists in the electricity market.

Case 2: The P2G system is utilized for the wind farm, and is described with simplified model as given in [36]. The adoption of smart contracts is not considered.

Case 3: The P2G system is utilized for the wind farm, the non-linearity nature of electrolysis is considered. The adoption of smart contracts is not considered.

Case 4: The P2G system is utilized for the wind farm, while the non-linear nature of electrolysis is considered. the adoption of smart contracts among all the market operators and the wind farm owner is considered.

The parameters of P2G system obtained from [37] are shown in Table 1. Economic parameters for WTGs, electrolysis, methanation, and BoP devices are shown in Table 2, which are obtained from [30,31] and assumptions. The equilibrium price in the electricity market and its penalty in the smart contract are set as 0.4 ¥/kWh and 2.6 ¥/kWh, respectively. The equilibrium price of methane and its penalty are set as 2.56 ¥/Nm^3 and 16.9 ¥/Nm^3, respectively. The equilibrium price of carbon and its penalty are set as 0.59 ¥/Nm^3 and 19.5 ¥/Nm^3, respectively. The cost for curtailing wind power is set as 1.2 ¥/kWh. The scheduling period is 24-h with a 1-h time interval.

Table 1. Parameters of P2G system.

	Parameters			
P2G	$T = 335.15$ K	$R = 8.314$ J/mol·K	$F = 96{,}485$ C/mol	$p_{H2} = 29.8$ bar
	$p_{O2} = 2.8$ bar	$p_{H2O} = 1$ bar	$\alpha_a = 2$	$\alpha_c = 0.5$
	$i_a = 1 \times 10^{-6}$ A/cm^2	$i_c = 1 \times 10^{-3}$ A/cm^2	$R_{pem} + R_{con} = 0.12$ R·cm^2	$\mu_1 = 0.001$
	$\mu_2 = 3.6$	$\mu_3 = 3600$	$\mu_4 = \mu_5 = 4$	$N_{stack} = 3$
	$N_{cell} = 250$	$A_{cell} = 1100$ cm^2	$\rho_{CH4} = 0.7174$ kg/m^3	$\eta_f = 99\%$
	$\rho_{CO2} = 1.977$ kg/m^3	$i_{cell,\ min} = 0.15$ A/cm^2	$i_{cell,\ max} = 3$ A/cm^2	

Table 2. Economic parameters for WTGs, electrolysis, methanation, and BoP devices in the wind farm.

Component i	CAP_i (MW)	$INVE_i$ (¥/kW)	τ (%)	y_i (year)	$OPEX_i$ (% of CAP_iINVE_i)
WTGs	40	3500	7	20	2.75
Electrolysis	6	4000	7	20	2.75
Methanation	4.5	3500	7	20	2.75
BoP devices	3	3000	7	20	2.75

5.2. Analysis of Scheduling Results among Different Cases

Before activating the smart contract protocol (i.e., OVN), potential contract items (i.e., multiple energy volumes that will be trade and record in real-time energy markets) are calculated off-chain by the wind farm owner are obtained, which are shown in Figure 4. Considering the infeasibility of complex calculations in smart contract, the on-chain OVN only needs to claim some typical wind power curves k_0 based on (12) and its corresponding scheduling plans among multiple market operators and the wind farm owner. In this paper, the claimed typical wind power curve k_0, which satisfies Equation (12), is $k_0 = 3$.

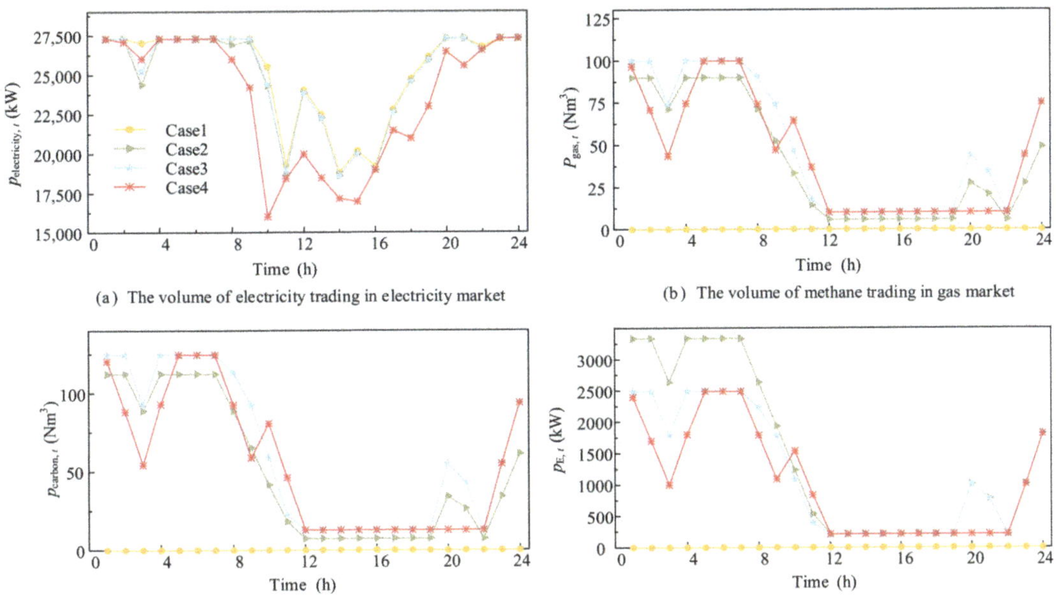

(a) The volume of electricity trading in electricity market

(b) The volume of methane trading in gas market

(c) The volume of carbon trading in carbon market

(d) The wind power consumption of the P2G facility

Figure 4. Day-ahead optimal scheduling results of P2G in remote wind farms considering multiple energy markets.

In multiple energy markets, the day-ahead scheduling results of the wind farm with a P2G system embedded are calculated off the chain in above four cases, respectively. The optimal scheduling results of P2G in remote wind farms in each time period are shown in Figure 4. There are some differences between the scheduling results of different cases. Based on the deviation, this section has detailed analyses on the capability of P2G system, the non-linearity nature of electrolysis and the performance of smart contract.

5.2.1. Analysis on the Capability of P2G System

The revenue of the remote wind farm equipped with a P2G system from multiple energy markets and wind power curtailment rate are shown in Table 3. In Case 1, due to the lack of a P2G system, the wind farm only trades with electricity market, and the surplus wind power cannot be absorbed to generate methane and reduce carbon emissions. In Cases 2, 3, and 4, due to the embedding of the P2G system, surplus wind power has a way to be absorbed. The steps of electrolysis and methanation are carried out in turn to generate methane and consume carbon by absorbing surplus wind power. Thus, the wind farm with a P2G system embedded can have energy trading with electricity, gas, and carbon markets, respectively, which has obvious economic benefits and carbon consumption effect.

Table 3. Revenues from multiple energy markets and effect on wind power curtailment reduction rate.

Case	Revenue from Electricity Market (¥)	Revenue from Gas Market (¥)	Revenue from Carbon Market (¥)	Wind Power Curtailment Reduction Rate (%)
1	241,900	/	/	9.56
2	239,152	2455	706	5.28
3	239,824	3056	879	6.03
4	225,865	2664	766	11.84

5.2.2. Analysis on the Non-Linearity Nature of Electrolysis

Operation situations of a P2G system which consist of power input, hydrogen production, methane production, and carbon consumption are shown in Table 4. In Case 2, the P2G system obviously assumes more surplus wind power than Case 3, while its hydrogen production, methane production, and carbon consumption are less than Case 3. The reason is that Case 2 crudely considers that the relationship between hydrogen production and power consumption is simple and linear, and it ignores the non-linearity nature of electrolysis. The hydrogen production efficiency and the hydrogen output deviation between the model used in this paper and the crude linear model are shown in Figure 5. Although the hydrogen production efficiency of the two models is equal when the power input of the electrolysis is at a certain value, the overall hydrogen production efficiency of the two models is still obviously different. With the increase of power input, the hydrogen production deviation shows a trend of first increasing, then decreasing, and finally increasing, which exactly reflects the complexity of the electrolysis model. The crude model is insufficient to reflect the complex and non-linear nature of P2G facilities, consequently leading to inaccurate scheduling results.

Table 4. Operation situations of a P2G system.

Case	P_E (MW)	$w_{E,H2}$ (Nm^3)	$w_{M,CH4}$ (Nm^3)	$w_{M,CO2}$ (Nm^3)
1	/	/	/	/
2	35.52	4776.96	958.85	1196.05
3	28.78	5947.19	1193.75	1489.05
4	24.82	5183.44	1040.45	1297.83

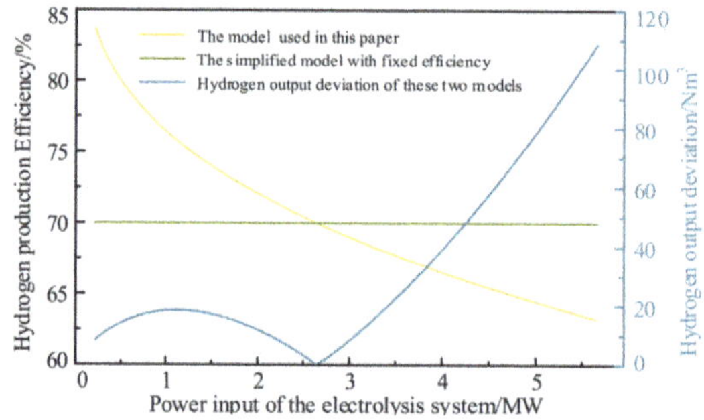

Figure 5. Hydrogen production efficiency and the hydrogen output deviation between the non-linear model in this paper and the simplified model.

5.2.3. Analysis of the Performance of Adopting Smart Contract

The typical wind power curve k_0 and the scheduling results under voted scheduling plan in Case 4 are shown in Figure 6. Caused by the differences between the typical wind power curve as the contract items and the predicted wind power curve at each hour, the scheduling results can be slightly different. The performance of adopting smart contract is shown in Table 5. When any participator violates the voted scheduling plan, a penalty is costed as the characteristics of a smart contract. With this type of smart-contract-enabled penalty mechanism in Case 4, the total penalized cost C_{AC} is much lower than Cases 1, 2, and 3, while the wind power curtailment rate is higher than Cases 2 and 3. According to the economic parameters in Table 2, the cost of the wind farm in the scheduling day is SIO = ¥76,691. The total net revenue Y thus can be obtained as shown in the last column in Table 5.

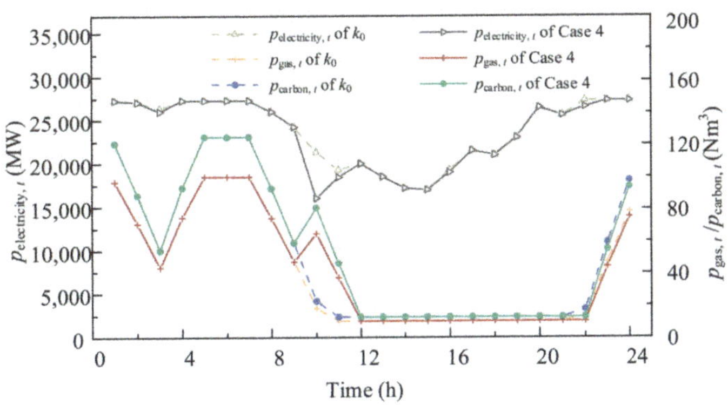

Figure 6. Scheduling plans of typical wind power curve k_0 and the scheduling results of Case 4.

Table 5. Performance of adopting smart contract in different cases.

Case	Penalization of Wind Farm (¥)	Violation of Electricity (MW)	Violation of Methane (Nm^3)	Violation of Carbon (Nm^3)	C_{AC} (¥)	Y (¥)
1	76,710	32.58	977	1219.27	125,000	−36,501
2	42,337	35.03	218.43	272.47	100,073	23,210
3	48,401	32.52	237.33	296.04	94,342	24,324
4	95,035	7.51	84.03	104.81	22,993	34,576

To illustrate the impact of penalizations for violating the voted scheduling plan in the smart contract, a comparison between the values of penalty C_{AC} and Y between Case 3 and Case 4 are shown in Table 6. The first column shows how much the penalties scale compared with the three original values defined in Section 5.1. From Table 5, it is observed that a penalty mechanism impacts the performance of adopting smart contract significantly. From Table 6, it is further observed that when this penalty for violating contract items is small, the effect of considering the performance of adopting smart contract during scheduling on Y is not significant, or even harms its effectiveness. In practical uses, the values of penalizations in the smart contract can be selected by experience and more tests.

Table 6. Performance under different penalties of violating contract items.

Scale with the Reference Value of Penalities c_j	Cost Item	Case 3	Case 4
×0.8	C_{AC} (¥)	75,474	55,936
	Y (¥)	43,192	41,801
×1	C_{AC} (¥)	94,342	22,993
	Y (¥)	24,324	34,576
×1.2	C_{AC} (¥)	113,211	83,097
	Y (¥)	54,550	56,673
×1.5	C_{AC} (¥)	141,514	69,029
	Y (¥)	−22,848	57,745
×2	C_{AC} (¥)	188,685	92,038
	Y (¥)	−70,019	55,444

5.2.4. Analysis on Investment and Return

Taking Case 4 as an example, the total net revenue in an available day as well as the sum of investment cost and operational cost for wind farm, electrolysis, methanation, and the BoP devices are shown in Figure 7. From the perspective of investment return, the income of the wind farm with a P2G system embedded is sufficient to pay the cost of the system in a day, which is economically feasible. It can be calculated that it takes 14.95 years to recover the investment cost of WTGs and the P2G system. Although the payback years are not few, different researchers have shown that the situation can change in 10–20 years [38,39]. Considering the advantages of P2G in enhancing the economic and decarbonizing benefits, wind farms with P2G is worthy to invest.

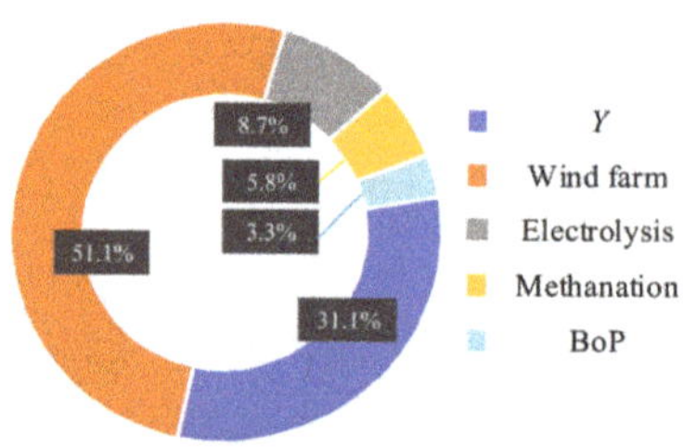

Figure 7. Total net revenue as well as the sum of investment cost and operational cost for WTGs, electrolysis, methanation, and BoP devices.

6. Conclusions

Predictions of uncontrollable wind power outputs are often not accurate enough, and the curtailment affects the economics of wind farms. By deploying P2G, wind farms can not only benefit from participating in multiple energy markets, but also contribute more for carbon reduction. An automated scheduling approach for remote wind farms equipped with P2G systems considering multiple energy markets is proposed in this paper in the presence of instable and unreliable information. Moreover, considering the insufficient calculation capability of smart contracts, a structure of off-chain solving and on-chain transaction is further developed. According to the simulation results, the main conclusions are summarized as follows:

1. The results verify the effectiveness of the non-linear model of the P2G system. The electrolysis process is full of complexity and non-linearity, which should be taken into account when constructing the P2G model to improve accuracy of scheduling results.
2. The proposed framework can cope with the limited complexity of smart contracts and insufficient computation. Specifically, off-chain solving is able to use a non-linear P2G model to obtain more accurate results, while the on-chain protocol only needs to consider a small set of potential scheduling plans.

3. The proposed approach can effectively make full use of remote wind farms with P2G equipped—i.e., improve the economics of scheduling while reducing wind curtailment and decarbonization—while the execution of real-time scheduling can be ensured by smart contract items agreed a day ahead.

This paper is an exploration of adapting the fast-developing blockchain technology in the field of energy trading in multiple energy markets. For future research, the market behaviors from the multiple energy markets will be considered. Further verification will be done on blockchain-based platforms to capitalize energy trading. In addition, more market-realistic situations, such as more than one remote wind farms equipped with P2G systems, will be studied.

Author Contributions: Conceptualization, Z.G. and Z.J.; Data curation, Z.G. and Z.J.; Formal analysis, Z.J. and H.L.; Funding acquisition, Z.J.; Methodology, Z.G. and Z.J.; Validation, Z.G. and H.L.; Investigation, Z.G., and Z.J.; Writing—original draft, Z.G. and Z.J.; Writing—review and editing, Q.W. and H.L.; Supervision, Q.W.; Visualization, Z.G. and H.L. All authors have read and agreed to the published version of the manuscript.

Funding: This research was funded by National Natural Science Foundation of China (52107100), Natural Science Foundation of Jiangsu Province (BK20190710), General Project of Natural Science Research in Colleges and Universities of Jiangsu Province (19KJD470004) and Key Research and Development Program of Jiangsu Province (BE2020081-4).

Institutional Review Board Statement: Not applicable.

Informed Consent Statement: Not applicable.

Data Availability Statement: Not applicable.

Conflicts of Interest: The authors declare no conflict of interest.

Appendix A

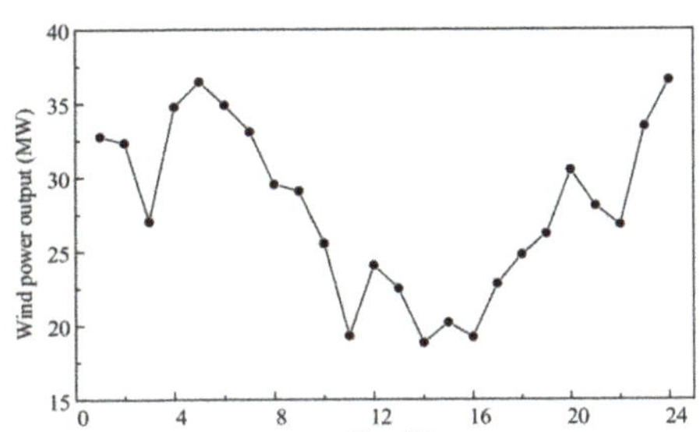

Figure A1. Day-ahead wind power output curve.

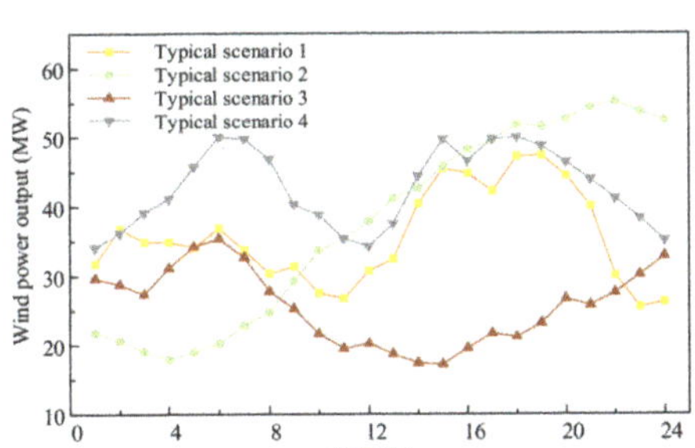

Figure A2. Several typical day-ahead predicted wind power curves.

References

1. Fragkos, P.; Fragkiadakis, K.; Paroussos, L. Reducing the decarbonisation cost burden for EU energy-intensive industries. *Energies* **2021**, *14*, 236. [CrossRef]
2. Ge, P.; Hu, Q.; Wu, Q.; Dou, X.; Wu, Z.; Ding, Y. Increasing operational flexibility of integrated energy systems by introducing power to hydrogen. *IET Renew. Power Gener.* **2020**, *14*, 372–380. [CrossRef]
3. Jiang, Y.; Guo, L. Research on wind power accommodation for an electricity-heat-gas integrated microgrid system with power-to-gas. *IEEE Access* **2019**, *7*, 87118–87126. [CrossRef]
4. Yang, Z.; Gao, C.; Zhao, M. The optimal investment strategy of P2G based on real option theory. *IEEE Access* **2020**, *8*, 127156–127166. [CrossRef]
5. Zhang, Z.; Zhang, Y.; Huang, Q.; Lee, W. Market-oriented optimal dispatching strategy for a wind farm with a multiple stage hybrid energy storage system. *CSEE J. Power Energy Syst.* **2018**, *4*, 417–424. [CrossRef]
6. Xu, D.; Wu, Q.; Zhou, B.; Li, C.; Bai, L.; Huang, S. Distributed multi-energy operation of coupled electricity, heating and natural gas networks. *IEEE Trans. Sustain. Energy* **2020**, *11*, 2457–2469. [CrossRef]
7. Li, Y.; Liu, W.; Shahidehpour, M.; Wen, F.; Wang, K.; Huang, Y. Optimal operation strategy for integrated natural gas generating unit and power-to-gas conversion facilities. *IEEE Trans. Sustain. Energy* **2018**, *9*, 1870–1879. [CrossRef]
8. Chen, Z.; Zhang, Y.; Ji, T.; Cai, Z.; Li, L.; Xu, Z. Coordinated optimal dispatch and market equilibrium of integrated electric power and natural gas networks with P2G embedded. *J. Mod. Power Syst. Clean Energy* **2018**, *6*, 495–508. [CrossRef]
9. Yang, J.; Zhang, N.; Cheng, Y.; Kang, C.; Xia, Q. Modeling the operation mechanism of combined P2G and gas-fired plant with CO2 recycling. *IEEE Trans. Smart Grid* **2019**, *10*, 1111–1121. [CrossRef]
10. Liu, J.; Sun, W.; Yan, J. Effect of P2G on flexibility in integrated power-natural gas-heating energy systems with gas storage. *Energies* **2021**, *14*, 196. [CrossRef]
11. Wang, C.; Dong, S.; Xu, S.; Yang, M.; He, S.; Dong, X.; Liang, J. Impact of power-to-gas cost characteristics on power-gas-heating integrated system scheduling. *IEEE Access* **2019**, *7*, 17654–17662. [CrossRef]
12. Yan, B.; Fan, H.; Luh, P.B.; Moslehi, K.; Feng, X.; Yu, C.; Bragin, M.; Yu, Y. Grid integration of wind generation considering remote wind farms: Hybrid markovian and interval unit commitment. *IEEE/CAA J. Autom. Sin.* **2017**, *4*, 205–215. [CrossRef]
13. Naz, M.N.; Imtiaz, S.; Bhatti, M.K.L.; Awan, W.Q.; Siddique, M.; Riaz, A. Dynamic stability improvement of decentralized wind farms by effective distribution static compensator. *J. Mod. Power Syst. Clean Energy* **2020**. [CrossRef]
14. Zhang, N.; Wang, Y.; Kang, C.; Cheng, J.; He, D. Blockchain technique in the energy internet: Preliminary research framework and typical application. *Proc. CSEE* **2016**, *36*, 4011–4023.
15. Andoni, M.; Robu, V.; Flynn, D.; Abram, S.; Geach, D.; Jenkins, D.; McCallum, P.; Peacock, A. Blockchain technology in the energy sector: A systematic review of challenges and opportunities. *Renew. Sustain. Energy Rev.* **2019**, *100*, 143–174. [CrossRef]
16. Wang, B.; Dabbaghjamanesh, M.; Kavousi-Fard, A.; Mehraeen, S. Cybersecurity enhancement of power trading within the networked microgrids based on blockchain and directed acyclic graph approach. *IEEE Trans. Ind. Appl.* **2019**, *55*, 7300–7309. [CrossRef]
17. Tschorsch, F.; Scheuermann, B. Bitcoin and beyond: A technical survey on decentralized digital currencies. *IEEE Commun. Surv. Tutor.* **2016**, *18*, 2084–2123. [CrossRef]
18. Ding, W.; Wang, G.; Xu, A.; Chen, H.; Hong, C. Research on key technologies and information security issues of energy blockchain. *Proc. CSEE* **2018**, *38*, 1026–1034.
19. di Silvestre, M.L.; Gallo, P.; Guerrero, J.M.; Musca, R.; Sanseverino, E.R.; Sciumè, G.; Vásquez, J.C.; Zizzo, G. Blockchain for power systems: Current trends and future applications. *Renew. Sustain. Energy Rev.* **2020**, *119*, 109585. [CrossRef]
20. Ahl, A.; Yarime, M.; Goto, M.; Chopra, S.S.; Kumar, N.M.; Tanaka, K.; Sagawa, D. Exploring blockchain for the energy transition: Opportunities and challenges based on a case study in Japan. *Renew. Sustain. Energy Rev.* **2020**, *117*, 109488. [CrossRef]
21. Mengelkamp, E.; Gärttner, J.; Rock, K.; Kessler, S.; Orsini, L.; Weinhardt, C. Designing microgrid energy markets a case study: The Brooklyn microgrid. *Appl. Energy* **2018**, *210*, 870–880. [CrossRef]
22. Guo, Z.; Ji, Z.; Wang, Q. Blockchain-enabled demand response scheme with individualized incentive pricing mode. *Energies* **2020**, *13*, 5213. [CrossRef]
23. Hou, W.; Guo, L.; Ning, Z. Local electricity storage for blockchain-based energy trading in industrial internet of things. *IEEE Trans. Ind. Inform.* **2019**, *15*, 3610–3619. [CrossRef]
24. Zou, W.; Lo, D.; Kochhar, P.S.; Le, X.D.; Xia, X.; Feng, Y.; Chen, Z.; Xu, B. Smart contract development: Challenges and opportunities. *IEEE Trans. Softw. Eng.* **2019**, *47*, 2084–2106. [CrossRef]
25. Zhou, Z.; Wang, B.; Guo, Y.; Zhang, Y. Blockchain and computational intelligence inspired incentive-compatible demand response in internet of electric vehicles. *IEEE Trans. Emerg. Top. Comput. Intell. 2019. 3*, 205–216. [CrossRef]
26. van Cutsem, O.; Dac, D.H.; Boudou, P.; Kayal, M. Cooperative energy management of a community of smart-buildings: A blockchain approach. *Int. J. Electr. Power Energy Syst.* **2020**, *117*, 105643. [CrossRef]
27. Shahzad, B.; Crowcroft, J. Trustworthy electronic voting using adjusted blockchain technology. *IEEE Access* **2019**, *7*, 24477–24488. [CrossRef]
28. Zhao, D.; He, Q.; Yu, J.; Jiang, J.; Li, X.; Ni, M. Dynamic behaviour and control strategy of high temperature proton exchange membrane electrolyzer cells (HT-PEMECs) for hydrogen production. *Int. J. Hydrogen Energy* **2020**, *45*, 26613–26622. [CrossRef]

29. Zhang, Z.; Xing, X. Simulation and experiment of heat and mass transfer in a proton exchange membrane electrolysis cell. *Int. J. Hydrogen Energy* **2020**, *45*, 20184–20193. [CrossRef]
30. Gorre, J.; Ruoss, F.; Karjunen, H.; Schaffert, J.; Tynjälä, T. Cost benefits of optimizing hydrogen storage and methanation capacities for Power-to-Gas plants in dynamic operation. *Appl. Energy* **2020**, *257*, 113967. [CrossRef]
31. Parra, D.; Zhang, X.; Bauer, C.; Patel, M.K. An integrated techno-economic and life cycle environmental assessment of power-to-gas systems. *Appl. Energy* **2017**, *193*, 440–454. [CrossRef]
32. Song, J.G.; Moon, S.J.; Jang, J.W. A scalable implementation of anonymous voting over Ethereum blockchain. *Sensors* **2021**, *21*, 3958. [CrossRef] [PubMed]
33. Seifelnasr, M.; Galal, H.S.; Youssef, A.M. Scalable Open Vote Network on Ethereum. Available online: http://fc20.ifca.ai/wtsc/WTSC2020/WTSC20_paper_10.pdf (accessed on 12 September 2021).
34. Sun, H.; Meng, J.; Peng, C. Coordinated Optimization Scheduling of Multi-region Virtual Power Plant with Wind-power/Photovoltaic/Hydropower/Carbon-Capture Units. *Power Syst. Technol.* **2019**, *43*, 4040–4051.
35. Yang, Z.; Zhang, F.; Liang, J.; Han, X.; Xu, Z. Economic Generation Scheduling of CCHP Microgrid with Heat Pump and Energy Storage. *Power Syst. Technol.* **2018**, *42*, 1735–1743.
36. Chen, Z.; Zhang, Y.; Ji, T.; Li, C.; Xu, Z. Economic dispatch model for wind power integrated system considering the dispatchability of power to gas. *IET Gener. Trans. Distrib.* **2019**, *13*, 1535–1544. [CrossRef]
37. Yu, J.; Shi, Q.; Yang, Z.; Dai, W.; Wang, X. Day-ahead Scheduling Method of Power-to-gas System Considering Operation Characteristics of Water Electrolysis and Methanation. *Autom. Electr. Power Syst.* **2019**, *43*, 18–25.
38. Zhang, X.; Bauer, C.; Mutel, C.L.; Volkart, K. Life cycle assessment of power-to-gas: Approaches, system variations and their environmental implications. *Appl. Energy* **2017**, *190*, 326–338. [CrossRef]
39. Aminu, M.D.; Nabavi, S.; Rochelle, C.A.; Manovic, V. A review of developments in carbon dioxide storage. *Appl. Energy* **2017**, *208*, 1389–1419. [CrossRef]

energies

MDPI

Article

Is Green Recovery Enough? Analysing the Impacts of Post-COVID-19 Economic Packages

Pedro R. R. Rochedo [1], Panagiotis Fragkos [2], Rafael Garaffa [1,3], Lilia Caiado Couto [4], Luiz Bernardo Baptista [1,*], Bruno S. L. Cunha [1], Roberto Schaeffer [1] and Alexandre Szklo [1]

[1] Centre for Energy and Environmental Economics (Cenergia), Energy Planning Programme (PPE), COPPE, Universidade Federal do Rio de Janeiro, Brazil-Centro de Tecnologia, Sala I-034, Cidade Universitária, Rio de Janeiro 21941-972, Brazil; pedrorochedo@ppe.ufrj.br (P.R.R.R.); rafael.garaffa@ec.europa.eu (R.G.); slcunha.bruno@ppe.ufrj.br (B.S.L.C.); roberto@ppe.ufrj.br (R.S.); szklo@ppe.ufrj.br (A.S.)

[2] E3 Modelling, 70-72 Panormou Street, PO 11523 Athens, Greece; fragkos@e3modelling.com

[3] European Commission, Joint Research Centre-JRC, 3-41092 Seville, Spain

[4] The Bartlett School of Environment, Energy and Resources, University College London, Central House, 14 Upper Woburn Place, London WC1H 0NN, UK; lilia.couto@ucl.ac.uk

* Correspondence: luizbernardo@ppe.ufrj.br

Abstract: Emissions pathways after COVID-19 will be shaped by how governments' economic responses translate into infrastructure expansion, energy use, investment planning and societal changes. As a response to the COVID-19 crisis, most governments worldwide launched recovery packages aiming to boost their economies, support employment and enhance their competitiveness. Climate action is pledged to be embedded in most of these packages, but with sharp differences across countries. This paper provides novel evidence on the energy system and greenhouse gas (GHG) emissions implications of post-COVID-19 recovery packages by assessing the gap between pledged recovery packages and the actual investment needs of the energy transition to reach the Paris Agreement goals. Using two well-established Integrated Assessment Models (IAMs) and analysing various scenarios combining recovery packages and climate policies, we conclude that currently planned recovery from COVID-19 is not enough to enhance societal responses to climate urgency and that it should be significantly upscaled and prolonged to ensure compatibility with the Paris Agreement goals.

Keywords: COVID-19; economic recovery; stimulus packages; climate scenarios; integrated assessment modelling

Citation: Rochedo, P.R.R.; Fragkos, P.; Garaffa, R.; Couto, L.C.; Baptista, L.B.; Cunha, B.S.L.; Schaeffer, R.; Szklo, A. Is Green Recovery Enough? Analysing the Impacts of Post-COVID-19 Economic Packages. *Energies* **2021**, *14*, 5567. https://doi.org/10.3390/en14175567

Academic Editor: Luigi Aldieri

Received: 25 July 2021
Accepted: 1 September 2021
Published: 6 September 2021

Publisher's Note: MDPI stays neutral with regard to jurisdictional claims in published maps and institutional affiliations.

1. Introduction

The impact of the COVID-19 pandemic on climate change mitigation will ultimately depend on long-term trajectory shifts caused by economic recovery [1]. The emission reduction rate observed during the restrictive confinement period in the first half of 2020 is broadly comparable to the annual emission reduction rate needed to achieve the 1.5 °C target [2]. However, the sharp 7% drop in emissions experienced during 2020 is likely to reflect only the very short term, not causing any lasting effect since the previous fossil fuel-based infrastructure is still in place and could rapidly return to full capacity [3,4]. IEA [5] has predicted a major surge in CO_2 emissions from the energy sector in 2021, as the world rebounds from the pandemic via accelerating rollouts of COVID-19 vaccinations in several countries and extensive fiscal responses to the economic crisis. Emissions pathways after COVID-19 will be shaped by how economic responses translate into infrastructure expansion, energy use, investment planning and societal changes.

The urgency to curb greenhouse-gas (GHG) emissions and attain the Paris Agreement temperature goals is now at risk of being overlooked by the need for an economic response to the COVID-19 pandemic crisis. The economy-wide recession has led to a steep decrease in oil and gas prices and a widely agreed need for governments to intervene with substantial economic stimulus [6], which could propel or undermine the energy transition, depending on future investment profiles [7].

Arguably, both the climate crisis and the pandemic-related crisis should be tackled at once through a low-carbon economic response, by ensuring that large funding is directed to clean energy [1]. As a response to the COVID-19 crisis, most governments worldwide have launched recovery packages aiming to boost their economies, support employment and enhance their competitiveness [8]. Climate action is embedded in most of these packages, but with sharp differences at regional level.

The European Union has launched a EUR 750 billion recovery package, from which at least 30% of expenditure is committed to mainstreaming climate action [9]. The United States Biden administration, similarly, has launched a "Build Back Better plan" which aims at canalising USD two trillion into low-carbon investment, including USD 400 billion directly to clean energy over the next ten years [10]. In contrast, an economic recovery based on low oil prices, such as the stimulus announced by Indonesia, Turkey and Russia [11], and investment in traditional infrastructure would hinder progress towards limiting global temperature rise and would increase the risk of locking our economies into a high-emission trajectory.

Climate change research addresses long-term impacts of current and mid-term decision-making through modelling to respond to "what if" questions. It assesses the long-term impacts of policies and societal changes over emissions and consequent temperature changes. Scenarios play a key role as long-term research tools for the transition to a low-carbon world. The analysis of common scenarios using multiple modelling frameworks allows the research community to produce integrated and comparable analyses of climate change impacts, adaptation and mitigation [12,13]. Providing shared scenarios is crucial to promoting interactions among disciplines and research interests, in order to make conclusions compatible and consistent across the literature, thus allowing easier communication of modelling results, as well as reducing scattered individual efforts towards elaborating consistent assumptions for their own scenarios [13].

While the scenario framework used by the Intergovernmental Panel on Climate Change (IPCC) and other authors [12,14–18] still serves as the basis for future narratives and mitigation pathways, COVID-19 raises a substantial policy shift, which impacts mitigation in the long-run, as it changes the core socio-economic assumptions underpinning these scenarios, the investment planning in various countries and (potentially) consumer behaviour (e.g., through reduced air transport and increased home working). The climate research community will therefore have to update scenarios reflecting such trade-offs in order to analyse future pathways from the COVID-19 pandemic onwards and inform policy debate on appropriate ways of allocating recovery funds.

This paper draws on the existing IPCC scenario framework [14,15] to advance the field by including potential long-term impacts of policy responses to what is plausibly believed to be the harshest societal crisis of the century: the COVID-19 pandemic. We provide novel evidence on the energy system and emission implications of post-COVID-19 recovery packages by revealing the wide gap between pledged recovery packages and the actual investment needs of the energy transition. We test the hypothesis that currently planned recovery from COVID-19 will undermine the response to climate urgency by modelling post-COVID-19 scenarios until 2050 through two different modelling frameworks: the COFFEE-TEA and the PROMETHEUS IAMs [19].

2. Materials and Methods

The following sections describe the modelling frameworks and the scenarios designed for this study, together with our analyses on current recovery packages.

2.1. Modelling Frameworks

This study uses two different modelling frameworks to assess the impacts of green recovery packages. The COFFEE-TEA IAM suite of models [20] comprehends a bottom-up, partial equilibrium, global model for the energy and land systems (COFFEEComputable Framework for Energy and the Environment) soft-linked to a global Computable General Equilibrium (CGE) model; the Total Economy Assessment (TEA) model. COFFEE represents the optimal pathway for the interaction and uptake of technologies and energy sources to meet a given demand for energy services, by minimising the total cost of the system from pre-established policy restrictions (Rochedo, 2016). The model captures the evolution of sectors such as energy, industrial processes, AFOLU, waste and others and their respective GHG emissions until 2100, including a detailed representation of energy resources, extraction and conversion technologies for each region, both in terms of volume and costs. TEA is a multi-regional and multi-sectoral model that represents the production and trade of goods, capturing industry-to-industry linkages, in the global economy [21]. TEA follows the standard microeconomic optimisation framework, assuming total market clearance and perfect competition. The TEA model provides consistent macroeconomic pathways, projecting future economic activities' demands to COFFEE, while COFFEE improves the representation of energy markets in TEA, given their compatibility in terms of base year data, sectoral and regional disaggregation.

PROMETHEUS is a comprehensive energy system model focusing on technology uptake analysis, energy price projections, and assessment of climate policies [22,23]. It captures the interactions between energy demand and supply at regional and global level and provides detailed projections of fuel mix in energy consumption, electricity production mix by technology, carbon emissions, energy prices and investment to the future. PROMETHEUS can provide medium and long term energy system projections up to 2050, in both the demand and the supply sides, under different policy and technology scenarios.

Most importantly, the modelling frameworks can be used for the impact assessment of energy and environment policies at regional and global levels, including price signals, such as carbon or energy taxation, subsidies, technology and energy efficiency promoting policies, Renewable Energy Systems (RES) supporting policies, and technology standards [22,24]. The modelling frameworks are therefore designed to address the questions about the short-, medium- and long-term effects of post-COVID-19 economic recovery based on long-term scenarios for global GHG emissions, capturing the extent to which pledged recovery packages manage to avoid carbon lock-in given key assumptions that drive investment in the energy system (e.g., oil prices, cost of technologies, efficiency, lifespan). For detailed information about the modelling frameworks, see Appendix A.

2.2. Scenario Design: COVID-19 Economic Recovery Packages Screening and Modelling

We depart from a baseline (CurPol) scenario framed within the Shared Socioeconomic Pathway—SSP2 "middle of the road" [25] rationale, but applying short-term regional GDP growth shocks due to COVID-19. We use short-term projections of the COVID-19 pandemic impact from the International Monetary Fund World Economic Outlook updated in October 2020 [26] and the OECD Economic Outlook of December 2020 [27]. The CurPol scenario does not comprise any additional economic recovery policy or climate policy apart from the policy framework currently in place, which is described in detail in [28]. From 2025 onwards, the SSP2 GDP growth rates are applied.

In order to design scenarios reflecting policies launched as a response to the COVID-19 economic crisis, we screened policy packages announced up to May 2021 for investment in three main technology groups related to low-carbon transition: Power generation, Energy Efficiency and Transport. For this purpose, we assessed government plans and tools created specifically to analyse the greenness and brownness of post-COVID-19 stimulus, namely: the Green Recovery Plan Tracker [29], the Energy Policy Tracker [11], the Climate Action Tracker [30] and the Greenness Stimulus Index [31]. When regional trend data are needed, the IEA Country Statistics [30] are used.

Markedly, the European Union and the United Kingdom led in terms of launching green recovery plans still in 2020. The Next Generation EU Recovery Plan, consistent with the European Green Deal, commits at least 30% of its EUR 750 billion budget to climate action, while the remaining 70% should follow the principle to "do no harm" to the environment (European Council Conclusions, 17–21 July 2020—Consilium, n.d.). At the same time, countries such as France and Germany, as well as the UK, outperform in the greenness of their stimulus packages, with a net positive impact towards climate action [31]. Additionally, the Energy Policy Tracker traced no commitment to direct fossil fuel support from the European Commission, in contrast with a USD 385.36 commitment to clean energy investment [11].

China still faces major uncertainties regarding the emission profile of its economic recovery plans. While China has announced a target for net zero carbon emissions by 2060, and committed additional USD 22 billion to clean energy investment when compared to fossil fuel energy [11], it still plans to install as many new GW of coal power plants as its previous trajectory [31].

The US has notably the largest economic stimulus package in the world. In the early stages of the pandemic, the US administration pledged USD 2.98 trillion of public expenditure, which included environmental measures for the power, industry, manufacturing and transport sectors, involving, for example, penalty exemptions. The US overall energy investment commitment originally included USD 72.35 billion to oil, oil products and coal, and USD 27.27 billion to support clean energy, mostly directed to biofuels and wind power [11]. A clear shift took place when, in 2021, the Biden administration committed to "Make a historic investment in clean energy and innovation", pledging an additional USD 400 billion to renewable energy investment [10]

Economies that heavily rely on fossil fuel exploitation such as Russia and Middle Eastern countries unsurprisingly indicate a fully brown recovery [11,31]. The remaining world regions seem to show rather dubious stimulus profiles, with investment directed both ways, but mostly showing brown net impacts [32,33].

Having screened national and regional policy packages for the post-COVID-19 pandemic economic responses, we translate them into assumptions for each of the scenarios and their main policy instruments (Table 1). The Recovery Packages scenario (RecPac) assumes the implementation of plans for investments on a portfolio of green energy options in different countries, amounting USD 1 trillion over the 2020–2025 period. In both IAMs, green recovery packages are implemented as investment subsidies to low-carbon technologies, including solar PV, wind, electric vehicles, biofuels, heat pumps and efficiency measures. The implementation of subsidies incentivises the uptake of clean energy technologies in power production, transport and buildings sectors.

Given that economic recovery packages comprise broader sectoral coverage than solely green energy and that investment in infrastructure requires longer maturity, we further assess the implications of a 5-fold increase in green energy investments as compared to the RecPac scenario. We call it the Enhanced Recovery scenario (EnhRec), where the total amount invested in green energy reaches approximately USD 5 trillion over the 2020–2025 period, which is in line with the 3-year extension of the recovery packages found in [8]. The scenario conceptualises a situation of prolonged needs for recovery packages, given that most countries face challenges to fight new COVID-19 variants, upscale vaccination rates and boost their economies. It gives an indication of how much investment in green energy is required in order to support the energy transition.

To assess the ambition gap of the recovery packages in previous scenarios we simulate a Climate Ambition scenario (CliAmb) that is based on a remaining carbon budget of 600 $GtCO_2$ over 2018–2100, considered compatible with a 1.5 °C average global warming by 2100 without temperature overshoot [34]. In this scenario, we simulate an economy-wide, global carbon market, in the form of an emission trading system in TEA, with the resulting carbon prices taken as input to COFFEE.

Table 1. Summary of policy scenarios.

Scenario	Tag	Policy Instruments	Description
Baseline	CurPol	Current policies	Current energy and climate policies. Short-term COVID-19 socio-economic impacts are included, but recovery packages are not.
Recovery Packages	RecPac	CurPol + Direct investment, subsidies	Recovery packages implemented as investment in green energy technologies reflecting national policies announced up to May 2021
Enhanced Recovery	EnhRec	CurPol + Direct investment, subsidies	Green energy investments are increased by 5 times as compared to the RecPac scenario to cover the 2021–2025 period.
Climate Ambition	CliAmb	Carbon pricing	Long-term pathways consistent with a well below 2 °C average global warming by 2100 based on a carbon budget of 600 GtCO2 over 2018–2100 without temperature overshoot.
Global Governance	GloGov	CurPol + Direct investment, subsidies	Total amount of recovery packages announced up to May 2021 implemented as investment in green energy technologies globally (modelling framework optimal choice).

Finally, we also account for inter-regional disparities by simulating a Global Governance scenario (GloGov) in which the total amount of green recovery funds is allocated globally (i.e., investments are not restricted to each region). We acknowledge the fact that a mechanism of global governance is extremely difficult to be implemented in the context of the COVID-19 pandemic. Therefore, the results of the GloGov scenario should be interpreted as a hypothetical exercise, reflecting the global least-cost optimal solution of the modelling framework, given the green energy technological portfolio included in the two IAMs. The models therefore allocate the total sum of each pledged recovery package to choose the optimal set of technologies and their locations.

We translate the green recovery packages into variables and parameters to be simulated in the modelling framework. The recovery packages were inserted in the modelling tools by changing specific parameters depending on model formulation, in particular by imposing additional investment in low-carbon technologies exogenously or by inserting subsidy rates in the capital costs to reduce the purchase price and accelerate the deployment of mitigation options. We start by allocating the amount of packages to sectors following the allocation proposed by the IEA (2020) [7]. In particular:

- 33% of the total amount goes to power generation, mostly in renewable energy technologies (wind and solar) but also to grid enhancements to support the increased uptake of variable renewable sources;
- 30% of the total amount is directed to low-emission transport modes, mostly in the purchase of electric cars;
- 30% of the total amount goes to increase energy efficiency and electrification of buildings; and
- The remaining 7% is directed to increase energy efficiency in industrial sectors.

After setting the sectoral allocation, we define what instruments are used in each sector. Here, our choice is somehow limited by the modelling framework—typically, bottom-up models with rich technological details—so we mainly explore supply-side instruments, not including demand-side instruments that could play a role in a green recovery context (e.g., consumer behaviour, digital services, lifestyle changes).

On the supply-side instruments, we therefore rely on direct investments for the expansion of renewable energy, mostly to wind and solar PV, as well as to grid enhancements to support the increased penetration of variable RES; subsidies on the purchase of electric vehicles and other zero-emission alternatives in the transport sector; and direct incentives through reduced prices of efficient equipment purchases and subsidise costs to increase renovation rates and accelerate the deployment of heat pumps and other low-emission options in buildings.

3. Results

In this section, we present the results of the different policy scenarios. The IAMs depart from similar/comparable but different baselines (CurPol), and so the modelling results should be interpreted in relative terms when comparing them across the modelling frameworks.

3.1. Policy Scenarios (National Pledges)

Figure 1 describes the global CO_2 emissions pathway of each modelling framework by scenario from 2020 to 2050. In the RecPac scenario, COFFEE shows a small decrease in global emissions between 2020 and 2025, mostly reflecting short-term effects of the investment in green energy. In the absence of additional stimulus to green energy, this trend is, however, reversed from 2025 onwards, with emissions returning to the original pathway of the CurPol scenario and achieving 34.7 $MtCO_2$ in 2050. The emissions trajectory in PROMETHEUS presents similar behaviour as COFFEE in RecPac, particularly after 2025, with the model reaching 40.9 $MtCO_2$ in 2050, showing a decline of 1–2 Gt annually from CurPol over 2020–2050. PROMETHEUS shows a larger reduction in global emissions from CurPol levels in the short-term (by 2025), induced by the implementation of green recovery measures as investment subsidies stimulating the increased uptake of renewable energy, electric vehicles, and energy efficiency.

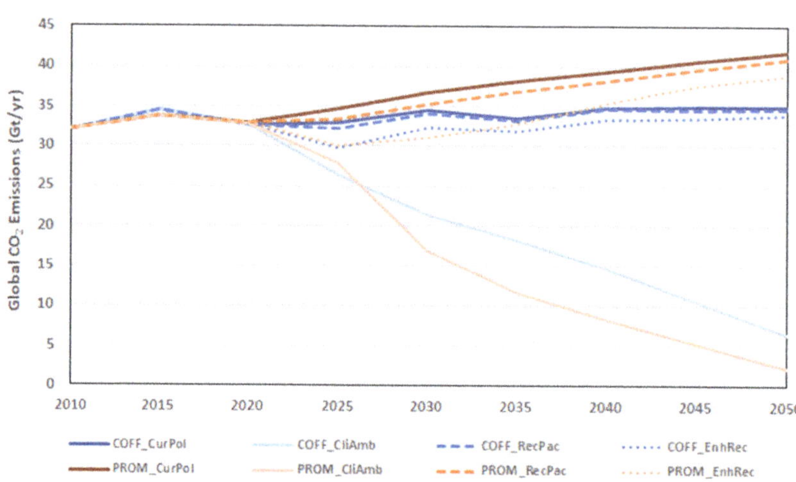

Figure 1. Global CO_2 emissions pathway over 2010–2050.

In the EnhPac scenario, the additional investment in green energy leads to larger mid-term effects in terms of emissions mitigation in both models—with global emissions declining by 9.6–13.2% in 2025 and 6.2–15.3% in 2030 from Cur Pol levels. This shows that the prolongation of green recovery packages can support further emission reductions and partly close the emissions gap with the cost-optimal pathway to 1.5 °C in 2030. However, if not combined with ambitious climate policy, alone they are not sufficient to trigger structural changes towards net zero by mid-century, with global emissions amounting to 34.0–38.9 $GtCO_2$ across models in 2050, which is clearly not compatible with the goal of carbon neutrality by 2050.

Figure 2 presents the ambition gap for different scenarios in 2030. The ambition gap accounts for the difference in global CO_2 emissions between the policy scenarios (RecPac and EnhPac) and the more ambitious mitigation scenario compatible with the Paris Agreement goal of 1.5 °C (CliAmb). The implementation of recovery packages results in limited emission reductions, thus closing only a small part of the emission gap from the 1.5 °C cost-optimal pathway in 2030 (3–7% across models in the RecPac scenario). The Enhanced Recovery scenario leads to larger mitigation, closing 16–29% (across models) of the ambition gap in 2030.

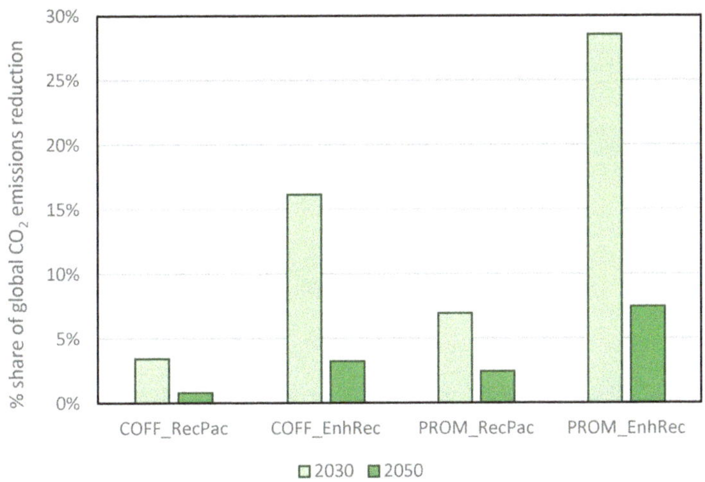

Figure 2. Closing the ambition gap—Share of global CO_2 emissions reduction from CurPol levels achieved in RecPac and EnhRec scenarios compared to reductions required to achieve the 1.5 degree target in a cost-optimal way in 2030 and 2050.

The impacts by 2050 are even smaller, with recovery packages representing about 1–7% of the overall effort towards the Paris Agreement goal of 1.5 °C. The impacts of green recovery packages vanish in the longer term, as in the absence of strong climate policy signals for investment in green energy and reducing fossil fuel use beyond 2025, emission pathways return to their CurPol trends with limited reductions until 2050.

Mitigation in policy scenarios comes as a consequence of changes in the energy system, triggered by the increased deployment of renewable energy, energy efficiency, low-carbon fuels and electrification of energy services [22]. Figure 3 presents the results of both modelling frameworks under alternative policy scenarios in 2030 and 2050 for: (a) final energy consumption; (b) changes in total energy use of the transport sector; (c) the share of renewables in electricity generation; and (d) the global emission factor of electricity generation (CO_2 emissions per MWh produced). The bars in Figure 3 describe the results for the ambition gap—i.e., the difference between CliAmb and CurPol scenarios—while the empty dot and the filled dot represent the levels achieved in RecPac and EncRec scenarios, respectively.

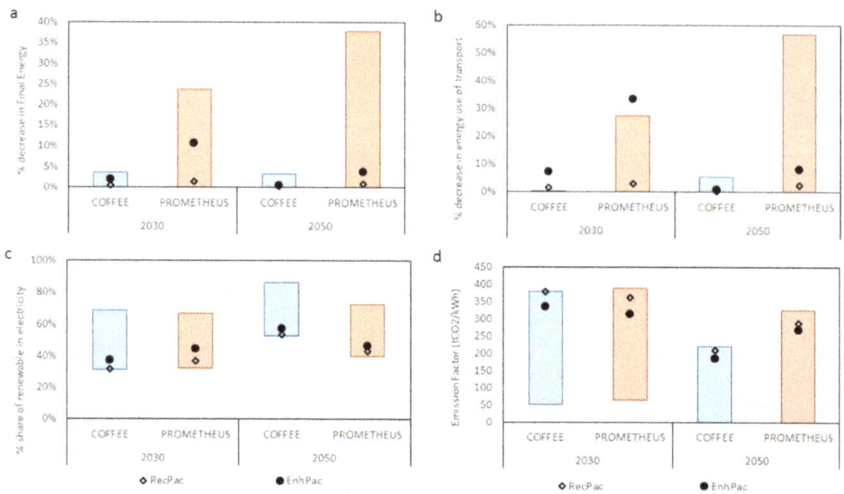

Figure 3. Energy system transformation—decrease in final energy consumption (**a**); decrease in energy use of transport (**b**); share of renewables in electricity generation (**c**); and global emission factor of electricity generation (**d**) in 2030 and 2050.

Figure 3a shows low to moderate changes in final energy consumption in RecPac and EnhPac scenarios (−0.2% to 10.6%). Final energy use in COFFEE shows a more similar trajectory in these scenarios than in CliAmb, while PROMETHEUS presents a more substantial reduction in final energy consumption in the long-term, particularly due to energy efficiency measures and to a more rapid electrification of the transport sector.

As illustrated in Figure 3b, PROMETHEUS shows a reduction in transport-related energy consumption of nearly −33% in 2030, while the penetration of electric vehicles in COFFEE is more moderate, which, combined with a greater use of biofuels, leads to a reduction of around −7% in 2030. Given the lack of long-term climate policies to increase the uptake of low and zero-emission vehicles, the projected reduction in energy consumption in transport declines over time, ranging from −0.1% to −8.1% in 2050.

Nonetheless, results suggest that the green recovery packages promote a greater transformation in the power sector, in particular due to a fast increase in wind and solar PV electricity generation. In both RecPac and EnhRec scenarios, the share of renewable energy in electricity production reaches substantial levels in 2030 (32–37% in RecPac and 38–44% in EnhPac), lying within the projected range of the CliAmb scenario. Although pushed by the green recovery packages, results confirm that the penetration of renewables in electricity generation is not solely driven by the packages, and a greater share than in 2030 is reached by mid-century driven by technology cost reduction and increased adoption of renewable energy technologies (43–54% in RecPac and 47–57% in EnhRec).

The transformation of the energy system can also be illustrated by the global emission factors of electricity generation (Figure 3d). Over 2030–2050, emission factors decrease from a range of 313–380 $MtCO_2/MWh$ to 185–287 $MtCO_2/MWh$ as a result of the decarbonisation of the power system, showing substantial decrease as compared to 2015 (485–576 $MtCO_2/MWh$). However, although they are in the upper ranges of the CliAmb scenario, these levels are far from meeting the lower bounds in 2030 (52–64 $MtCO_2/MWh$) or even zeroing emissions in 2050 as required to meet the Paris Agreement goals of 1.5 °C.

Closing the ambition gap comes at different costs across the modelling frameworks. Figure 4 presents the results for the green energy investment required to close the gap in 2030 and 2050. In the horizontal axis, the investment gap is the level of cumulative investment (present value (PV) in 2020 of the level of investment over 2020–2030, discounted at a 5% p.y. rate) in policy scenarios compared to the level of the more ambitious mitigation scenario (CliAct). In the vertical axis, the emission gap is shown. In Figure 4, numbers for

RecPac and EnhRec scenarios are summarised as COFFEE and PROMETHEUS Standard, while results for the GloGov scenario appear as COFFEE Global Optimal.

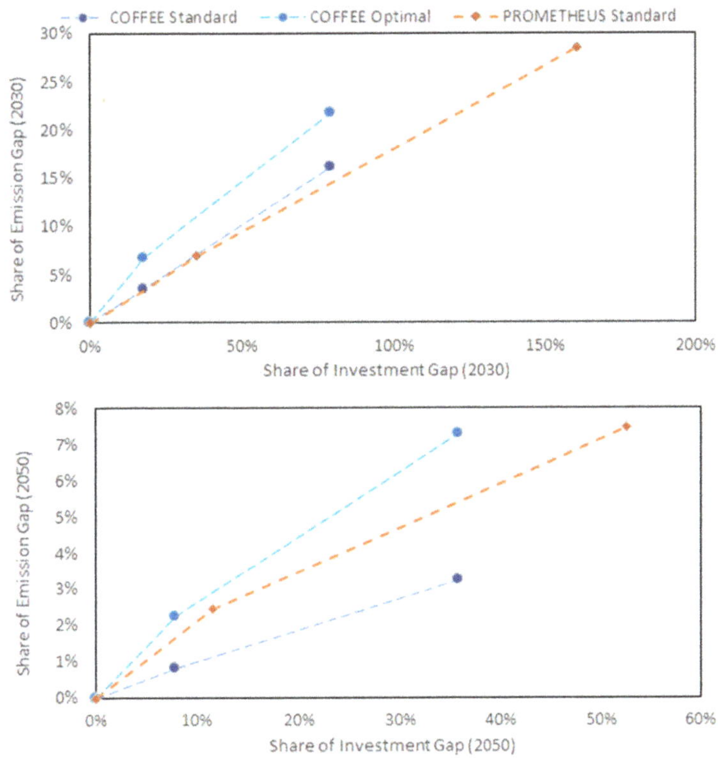

Figure 4. Ambition gap and investment gap in 2030 and 2050.

Green recovery packages close a high fraction of the investment gap in 2030 (17–35% in RecPac and 79–116% in EnhRec), but a relatively smaller part of the emission gap (3–7% in RecPac and 16–29% in EnhRec). This result suggests that other policy instruments that incentivise changes not only in the investment patterns, but also in the use patterns of energy infrastructure, vehicles, appliances and equipment, are required to achieve greater levels of mitigation (e.g., carbon pricing that penalises the use of fossil fuels).

We also note that, in the CliAmb scenario, carbon prices in the global emission trading system rise from USD 29/tCO_2 to USD 55/tCO_2 over the 2025–2050 period, corresponding to a total revenue of USD 643 billion in 2025 and USD 1082 billion in 2050. The simulation of a comprehensive carbon pricing instrument adds to our analysis of a green economic recovery by providing a few insights. First, the total revenue of the carbon market serves as a proxy of what the figures at play are and how they compare to the amount of the green recovery packages announced. For instance, green recovery amounts to USD 1 trillion, while our simulations suggest a global carbon market of USD 2.3 trillion over the 2020–2025 period. Second, carbon pricing is widely regarded as a cost-effective instrument by internalising the cost of the pollutants in the prices of goods and services, therefore reducing the costs of the climate policy. As illustrated in Figure 4, despite the substantial effort in closing the investment gap, policy instruments included in the green recovery packages are less efficient in terms of closing the ambition gap. Third, in CliAmb, a global carbon pricing is in place over the full period, highlighting the relevance of long-term signals to abate emissions, in contrast to the instruments included in the green recovery packages that do not provide long-term signals and are discontinued after 2025.

3.2. Global Governance

Differences across the RecPac/EncRec and GloGov results also reveal that going global achieves a greater reduction in worldwide emissions than implementing national green recovery strategies independently. By simulating a hypothetical mechanism of global governance (GloGov scenarios), results suggest a further reduction of 4–6 percentage points in global emissions relative to the scenarios where recovery packages are simulated following national pledges, meaning that a greater share of the emission gap is achieved with the same amount of money invested, thus increasing the overall cost-efficiency of recovery packages. However, recovery packages announced up to May 2021 are highly concentrated in developed regions, particularly in Europe and North America. Investment allocation by regions and sectors in the Global Governance scenario in 2030 is presented in Figure 5. The modelling framework optimal choice leads to a different allocation of recovery funds as compared to the RecPac scenarios, both in terms of sectors and regions of where investment is directed to.

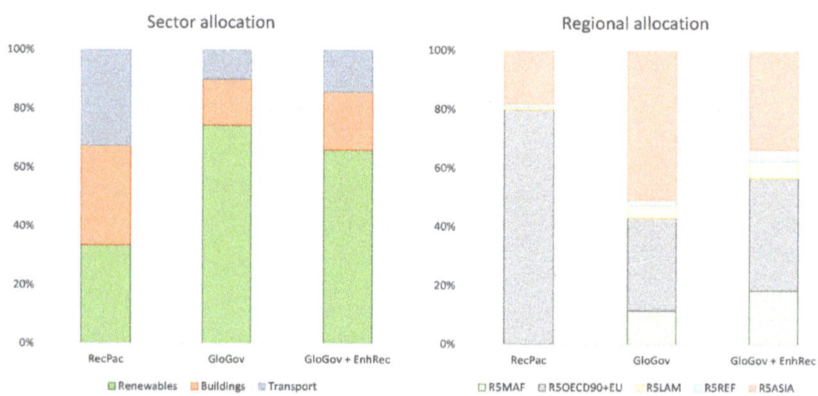

Figure 5. Investment allocation by regions and sectors in Global Governance scenario. Note: R5MAF (Middle East and Africa), R5OECD90 + EU (OECD countries), R5LAM (Latin America), R5REF (Rest of the World), R5ASIA (Asia).

Results indicate that a greater share of green investments would flow to renewables (60–70%) in GloGov and GloGov + EnhRec scenarios (GloGov + EnhRec scenario simulates the Global Governance scenario with the green energy investments being increased by 5 times as compared to the amount of RecPac scenario to cover the 2021–2025 period), decreasing the amount directed to buildings and transport sectors. The optimal solution of the modelling framework is chosen due to having the least-cost abatement opportunities, which are found in the renewable energy sector, especially as solar PV and wind technologies are already cost competitive to fossil fuels in many parts of the world (IEA, 2021) [7].

Most interestingly, the choice is not restricted to sectors, but also includes the regional dimension. The results suggest that the optimal allocation of investments would differ substantially from the initial one where Europe and North America are protagonists; the joint share of OECD economies declines from 80% in the RecPack scenario to 31% in the GloGov case. In both scenarios (GloGov and GloGov + EnhRec), Asia stands as the strongest candidate to where investments should be directed to, given that the least-cost abatement opportunities are placed within this region, resulting in a 35–55% share of total investment. Africa takes a greater share of investments as compared to RecPac scenario, reaching up to 20% of the total investment, while other regions also emerge in the scenarios' results (Latin America and Rest of the World) with lower shares.

4. Discussion

Scenario results show that even an enhanced green recovery strategy would not be enough to close the emission or investment gap in order to shift the global emission pathway consistently with the Paris Agreement temperature goals. A larger and fully green stimulus should be implemented; it is clear that a fossil-based recovery would cause an unaffordable delay to climate action. The IEA [5] projects a sharp rebound in electricity demand of nearly 5% in 2021 and 4% in 2022, with an inevitable rebound in fossil-fuel generation since renewable investments have been postponed by the pandemic. Despite low investment attractiveness and the stranded-asset threat, countries may seek to accelerate fossil fuel production in the context of moderate crude oil prices. The critical post-COVID-19 situation in emerging countries may generate relatively predatory strategies based on mineral extraction and agricultural production [33] with long-term repercussions on land use, fossil fuel use and GHG emissions. A fossil-based post-COVID-19 recovery would create a carbon lock-in, which would delay climate compatible development in those economies.

Our model-based analysis shows that recovery packages stimulating investment in clean energy and energy efficiency can reduce global emissions by 10–13% in 2025 and 6–15% in 2030 relative to the CurPol scenario. So, they can close less than 7% of the emission gap to Paris-compatible pathways in 2030 [35] (and up to 30% if they are enhanced and prolonged for 5 years), but cannot induce the structural changes required to reach global net-zero energy systems by 2050. Current green recovery packages are not enough to deal with climate urgency, but (if upscaled and combined with ambitious climate policies) they can potentially catalyse the transition to net-zero energy emissions by mid-century. A green recovery should therefore include considerably more ambitious climate policies.

Interestingly, results have shown that green recovery packages provide more of an investment gap closure than an emission gap closure (Figure 4). In the enhanced recovery case (EnhRec), in 2030, the resulting level of investment can meet or even exceed projected requirements (in the case of the PROMETHEUS model), while the emission gap closure could reach a maximum of 29%. This could mean that chosen technologies need a large upfront investment to reach a minimum scale, or that infrastructure should be put in place beforehand. It can also mean that combined policies are necessary as demand drivers. As proposed in the CliAmb scenario, a global carbon pricing mechanism, namely an ETS, should be effective as an incentive for such shifts.

Combining green recovery packages (in the form of investment subsidies to low-carbon technologies) with carbon pricing schemes may drive the required medium and long-term system transformations towards net zero by mid-century. Currently pledged recovery packages, if fully green, can propel the post-pandemic economic recovery "doing no harm" to climate ambition. Enhanced packages could probably accelerate economic recovery, and be more successful in closing the emission gap. However, ultimately, combining strengths of recovery packages with carbon pricing could accelerate the technological transition while ensuring post-pandemic economic stimulus. Green recovery packages would avoid redundancies through the creation of green jobs, while carbon pricing sustains mitigation in the longer, necessary, time frame. Therefore, this combination could be a successful way of closing the gap between RecPac/EnhPac and CliAmb scenarios, not only until 2030, but also in the long run. On the one hand, mitigation achieved through green recovery packages can increase the social acceptance of climate policy by reducing the need for high carbon pricing. On the other hand, the introduction of (mild) carbon pricing schemes can increase the effectiveness of green recovery packages in terms of emissions reduction by penalising also the use of fossil fuels and not only investment decisions taken by energy consumers and producers.

Overall, our model-based analysis shows that green recovery packages can accelerate energy system transformation with higher uptake of renewable energy, electric vehicles and energy efficiency until 2030, but cannot deliver the systemic long-term restructuring to pave the way towards carbon neutrality by 2050. Additionally, our analysis makes the case for a hypothetical mechanism of global governance for green stimulus packages. Institutionally challenging as it may be, global optimal allocation of recovery packages yields a larger level of mitigation through larger shares of wind and PV power generation. It could also potentially lead to reducing inequalities, since resources would migrate from Europe and North America to less developed regions, mostly Africa, Latin America and parts of Asia.

5. Conclusions

Investment choices for the post-pandemic recovery will strongly affect the climate trajectory in this century. While most policy packages launched can potentially undermine the response to climate urgency, pursuing a green recovery is the minimum to set the world on track for keeping the Paris Agreement temperature goals within sight.

Emission pathways after COVID-19 will be shaped by how governments' economic response translates into infrastructure expansion, energy use, investment planning and societal changes. As a response to the COVID-19 crisis, most governments worldwide launched recovery packages aiming to boost their economies, support employment and enhance their competitiveness. Climate action is pledged to be embedded in most of these packages, but with substantial geographical heterogeneity. In this paper, we provide novel evidence on the energy system and emission implications of post-COVID-19 recovery packages by assessing the gap between pledged recovery packages and the actual investment needs of the energy transition to reach Paris goals. Using two well-established IAMs and analysing various scenarios combining recovery packages and climate policies, we conclude that the currently planned recovery from COVID-19 is not enough to enhance societal responses to climate urgency and should be significantly upscaled and prolonged to ensure compatibility with the Paris Agreement goals.

We point out that our impact assessment does not account for economy-wide impacts of economic stimulus, sectoral feedbacks, or the effects of money creation through discretionary fiscal policy [36]. Shifts in energy demand caused by societal changes resulting from the pandemic are not considered either, or those related to furlough schemes, which are noticeably concentrated in the very short term. Additionally, many of the policy instruments assessed in our simulations imply structural changes across supply chains (e.g., electrification of road transport). Although these changes are explicitly or implicitly represented in our modelling frameworks, we acknowledge that they are often represented in an aggregated way, which can lead to optimistic assumptions about the penetration rates of technologies. Finally, despite having explored five different scenarios, we have not analysed the combination of recovery packages with a global carbon pricing mechanism to sustain emission reductions in the longer run [37], which would probably represent the next step to expand this study.

The analysis can be significantly expanded in various dimensions that were not fully captured in this paper and could be the source of future works. As observed, recovery packages cover a wider range of measures other than climate policies, such jobs and firms direct support, which are of high relevance for political decision making [38–42]. Assessing the overall socio-economic impacts of recovery packages and possible policy measures to boost the economy and create jobs (e.g., VAT reduction, investment tax reduction, lower social security contributions, etc.), is one dimension to be explored. Other ways to use green recovery packages related to energy transition (e.g., subsidies, grants/loans, low-carbon R&D, procurement, fuel mandates, regulation) could be explored, also considering a broader set of technological options, particularly in sectors where emissions are harder to abate, due to high costs or other barriers. Finally, further improvements can be driven by including additional modelling tools towards a multi-model scenario comparison study, such as in [37], to derive more robust policy recommendations and by including real-world data and estimations on technology allocation of green recovery packages (which differ by country).

Author Contributions: Conceptualization, P.F., R.S. and A.S.; Data curation, L.C.C. and L.B.B.; Formal analysis, P.F., R.G. and L.C.C.; Funding acquisition, P.F. and R.S.; Investigation, L.C.C. and B.S.L.C.; Methodology, P.R.R.R., P.F., R.G., L.B.B. and B.S.L.C.; Resources, P.F. and R.S.; Supervision, P.R.R.R., P.F. and R.S.; Validation, P.F., R.S. and A.S.; Visualization, P.R.R.R., R.G. and L.B.B.; Writing—original draft, P.F., R.G. and L.C.C.; Writing—review and editing, P.F., R.G., L.C.C. and B.S.L.C. All authors have read and agreed to the published version of the manuscript.

Funding: The research leading to this study has received funding from the European Union Horizon 2020 research and innovation program under grant agreement No 821124 (NAVIGATE) and No 101003866 (NDC ASPECTS). Lilia Caiado Couto is funded by the Brazilian Federal Agency for Support and Evaluation of Graduate Education (CAPES) through the PhD scholarship Programa de Doutorado Pleno no Exterior Processo no 88881.129207/2016-01. Bruno Scola Lopes da Cunha would like to express his gratitude to the financial support from the Human Resources Program of the National Agency of Petroleum, Natural Gas, and Biofuels—PRH-41/ANP/Finep (in Portuguese). Luiz Bernardo Baptista acknowledges the Brazilian National Council for Scientific and Technological Development (CNPq) for funding provided.

Institutional Review Board Statement: Not applicable.

Informed Consent Statement: Not applicable.

Data Availability Statement: Data can be provided upon request.

Acknowledgments: The authors thank André Lucena (COPPE-UFRJ) and Angelo Gurgel (FGV/SP) for their valuable comments on earlier versions of the paper.

Conflicts of Interest: The authors declare no conflict of interest. The funders had no role in the design of the study; in the collection, analyses, or interpretation of data; in the writing of the manuscript, or in the decision to publish the results.

Appendix A

Appendix A.1 COFFEE-TEA Integrated Assessment Modelling Suite

The COFFEE-TEA is an integrated assessment modelling suite. The COmputable Framework For Energy and Environment (COFFEE) model is a global perfect-foresight, least-cost optimisation, and partial equilibrium model that is based on the Model for Energy Supply Strategy Alternatives and their General Environmental impacts (MESSAGE) platform, a linear programming optimisation platform for energy systems and physical balances (mass, energy, exergy and land) developed by the International Institute for Applied System Analysis (IIASA).

COFFEE was developed to assess long-term energy supply strategies, based on technological deployment and resource availability, given constraints on GHG emissions and other air pollutants from the energy and land-use systems. Each of the model's regions has a detailed representation of energy extraction and conversion technologies, and individualised estimates of energy resources (both in terms of volume and costs), which are mostly reported as cost supply curves. The model accounts for all primary energy produced by the energy systems and its later transformation into secondary and, further, into final energy. The international trade of the energy commodities is also captured by the model. Final energy is consumed by end users to fulfil the energy service demands.

Regarding sectoral coverage, COFFEE is divided into five main sectors: Energy, Industry, Transportation, Services/Residential (Buildings) and Agriculture (Table A1). Industry is divided into four subsectors: cement, iron and steel, chemical and other industries. The model includes explicit demand for clinker, steel, and non-energy products such as plastics and ammonia. Furthermore, there is demand for industry energy services, such as: direct heat; steam; HVAC; lighting; drive; and other uses.

The transport sector is divided into freight and passenger transport, measured in ton/kilometer (tkm) and passengers/kilometer (pkm), respectively. In COFFEE, the transport service is represented by different technologies of private transport (light duty vehicles, motorcycles, three-wheelers) and public transport (buses, trains, ships, airplanes). Each of these has a set of technologies varying from energy vector and efficiency levels, including variations of conventional vehicles, flex vehicles, hybrid vehicles, battery electric vehicles and fuel cell vehicles. Freight transport includes transport technologies such as trucks (Light, Middle and Heavy Duty), trains and ships, with all technologies previously listed for passenger vehicles also applying for trucks. Additionally, COFFEE relies on the production of drop-in synthetic fuels (such as diesel, jet fuel and marine bunker) as mitigation options for the freight transport sector.

The buildings sector includes both the residential and the commercial/public sectors. Each subsector has regional energy services demand for: space heating, water heating, cooking, lighting, appliances (electrical) and space cooling. To meet demand, the model has a range of technology options, ranging from low-efficiency lamps and cookers (non-commercial wood and kerosene), mid-range commercial options of appliances and heaters/air conditioning, up to more advanced options, such as LED lamps and highly efficient appliances. This sector also presents Distributed Generation (DG) options, either through photovoltaic (PV) or solar water heating.

Residues and agriculture sectors have a lesser impact on the energy consumption, despite being significant socioeconomic and environmental sectors. As for residues, they include the water management and municipal solid wastes. This sector has a low energy (mostly electric) consumption, but its mitigation options have a great impact on non-CO_2 emissions, including options for renewable energy, such as landfill gas and incineration. Regarding agriculture, the energy consumption for agricultural practices and crop processing is accounted in COFFEE.

The land-use system also presents several mitigation options through the adoption of sustainable practices and production of bioenergy, all of which are fundamental in long-term climate stabilisation scenarios. COFFEE derives from most global integrated assessment models in two manners: spatial resolution and integration with other sectors. Firstly, COFFEE does not have a spatial explicit representation of the land system. The model includes cost categories of each land cover to represent a cost supply curve of available land for use and land use change. As such, the cost supply curve for bioenergy, for instance, is completely endogenous and subject to competition for other land uses, such as crop and livestock production. Nonetheless, COFFEE also differs from most IAMs in the sense that the integration between the energy and the land-use systems is hard linked, meaning that its optimal solution accounts for the constraints and costs of both sectors simultaneously, including any potential trade-offs and synergies.

The Total-Economy Assessment (TEA) is a global top-down, recursive dynamic, Computable General Equilibrium (CGE) model. TEA uses the general equilibrium microeconomic theory as an operational tool in empirical analyses. The model simulates the evolution of the global economy, capturing industry-to-industry linkages, to assess policies on issues related to climate change, energy transitions, resource allocation, trade flows, technological change, income distribution, among others.

TEA is built as a mixed (non-linear) complementary problem on Mathematical Programming System for General Equilibrium (MPSGE), a tool written in the General Algebraic Modelling System (GAMS) software. To reach the general equilibrium of the economy, the TEA model assumes total market clearance (supply equals demand through commodity price equilibrium), zero profit condition for producers and perfect competition. The equilibrium is obtained when prices and quantities (endogenous variables) are balanced so that agents cannot improve their situation (welfare) by changing their behaviour, nor making other agents worse-off (Pareto optimal condition).

Production in each sector is represented by multi-level nested Constant Elasticity of Substitution (CES) functions, which use intermediate goods, labour, capital, land and energy as their input. The CES functions describe the substitution possibilities between factors of production and intermediate inputs in the production process, based on a least-cost approach. International trade follows Armington's aggregation [43], in which a composite CES function differentiates consumer's preferences between imported and domestic goods. Consumer preferences (household sector) are expressed by a CES utility function. Firms maximise their profits and the household sector maximises its welfare (utility) under budget constraints. Such choices are determined by the parameters of substitution and transformation elasticities in the utility and production functions.

In the TEA model, the macroeconomic closure assumes full employment of the factors of production (capital and labour). Savings equal investment in the general equilibrium, but regionally the imbalances are closed by a surplus (or deficit) in the current account. An endogenous real exchange rate clears the current accounts and the capital account decreases exogenously in the long-run. Capital stock evolves at each period with the formation of new capital that depends on the investment level in that period and the capital depreciation rate [44].

COFFEE and TEA models are long-term global models suitable for policies and climate aspects evaluation. They are integrated and have perfect compatibility in terms of base year data, sectoral and regional disaggregation. The COFFEE model provides data inputs for electricity generation and production shares to the TEA model, which accounts for the transformation of primary energy (coal, natural gas and crude oil) to secondary energy (oil products and electricity) to be consumed by end-use sectors, such as transport sectors (land, air and waterway) and energy-intensive industries (iron and steel, chemical, non-metallic minerals and other manufactures). The food production and land-use systems, which include the agricultural and livestock sectors, are also represented in the TEA model. The sectoral coverage comprehends 21 sectors that can be grouped into the five main sectors represented in the COFFEE model (Table A1).

Table A1. Sectoral coverage in COFFEE-TEA IAM suite.

COFFEE	TEA	Description
Agriculture	AGR	Agriculture
	CTL	Cattle
	OAP	Other animal products
	FSH	Fishing
Energy	COL	Coal
	CRU	Crude oil
	ELE	Electricity
	GAS	Natural gas
	OIL	Oil products
Industry	I_S	Iron and steel
	CRP	Chemical rubber and plastic
	NMM	Manufacture of non-metallic mineral products
	MAN	Other manufacture
	OFD	Other food (except meat)
	OMT	Other meat products
Transportation	OTP	Land transport
	WTP	Water transport
	ATP	Air transport
Services	SER	Services
Residential	DWE	Dwellings

TEA and COFFEE are divided into the same 18 regions, including large representative economic regions/countries, such as Europe, China, USA and Japan, while putting emphasis on developing countries in which energy and environmental issues are relevant globally, such as India and Brazil (Figure A1). In addition, COFFEE has a global region represented as the 19th region of the model for the assessment of global climate policies.

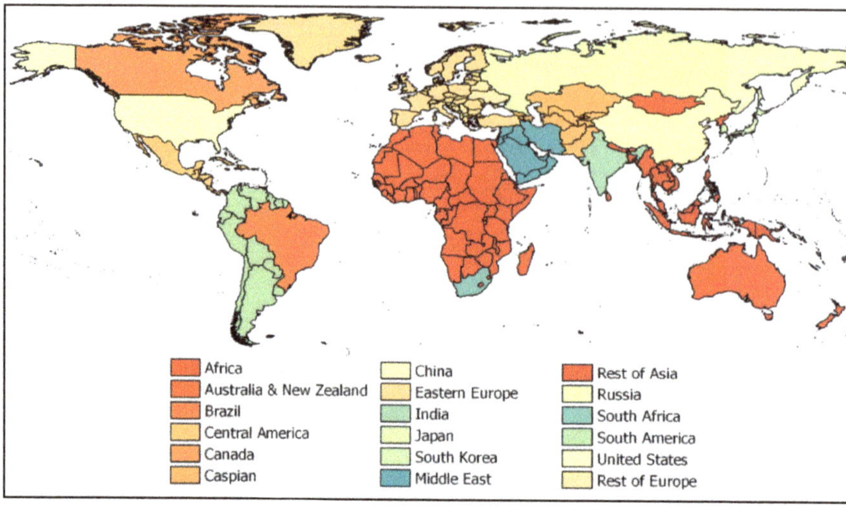

Figure A1. Regional breakdown of the COFFEE-TEA IAM suite.

The COFFEE-TEA suite is included in the category of IAMs that combines techno-economic and environmental variables to generate a cost-optimal solution in a hybrid approach; bottom-up technological solution with top-down macroeconomic consistency. COFFEE fully represents energy markets, while TEA projects future economic activities' demands based on macroeconomic drivers, such as population and GDP growth. The COFFEE-TEA IAM suite accounts for the three main GHG gases: CO_2, CH_4 and N_2O. These emissions are associated with the main sectors of land-use, agriculture and livestock, fugitive emissions, fuel combustion, industrial processes, and waste treatment. The model runs with a 5-year time step, from 2010 to 2100, with historical data (2010–2020) being used for calibration.

Appendix A.2 PROMETHEUS Model

PROMETHEUS is a global energy system model covering in detail the complex interactions between energy demand, supply and energy prices at the regional and global level. Its main objectives are: (1) to assess climate change mitigation pathways and low-emission development strategies for the medium and long-term; (2) to analyse the energy system, economic and emission implications of a wide spectrum of energy and climate policy measures, differentiated by region and sector); and (3) to explore the economics of fossil fuel production and quantify the impacts of climate policies on the evolution of global energy prices.

PROMETHEUS quantifies CO_2 emissions and incorporates environmentally oriented emission abatement technologies (such as RES, electric vehicles, CCS, energy efficiency) and policy instruments, such as carbon pricing schemes that may differentiate by region and economic activity. The model can be used to assess energy and climate policies, as it endogenously determines the international prices of fossil fuels through detailed world and regional supply/demand dynamics and technology dynamics mechanisms focusing on low-carbon technologies (e.g., wind, PV, electric cars, CCS, advanced biofuels, hydrogen).

PROMETHEUS is a recursive dynamic energy system simulation model. The economic decisions regarding the investment and operation of the energy system are based on the current state of knowledge of parameters (costs and performance of technologies, etc.) or with a myopic anticipation of future costs and constraints. Some foresight can be forced in the electricity production sector. The PROMETHEUS model assumes market equilibrium, where each representative agent (e.g., energy producer or consumer) uses information on prices and makes decisions about the allocation of resources. The interactions of representative agents are governed by market dynamics with market-derived prices to balance energy demand and supply in each sector (e.g., electricity production, transport and energy industries). The regional fuel markets are also integrated to form an international (global or regional) market equilibrium for crude oil, natural gas and coal. The model produces projections of global and regional fossil fuel prices, which depend on demand, supply, technology and resources. The model runs with a 1 year time step, usually from 2018 to 2050, the 2015–2018 period being entirely set by data and used for calibration.

References

1. Hepburn, C.; O'Callaghan, B.; Stern, N.; Stiglitz, J.; Zenghelis, D. Will COVID-19 fiscal recovery packages accelerate or retard progress on climate change? *Oxf. Rev. Econ. Policy* **2020**, *36*, S359–S381. [CrossRef]
2. Le Quéré, C.; Jackson, R.B.; Jones, M.W.; Smith, A.J.; Abernethy, S.; Andrew, R.M.; De-Gol, A.J.; Willis, D.R.; Shan, Y.; Canadell, J.G.; et al. Temporary reduction in daily global CO_2 emissions during the COVID-19 forced confinement. *Nat. Clim. Chang.* **2020**, *10*, 647–653. [CrossRef]
3. Forster, P.M.; Forster, H.I.; Evans, M.J.; Gidden, M.J.; Jones, C.D.; Keller, C.A.; Lamboll, R.D.; Quéré, C.; Le Rogelj, J.; Rosen, D.; et al. Current and future global climate impacts resulting from COVID-19. *Nat. Clim. Chang.* **2020**, *10*, 913–919. [CrossRef]
4. Le Quéré, C.; Peters, G.P.; Friedlingstein, P.; Andrew, R.M.; Canadell, J.G.; Davis, S.J.; Jackson, R.B.; Jones, M.W. Fossil CO_2 emissions in the post-COVID-19 era. *Nat. Clim. Chang.* **2021**, *11*, 197–199. [CrossRef]
5. IEA. *Global Energy Review 2021*; IEA: Paris, France, 2021. Available online: https://www.iea.org/reports/global-energy-review-2021 (accessed on 10 July 2021).
6. The World Bank. Global Economic Prospects (Issue June). 2020. Available online: https://www.worldbank.org/en/publication/global-economic-prospects (accessed on 10 July 2021).
7. IEA. *Sustainable Recovery*; IEA: Paris, France, 2021. Available online: https://www.iea.org/reports/sustainable-recovery (accessed on 15 June 2021).
8. IEA. *Sustainable Recovery Tracker*; IEA: Paris, France, 2021. Available online: https://www.iea.org/reports/sustainable-recovery-tracker (accessed on 13 June 2021).
9. European Council Conclusions, 17–21 July 2020—Consilium. Available online: https://www.consilium.europa.eu/en/press/press-releases/2020/07/21/european-council-conclusions-17-21-july-2020/ (accessed on 2 November 2020).
10. Biden, J. 9 Key Elements of Joe Biden's Plan for a Clean Energy Revolution | Joe Biden for President: Official Campaign Website. 2021. Available online: https://joebiden.com/9-key-elements-of-joe-bidens-plan-for-a-clean-energy-revolution/ (accessed on 10 July 2021).
11. Energy Policy Tracker-Track Funds for Energy in Recovery Packages. Available online: https://www.energypolicytracker.org/ (accessed on 1 November 2020).
12. O'Neill, B.C.; Kriegler, E.; Riahi, K.; Ebi, K.L.; Hallegatte, S.; Carter, T.R.; Mathur, R.; van Vuuren, D.P. A new scenario framework for climate change research: The concept of shared socioeconomic pathways. *Clim. Chang.* **2014**, *122*, 387–400. [CrossRef]
13. van Vuuren, D.P.; Kriegler, E.; O'Neill, B.C.; Ebi, K.L.; Riahi, K.; Carter, T.R.; Edmonds, J.; Hallegatte, S.; Kram, T.; Mathur, R.; et al. A new scenario framework for Climate Change Research: Scenario matrix architecture. *Clim. Chang.* **2014**, *122*, 373–386. [CrossRef]
14. IPCC. *Climate Change 2014: Mitigation of Climate Change Working Group III Contribution to the Fifth Assessment Report of the Intergovernmental Panel on Climate Change*; IPCC Report; Edenhofer, O., Pichs-Madruga, R., Sokona, Y., Farahani, E., Kadner, S., Zwickel, T., Seyboth, K., Adler, A., Baum, I., Brunner, S., et al., Eds.; Cambridge University Press: Cambridge, UK; New York, NY, USA, 2014.
15. IPCC. Global Warming of 1.5 °C an IPCC Special Report on the Impacts of Global Warming of 1.5 °C above Pre-Industrial Levels and Related Global Greenhouse Gas Emission Pathways, in the Context of Strengthening the Global Response to the Threat of Climate Change. 2018. Available online: http://www.ipcc.ch/report/sr15/ (accessed on 5 April 2021).
16. Kriegler, E.; Edmonds, J.; Hallegatte, S.; Ebi, K.L.; Kram, T.; Riahi, K.; Winkler, H.; van Vuuren, D.P. A new scenario framework for climate change research: The concept of shared climate policy assumptions. *Clim. Chang.* **2014**, *122*, 401–414. [CrossRef]
17. Moss, R.H.; Edmonds, J.A.; Hibbard, K.A.; Manning, M.R.; Rose, S.K.; Van Vuuren, D.P.; Carter, T.R.; Emori, S.; Kainuma, M.; Kram, T.; et al. The next generation of scenarios for climate change research and assessment. *Nature* **2010**, *463*, 747–756. [CrossRef]
18. Fragkos, P.; Kouvaritakis, N. Model-based analysis of Intended Nationally Determined Contributions and 2 °C pathways for major economies. *Energy* **2018**, *160*, 965–978. [CrossRef]
19. IAMC. The Common Integrated Assessment Model (IAM) Documentation. 2021. Available online: https://www.iamcdocumentation.eu/index.php/IAMC_wiki (accessed on 23 July 2021).
20. IAMC. Model Documentation-COFFEE-TEA-IAMC-Documentation. Integrated Assessment Modelling Consortium. 2020. Available online: https://www.iamcdocumentation.eu/index.php/Model_Documentation_-_COFFEE-TEA (accessed on 23 May 2021).

21. Cunha, B.; Garaffa, R.; Gurgel, A.; TEA Model Documentation. FGV/AGRO-N° 001 Working Paper Series. 2020. Available online: https://hdl.handle.net/10438/28756 (accessed on 23 May 2021).
22. Fragkos, P. Assessing the Role of Carbon Capture and Storage in Mitigation Pathways of Developing Economies. *Energies* **2021**, *14*, 1879. [CrossRef]
23. Fragkos, P.; Kouvaritakis, N.; Capros, P. Incorporating Uncertainty into World Energy Modelling: The PROMETHEUS Model. *Env. Model Assess* **2015**, *20*, 549–569. [CrossRef]
24. Rochedo, P.R.; Soares-Filho, B.; Schaeffer, R.; Viola, E.; Szklo, A.; Lucena, A.F.; Koberle, A.; Davis, J.L.; Rajão, R.; Rathmann, R. The threat of political bargaining to climate mitigation in Brazil. *Nat. Clim Chang.* **2018**, *8*, 695–698. [CrossRef]
25. Fricko, O.; Havlik, P.; Rogelj, J.; Klimont, Z.; Gusti, M.; Johnson, N.; Kolp, P.; Strubegger, M.; Valin, H.; Amann, M.; et al. The marker quantification of the Shared Socioeconomic Pathway 2: A middle-of-the-road scenario for the 21st century. *Glob. Environ. Chang.* **2017**, *42*, 251–267. [CrossRef]
26. IMF. World Economic Outlook—A Long and Difficult Ascent (Issue October). 2020. Available online: https://www.imf.org/en/Publications/WEO/Issues/2020/09/30/world-economic-outlook-october-2020 (accessed on 22 May 2021).
27. OECD. *OECD Economic Outlook*; OECD Publishing: Paris, France, 2020; Volume 2020. [CrossRef]
28. Roelfsema, M.; van Soest, H.L.; Harmsen, M.; van Vuuren, D.P.; Bertram, C.; den Elzen, M.; Höhne, N.; Iacobuta, G.; Krey, V.; Kriegler, E.; et al. Taking stock of national climate policies to evaluate implementation of the Paris Agreement. *Nat. Commun.* **2020**, *11*, 2096. [CrossRef] [PubMed]
29. Coronavirus: Tracking How the world's 'Green Recovery' Plans Aim to Cut Emissions. Available online: https://www.carbonbrief.org/coronavirus-tracking-how-the-worlds-green-recovery-plans-aim-to-cut-emissions (accessed on 1 November 2020).
30. Home | Climate Action Tracker. Available online: https://climateactiontracker.org/ (accessed on 1 November 2020).
31. Vivid Economics. Greenness of Stimulus Index-Vivid Economics. 2020. Available online: https://www.vivideconomics.com/casestudy/greenness-for-stimulus-index/ (accessed on 25 April 2021).
32. Data & Statistics-IEA. Available online: https://www.iea.org/data-and-statistics?country=BRAZIL&fuel=Energysupply&indicator=ElecGenByFuel (accessed on 27 October 2020).
33. Carbon Brief, 2021, Data & Statistics-IEA. 2021. Available online: https://www.iea.org/ (accessed on 25 March 2021).
34. Rogelj, J.; Popp, A.; Calvin, K.V.; Luderer, G.; Emmerling, J.; Gernaat, D.; Fujimori, S.; Strefler, J.; Hasegawa, T.; Marangoni, G.; et al. Scenarios towards limiting global mean temperature increase below 1.5 °C. Nat. *Clim Chang.* **2018**, *8*, 325–332. [CrossRef]
35. UNEP. Emissions Gap Report 2020. United Nations Environment Programme, Nairobi. 2020. Available online: https://www.unep.org/emissions-gap-report-2020 (accessed on 5 June 2021).
36. Emmerling, J.; Fragkiadakis, K.; Fragkos, P.; Gulde, R.; Kriegler, E.; Mercure, J.F.; van Ruijven, B.; Simsek, Y.; Tavoni, M.; Wilson, C. Impacts of COVID-19 and Recovery Packages on Climate Change Mitigation—First Results From NAVIGATE. 2021. Available online: https://www.navigate-h2020.eu/policy-brief-on-first-research-results-of-the-navigate-project-on-impacts-of-covid-19/ (accessed on 8 July 2021).
37. Fragkos, P.; van Soest, H.L.; Schaeffer, R.; Reedman, L.; Köberle, A.C.; Macaluso, N.; Evangelopoulou, S.; De Vita, A.; Sha, F.; Qimin, C.; et al. Energy system transitions and low-carbon pathways in Australia, Brazil, Canada, China, EU-28, India, Indonesia, Japan, Republic of Korea, Russia and the United States. *Energy* **2021**, *216*, 119385. [CrossRef]
38. AlKhars, M.; Miah, F.; Qudrat-Ullah, H.; Kayal, A. A Systematic Review of the Relationship between Energy Consumption and Economic Growth in GCC Countries. *Sustainability* **2020**, *12*, 3845. [CrossRef]
39. Nesticò, A.; Maselli, G. Declining discount rate estimate in the long-term economic evaluation of environmental projects. *J. Environ. Account. Manag.* **2020**, *8*, 93–110. [CrossRef]
40. Chen, Z.; Marin, G.; Popp, D.; Vona, F. Green Stimulus in a Post-pandemic Recovery: The Role of Skills for a Resilient Recovery. *Environ. Resour. Econ.* **2020**, *76*, 901–911. [CrossRef]
41. ILO. COVID-19 and the World of Work: Jump-Starting a Green Recovery with More and Better Jobs, Healthy and Resilient Societies. July 2020. Available online: https://www.ilo.org/global/topics/green-jobs/publications/WCMS_751217/lang--en/index.htm (accessed on 9 July 2021).
42. Galvin, R.; Healy, N. The Green New Deal in the United States: What it is and how to pay for it. *Energy Res. Soc. Sci.* **2020**, *67*, 101529. [CrossRef]
43. Armington, P.S. A theory of demand for products distinguished by place of production. *Staff Pap.* **1969**, *16*, 159–178. [CrossRef]
44. IAMC-Documentation. Available online: https://www.iamcdocumentation.eu/index.php/Capital_and_labour_markets_-_COFFEE-TEA (accessed on 1 September 2021).

energies

MDPI

Article

The Impact of Carbon Disclosure on Financial Performance under Low Carbon Constraints

Wenting Lu, Naiping Zhu * and Jing Zhang

School of Finance & Economics, JiangSu University, No. 301 Xuefu Road, Jingkou District, Zhenjiang 212013, China; 2112019002@stmail.ujs.edu.cn (W.L.); jzhang2015@163.com (J.Z.)
* Correspondence: npzhu@ujs.edu.cn

Abstract: In the context of low-carbon constrained development, in order to avoid the risk brought by climate change, more and more companies choose to disclose carbon information, respond to the national policy of carbon emission reduction and focus on the sustainable development of enterprises. This paper will investigate the impact of carbon disclosure on financial performance based on the 2011–2018 CDP report, taking the Fortune 500 companies as a sample. The study finds that for carbon-intensive industries, carbon disclosure cannot significantly contribute to the improvement of financial performance in the current period, but for carbon-non-intensive industries, carbon disclosure can significantly contribute to the improvement of financial performance in the current period, and the positive impact of carbon disclosure on financial performance in the current period can be extended to the next period. Finally, based on the findings of the empirical study, this paper puts forward policy recommendations for the construction of China's carbon disclosure system.

Keywords: fortune 500; carbon disclosure; financial performance

Citation: Lu, W.; Zhu, N.; Zhang, J. The Impact of Carbon Disclosure on Financial Performance under Low Carbon Constraints. *Energies* **2021**, *14*, 4126. https://doi.org/10.3390/en14144126

Academic Editor: Nuno Carlos Leitão

Received: 5 May 2021
Accepted: 5 July 2021
Published: 8 July 2021

Publisher's Note: MDPI stays neutral with regard to jurisdictional claims in published maps and institutional affiliations.

1. Introduction

Greenhouse gas (GHG) emissions have become one of the primary threats to the existence of life on earth. The excessive GHGs in the earth's atmosphere led to undesirable consequences in the ecosystem, creating global warming or climate change (Liu et al. Sand) [1,2]. Since 1880, the global average temperature has risen by 0.85 °C according to the data from Intergovernmental Panel on Climate Change (IPCC) [3]. As the Climate Change 2013 predicted, by 2100, the world will face a rise of the ground temperature and an increase in carbon dioxide concentration, and the sea level will rise by 26 to 81 cm. The main reason for climate warming is most likely (95%) that we did not aggressively reduce carbon emissions. By the end of this century, global warming will exceed 2 °C, or may exceed 4 °C according to the Climate Change 2013 report [3]. At present, according to the Intergovernmental Panel on Climate Change (IPCC) special report on the impacts of global warming of 1.5 °C, the control of greenhouse gas (GHG) emissions has become a necessary condition for companies to achieve sustainable development and it is important to limit the global warming of 1.5 °C [4] and therefore, there is an increasing demand for carbon disclosure in society.

More and more climate events dominate the headlines of the media, and they have a major impact on the economic development of various countries all over the world. On 15 January 2020, the World Economic Forum (WEF) released the Global Risks Report 2020. The report presents the main risks the world facing in the coming year, pointing out that all five major global risks in 2020 are environmental risks, while climate change is the biggest risk facing the world in 2020 [5]. In this context, a series of new concepts and policies such as "low-carbon economy", "low-carbon development" and "low-carbon city" have emerged. Low-Carbon Economy has attracted people's attention in recent years. It is a business model based on low energy consumption, low pollution, and low emissions, which will guide the "Fourth Industrial Revolution", protect the existence of non-renewable energy, prevent global warming, realize sustainable economic development, and at the same time create a healthy and green home for people.

On 12 December 2015, 195 countries and the European Union unanimously agreed to adopt the "Paris Agreement" at the Paris Climate Change Conference, opening a new chapter for taking action on global climate change. In order to control carbon emissions, countries around the world have committed to emission reduction targets and have formulated policies and regulations to regulate and guide the carbon emission reduction behavior of enterprises. It has become an inevitable choice to take a low-carbon economy. As a consequence of stakeholders starting exerting started exerting pressure on corporations to decrease their GHG emissions, firms are now likely to play a vital role in reducing their GHG emissions and contributing to stabilizing climate change (Luo) [6]. In recent years, firms have been asked to disseminate information about climate change related activities, also referred to as carbon disclosures, to satisfy the concerns of relevant stakeholders (Li et al., Meng et al.) [7,8]. Under the emission reduction pressures in the world, China should need to change the current high energy-consuming and high-pollution development model, accelerate the adjustment of economic structure, promote technological progress, and improve energy efficiency [9]. Although various countries have introduced various laws and regulations on carbon emissions and policies to encourage carbon emission reduction, and researchers have also realized the importance of carbon disclosure, there are still some people who question the authenticity of this information because of the inherent uncertainty between measurement of carbon emissions and carbon emissions reduction. In the case of internationally recognized standards, CDP (Carbon Disclosure Project) adopts a unique set of rules that all participating companies must follow, which greatly reduces the opportunity for managers to manipulate carbon data. Liu believes that carbon information disclosure by enterprises is an effective way to improve social awareness of low carbon and environmental protection, and a good carbon information disclosure mechanism helps internal and external stakeholders to have a better understanding of corporate low carbon governance and strengthen the carbon regulation of government departments [10]. The study of carbon disclosure has been gaining increasing importance in recent years to help firms communicate their climate change activities to their stakeholders through environment disclosures (Hahn et al., Uyar et al.) [11,12]. With the deepening of low-carbon economy in China, the demand for carbon information from the market and corporate stakeholders is increasing. The relationship between the quality of carbon information and the performance of enterprises can be explored by evaluating the carbon information disclosed by enterprises. The relationship between the quality of carbon information disclosed by enterprises and their own performance is particularly important to motivate enterprises to disclose their carbon information. To this end, this paper will investigate the impact of carbon disclosure on financial performance based on the 2011–2018 CDP report, taking the Fortune 500 companies as a sample. Based on existing research perspectives, this paper uses the scoring index and carbon emission data from the CDP report, as well as the financial data of enterprises to investigate the relationship between carbon disclosure and financial performance of carbon-intensive companies and carbon-non-intensive companies. This paper takes voluntary disclosure theory, legitimacy theory, stakeholder interest theory, signaling theory, and sustainable development theory as its pillars, and applies them to the research fields of carbon disclosure and financial performance, and conducts a profound analysis on the theoretical level.

The paper deepens the understanding of carbon disclosure by enterprises from the theoretical level, which has certain significance for the practice of carbon disclosure of Chinese enterprises, and also promotes the development of empirical research related to carbon disclosure. At the same time, the empirical research in this paper also makes up for the deficiencies of existing research as few scholars have studied whether the significant impact of carbon disclosure on financial performance is deferred to the next period and few scholars have conducted comparative analysis of for carbon-intensive industries and carbon-non-intensive industries. Therefore, this research will investigate the impact of carbon disclosure on current financial performance for carbon-intensive industries and carbon-non-intensive industries, and on this basis, the intertemporal impact of carbon disclosure on financial performance.

2. Literature Review and Theoretical Hypotheses

2.1. Literature Review

Many scholars have found that, as both sides of a transaction, companies disclose carbon information so that stakeholders can have sufficient information to facilitate the formation of the transaction and promote the enhancement of corporate value. The earliest environmental information study had been included in social responsibility information to conduct relevant studies and environmental information disclosure were mostly at the level of theoretical and descriptive analysis (Mobley; Gray et al.) [13,14]. Friedman conducted a study on the relationship between social responsibility and corporate performance [15]. With the research on the determinants of social responsibility and environmental information disclosure (Deegan, Gordon; Karim et al.) [16,17] and the continuous research on economic consequences and market reactions (Dhaliwal et al.; Lys et al.) [18,19]. In recent years, the research on social responsibility information and environmental information disclosure has also been refined. Scholars have been studying the disclosure of social responsibility information and environmental information from internal and external perspectives, and from the perspective of report forensics respectively [20].

Most of the early studies on the relationship between carbon disclosure and corporate financial performance showed a negative relationship, validating the views of traditional economists and neoclassical schools. Hassel combined research with the cost-related theory and pointed out that the cost of disclosing carbon information in the process of maintaining the legitimacy of enterprises is greater than the benefit, and the more detailed and comprehensive the carbon information disclosure, the more costly it is, the carbon information disclosure will negatively affect the enterprise's interest goal, and the quality of disclosure is negatively related to the enterprise's value. There is a negative relationship between disclosure quality and corporate value, and carbon information disclosure will reduce the financial performance of enterprises [21]. Chapple et al. also verified that the expensive cost of carbon disclosure by firms to gain legitimacy outweighs the benefits of the act, and that corporate carbon management practices can cause changes in firm value [22]. Griffin et al. found that due to the high cost of disclosure, carbon information disclosure does not bring economic benefits to firms or even reduces their profitability [23].

According to the theory of information asymmetry, carbon information is an important non-financial information for investors' decision making, and high-quality carbon information disclosure can effectively improve the situation of information disadvantage for investors. The disclosure of high-quality carbon information can effectively improve the situation that investors are at an information disadvantage, and to a certain extent can reduce the risk faced by investors and protect the interests of investors. At the same time, enterprises can get the necessary resources for production and operation. Proactive disclosure of carbon information is the most important way for companies to gain access to stakeholders and achieve their sustainable development. Saka and Oshika pointed out that carbon information disclosure has a positive impact on market-based financial performance [24]. Velte et al. found that carbon performance was significantly associated with carbon disclosure, and carbon information disclosure can reduce information asymmetry, while carbon information disclosure can increase financial performance [25]. Although carrying out carbon emission reduction and disclosing carbon information will incur certain costs, failure to fulfill carbon emission reduction obligations and disclose related information may save enterprises some costs in the short term, but in the long term development will generate more explicit or implicit costs, affect the efficiency of resource allocation of enterprises, cause implicit harm to enterprises, which in turn affects their competitiveness and hinders the improvement of enterprise value and performance. Lemma et al. stated that firms can meet consumer expectations by reducing their carbon footprint and can reasonably expect that after meeting consumer expectations company can obtain from existing or potential customers [26]. Lueg state that disclosure hardly affects financial performance by changing free cash flow, but the increase in transparency from high quality disclosure can reduce information asymmetry between stakeholders and impact on financial performance by reducing risk [27].

According to signaling theory, carbon information disclosure can alleviate the pressure that companies may face and increase stakeholders' corporate recognition and support, which in turn promotes financial performance. Companies that are aware of the environmental crisis, when they are aware of the environmental crisis, they will immediately take measures to minimize the risk of environmental pollution and protect their reputation and image by disclosing information. This can reduce financial risk to a certain extent. Wegener empirically examined the impact of corporate disclosure of carbon information on the stock market based on CDP reports of Canadian companies and found that voluntary disclosure of carbon information increased shareholder motivation and reduced transaction costs, which in turn had a positive impact on stock market value [28]. Ziegler et al. and Schiager found that for U.S. energy companies, in response to pressure from global warming agencies to gain legitimacy, companies actively disclose their efforts to respond to climate change, enhance their corporate image and thus significantly improve their stock performance [29,30]. Saka and Oshika studied the relationship between carbon disclosure and equity market value of more than 1000 companies in Japan based on circumventing sampling bias and endogeneity issues and found that their equity market value increases as the content of carbon disclosure increases [24]. Ganda selected a sample of South African companies as a study and examined the impact of carbon emission reporting on financial value and found that carbon disclosure was positively correlated with accounting-based indicator Return on Assets (ROA) in most cases but negatively correlated with market-based indicator Market Value Added (MVA) by conducting panel regressions on the data of the sample companies from 2010–2015 [31]. Iskandar and Fran found that carbon emissions disclosure significantly negatively affects firm value and corporate social responsibility disclosure significantly positively affects firm value [32]. Siddique et al. examined how carbon performance affects carbon disclosure and how carbon disclosure affects financial performance and the results showed that carbon disclosure positively affects carbon performance, consistent with the signaling theory. It also showed that carbon disclosure negatively affects financial performance in the short-term, and positively affects financial performance in the long-term [33].

Some scholars have studied the possible economic impact of carbon information disclosure from the perspective of corporate governance. Carbon information disclosure is a rational choice for firms, and Schiager and Haukvik selected Nordic listed companies in the CDP report as the research object and studied the mechanism of carbon information disclosure affecting company value from both accounting and marketing perspectives, respectively, and found that carbon information disclosure by listed companies can enhance corporate value [30]. Borghei et al. analyzed the annual reports of Australian firms and found that the return on corporate assets increased in the year following carbon information disclosure, noting that carbon information disclosure positively affects corporate financial performance [34]. Brouwers et al. stated that carbon performance and information disclosure have a positive impact on corporate financial performance in the long run [35].

Based on the perspective of environmental information disclosure, many scholars' studies show that there is a positive relationship between environmental information disclosure and financial performance of enterprises. Freedman and Jaggi found that environmental disclosure in the petroleum industry shows a significant positive relationship with financial performance [36]. Murray et al. separated environmental information from social responsibility information in a separate study and found a positive relationship between the quality of environmental disclosure of many UK firms and their contemporaneous returns [37]. Anderson et al. concluded similarly that companies with good business performance have a correspondingly higher quality of environmental information disclosure [38]. Stanwick and Stanwick studied 469 listed companies in the 1994 Forbes 500 and found that the financial performance of the different groupings of high medium, low had different effects on the response to the environment, with high financial performance companies having a higher incidence of environmental policies and/or environmental commitments compared to low financial performance companies, while medium financial performance companies had the highest level of environmental policies and/or environmental commitments [39]. Nor et al. showed through empirical studies that there is a mixed result between the behavior of environmental disclosure and financial performance, and that environmental disclosure is significantly related to profitability [40].

Based on carbon disclosure perspective, Luo et al. investigated the actions of Fortune 500 companies in terms of carbon disclosure strategies in response to climate change challenges based on a 2009 report provided by Fortune 500 companies to the Carbon Disclosure Project (CDP), and found that economic pressures were significantly associated with the decision to disclose carbon [41]. Luo et al. also selected a sample of 2045 large firms from 15 countries and representing different industries, based on the reports provided by these firms to the Carbon Disclosure Project (CDP) in 2009, using profitability, leverage and growth as indicators of resource availability and the extent of firm participation in the CDP as indicators of the propensity to disclose carbon and conducting an empirical study for developing and developed countries, respectively. The study found that the propensity to disclose carbon is correlated with the indicator of resource availability and that this relationship is more significant in developing countries, suggesting that one of the reasons for the lack of committed carbon reductions and disclosure in these countries is the shortage of resources [42]. Matsumura et al. examined the impact of carbon emissions and voluntary disclosure of carbon emissions on firm value based on data on carbon emissions voluntarily disclosed by S&P500 companies to CDP (Carbon Disclosure Project) from 2006 to 2008. The study showed a negative correlation between carbon emissions and firm value [43]. Zhao and Li based on the data of Chinese listed companies, scored the quality of carbon emission information content and concluded that the return on net assets was positively related to the quality of carbon information disclosure [44].Zhao and Yan took the listed companies in the heavy polluting industry in the 2008–2011 China CDP report as a sample and found that the carbon disclosure level of the selected sample companies was significantly positively correlated with financial performance [45]. Li and Shi divided the carbon information disclosure quality evaluation index into 10 dimensions to score and explore the correlation between the carbon disclosure quality and financial performance. The study found that the higher the quality of carbon information disclosure, the higher the financial performance and there is intertemporal in this impact, but the intertemporal impact has a downward trend year by year [46]. Ganda based on the annual carbon emission reports of South African companies from 2010–2015, using panel regression, the results of the study indicate that carbon disclosure is positively related to ROA (return on assets) but negatively related to MVA (market value added) [31]. Zhu conducted the study from the perspective of financial management, analyzed the possible influence of carbon tax policy on green financial performance, green financial activities, green financial accounting and financial information disclosure of microeconomic entity power companies based on specific cases [47]. Piesiewicz and Ciechan-Kujawa (2021) conducted the study on 57 published integrated reports of listed companies in Poland, contributed to the integrated reporting examination by identifying quantitative and qualitative gaps when applying Integrated Reporting standards and found insignificant differences in the analysis of completeness of disclosures in performance [48].

Most of the national scholars' explorations on the economic consequences of carbon information disclosure have gathered in the aspects of carbon information disclosure and corporate value, cost of capital, and decision usefulness. The discussion on how the level of carbon information disclosure affects financial performance is relatively lacking, and more often explores the relationship between the role of environmental information disclosure and social responsibility information disclosure on financial performance. In addition, the findings of the current available studies are widely divergent, with positive, negative, and uncorrelated results.

Frost found from an empirical analysis of 60 Australian firms, mainly in extractive services, that better performing firms are willing to disclose more environmental information than poorer performing firms [49]. Clakson et al. chose two different measures of firm performance, and the empirical results both indicate that environmental information disclosure has a positive effect on firm performance [50]. Al-Tuwaijri et al. found a positive relationship between environmental performance, environmental information disclosure and firm performance by taking environmental performance into account [51]. Jenkins and Yakovleva found a positive effect between the level of social responsibility disclosure and corporate value for a sample of ten global mining companies [52].

In contrast, there are some scholars have shown a negative or no correlation between environmental disclosure and financial performance of firms. Freedman and Jaggi explored the relationship between the level of pollution disclosure, pollution performance and economic performance of firms in highly polluting industries, and found that for the total sample, there was no correlation between their level of pollution disclosure and economic performance. While the results of the subsample showed that large companies with poor economic performance provided the most detailed pollution information, for smaller firms there was no correlation between economic performance and pollution disclosure [53]. Richardson and Welker used a sample of Canadian companies from 1990–1992, found a significant positive relationship between social disclosure and the cost of equity capital, a positive relationship that was mitigated among companies with better financial performance, where companies may will be financially penalized to some extent for having disclosed socially responsible information [54]. The findings of Johnson suggested that a firm's illegal or irresponsible attitude will hurt it financially and have a negative impact on the firm's financial performance, however, merely complying with legal requirements or undertaking sporadic social responsibility will not bring any financial advantage to the firm [55]. Stanny and Ely found that there was no significant correlation between carbon disclosure and investment, and carbon disclosure did not promote company performance [56]. Hsu and Wang find that because of the high cost of environmental responsibility, corporate disclosure of greenhouse gas emissions does not produce timely economic benefits and may reduce corporate competitiveness [57]. Plumlee et al. studied heavy polluting industries and general industries separately, and the empirical results showed that the polluting industries are negatively related to the level of environmental information disclosure and the financial performance of firms, while general industries show a positive relationship [58].

By sorting out and summarizing the findings of previous studies mentioned above, it can be found that although the number of literatures on financial performance research from the perspective of environmental information disclosure is relatively large, scholars still cannot reach a consistent research conclusion. On the one hand, it is because scholars choose different methods for environmental information disclosure indicators; on the other hand, given the different research objects selected, the research results also differ, for example, carbon intensive industries and carbon non-intensive industries cannot be confused due to different industry characteristics and different pressure in terms of regulation.

The impact on financial performance from the perspective of carbon information disclosure is rarely studied. There are some scholars pay more attention to carbon information disclosure, but most of the studies are on the impact factors related to carbon information disclosure, and few empirical studies involve the impact on current financial performance and whether it is deferred to the next period. Moreover, in previous studies, many scholars have scored carbon information disclosure based on whether to disclose or the number of disclosures, which is highly subjective. Given that the Carbon Disclosure Project (CDP) benefits from the guidance of PricewaterhouseCoopers, its scoring system is more comprehensive and more authoritative, and the data also has stronger depth and breadth.

Therefore, this paper will use the Carbon Disclosure Leader Index published by CDP as a proxy variable for carbon disclosure in CDP reports.

2.2. Theoretical Hypothesis

Based on the voluntary disclosure theory, the cost and the credibility of the disclosed information are worth more notice since they are the private information that the company chooses to publish. Management tends to reveal information that is beneficial to increase the corporate value (e.g., lower greenhouse gas emissions) and conceal unfavorable information that reduces the corporate value (e.g., the failure to meet emission reduction target), and rational managers tend to hide unfavorable information disclosure. The value growth effect of sustainability disclosure has been proposed in many scholars' studies (Dhaliwal et al.; Clarkson et al.; De Villiers, Marques) [59–61]. In addition, according to the signaling theory, good information promotes the development of the enterprise and will bring potential economic benefits, however, bad news will hinder the development of the enterprise and is likely to reduce the financial performance of the firm in the current period. Therefore, based on the perspective of voluntary disclosure theory and signaling theory, this paper argues that enterprises with high quality of carbon disclosure have better financial performance.

In addition, since ten industries are covered in the sample companies selected in this paper, including energy industry, industrial industry, materials industry, consumer discretionary industry, consumer staples industry, financial industry, health care industry, information technology industry, telecommunications service industry, and public utilities industry. Given that carbon-intensive industries such as energy, industry, and materials face higher risks in terms of energy costs, climate change response, energy saving and emission reduction, they are more cautious in disclosing relevant data and take a relatively conservative stance on carbon disclosure. While for industries such as information technology and finance, the risk associated with data disclosure are relatively low, and they are more proactive in their own carbon disclosure and more confident in carbon emission reduction. Therefore, this paper combines the approach of Clarkson (2008) [50] and He [62] to divide the industry categories into two categories, namely, carbon-intensive industries (energy, industry, material industries, utilities) and non-carbon-intensive industries (the remaining six industries). This leads to hypotheses 1 and 2.

Hypotheses 1 (H1). *In carbon-intensive industries, carbon disclosure can contribute to the improvement of the company's current financial performance.*

Hypotheses 2 (H2). *In non-carbon-intensive industries, carbon disclosure can contribute to the improvement of the company's current financial performance.*

According to the signaling theory, information utilizers such as management have a time process of absorbing, thinking, and reflecting on receiving information until deciding. During this period, the investor must measure whether the identified information is useful, and if it is, then the investor needs to distinguish whether the information conveys a signal that is beneficial to the investment or not, and finally make a corresponding response. In addition, according to the stakeholder theory, not only the resource environment of the enterprise itself must be considered in the process, but also the opinions of other stakeholders at the moment. Therefore, the significant impact of carbon disclosure on the current period's financial performance, regardless of the industry, is likely not to happen overnight and this significant impact will be deferred to the next period. So, based on signaling theory and stake-holder theory, this paper proposes the following hypothesis 3.

Hypotheses 3 (H3). *On the premise that carbon disclosure has a significant impact on the company's current financial performance, it will be extended to the next period.*

3. Materials and Model Design

3.1. Sample Selection and Data Sources

The samples selected for this study are the Fortune 500 companies involved in the CDP report from 2011–2018 because the publicly available data of Carbon Disclosure Leadership Index (CDLI) are only from 2011–2018, and the carbon disclosure behaviors of these companies are still at the forefront of the times as of now, which can help us to provide a basis for the empirical research by providing the Carbon Disclosure Leadership Index (CDLI) and carbon emissions date. In addition, in order to study the current and intertemporal impact of carbon disclosure on financial performance, the relevant financial performance data used in this article come from 2011–2018. The carbon disclosure data used in this article are all sourced from the Carbon Disclosure Project (CDP), which are aggregated and processed by the CDP project team based on the corporate response data collected online. The financial data used come from Wind Information, which are all collected and filtered manually. Companies without CDLI scores and those with incomplete or discontinuous financial data were excluded, there are 94 remaining companies with a total of 752 samples. The sample companies involve 10 different industries, and all industries are divided into two categories, namely, carbon-intensive industries (denoted as IND = 1) and non-carbon-intensive industries (denoted as IND = 0).

3.2. Variable Selection

3.2.1. Measurement of Financial Performance

There are many indicators of financial performance, such as return on assets (ROA), earnings per share (EPS), return on equity (ROE), Z-score proposed by Edward Altman, return on investment (ROI) and profit margin. Among them, ROA and ROE are financial performance indicators recognized by scholars all over the world. Based on previous research, this paper selects the return on assets (ROA) as a substitute variable for financial performance. This indicator can reflect the company's financial performance at an overall level, and its high or low level can directly reflect the company's financial status, will not be affected by the company's extraordinary events, and has the characteristics of being objective, universal, and easy to obtain. At the same time, ROA is also one of the most concerned financial indicators of corporate stakeholders and it has good comparability.

3.2.2. Measurement of Carbon Disclosure

Since the focus of our study is carbon, we use the Carbon Disclosure Leader Index (CDLI) to measure the degree or level of carbon disclosure based on content analysis. Most of the questions in the CDP questionnaire are two-choice questions. Participants who provided truthful information gave a "yes" (count as 1 point) or "no" (count as 0 point) answer, such as whether the company has an environmental committee.

The design and development of CDLI have received professional guidance from Pricewaterhouse Coopers, which is better than self-designed methods. Its scores can reflect the depth and breadth of corporate carbon information. By using a content analysis method that facilitates the analysis and interpretation of relevance, materiality, and substance of disclosure rather than simply the number of counts and the extend of disclosure, the CDLI score has also received widespread support at present.

3.2.3. Measurement of Financial Performance

Other control variables include net profit margin, debt to asset ratio, enterprise scale, growth rate of total operating income, etc. The specific variables are shown in Table 1:

Table 1. Variable design.

Variable Type	Variable Name	Symbol	Description
Dependent Variable	Return on Assets	ROA	Return on Assets = Net Income/Total Assets
Independent Variable	Carbon Disclosure	CDLI	Carbon Disclosure Leader Index
Control Variable	Net Profit Margin	NPM	Net Profit Margin = Net Profit/Sales Revenue
	Debt to Asset Ratio	LEV	Debt to Asset Ratio = Debt/Total Assets
	Enterprise Scale	SIZE	Enterprise Scale = $\ln_{\text{Total Assets}}$
	Growth Rate of Total Operating Income	GR	Growth Rate of Total Operating Income = (Current Period Gross Operating Income – Previous Period Gross Operating Income) /Current Period Gross Operating Income
Grouping Variable	Whether belongs to carbon intensive industry	IND	1 for carbon-intensive industries and 0 for non-carbon-intensive industries

3.3. Regression Model Setting

This paper investigates the impact of carbon disclosure on financial performance in carbon-intensive industries and non-carbon-intensive industries. The financial performance of the current period and the next period are used as explanatory variable and carbon disclosure is used as the explanatory variables. Stata16.0 is used as the multivariate statistical analysis software for this paper, and the following two multiple regression models are established:

$$ROAt = a0 + a1CDLIt + a2NPMt + a3LEVt + a4SIZEt + a5GRt + \varepsilon. \quad (1)$$

$$ROAt + 1 = a0 + a1CDLIt + a2NPMt + a3LEVt + a4SIZEt + a5GRt + \varepsilon \quad (2)$$

Among them, Model 1 is used to investigate the impact of carbon disclosure on financial performance in the current period, and Model 2 is used to study the impact of carbon disclosure on financial performance in the next period when carbon disclosure has a significant impact on the current period financial performance. Both models are suitable for empirical analysis of carbon-intensive industries, non-carbon-intensive industries, and full samples.

4. Empirical Results and Analysis

4.1. Descriptive Statistics

4.1.1. Descriptive Statistical Analysis of the Full Sample

There are 752 samples selected in this article, including 10 industry categories. The specific industry distribution is shown in Table 2 below:

Table 2. Classification of industries to which the full sample belongs.

Category	Quantity	Proportion
Energy	112	14.89%
Industrial	80	10.64%
Materials	88	11.70%
Public Utilities	40	5.32%
Consumer Discretionary	40	5.32%
Consumer staples	56	7.45%
Finance	160	21.28%
Health Care	80	10.64%
Information Technology	64	8.51%
Telecommunication services	32	4.26%
Total	752	100%

As can be seen from Table 2 that the financial industry, energy industry, material industry, industrial industry, and health care industry account for a large proportion of the entire sample. The proportions are 21.28%, 14.89%, 11.70%, 10.64%, and 10.64%, respectively. Among them, there are 320 sub-samples of carbon-intensive industries (energy, materials, industrial and public utilities), accounting for a total of 42.55%. Non-carbon-intensive industries (consumer discretionary, consumer staples, finance, health care, information technology, telecommunications services) has a total of 432 sub-samples, accounting for a total of 57.45%.

Table 3 presents the minimum, maximum, average, and standard deviation of all variables in the full sample of 752. The average return on assets (ROA) is 0.0883, and the standard deviation is 0.0741, indicating that the difference in the return on assets in the entire sample is not significant. The minimum value of carbon disclosure (CDLI) is 25, the maximum value is 100, the average value is 80.8546, and the standard deviation is 14.1486, indicating that there are large differences in carbon disclosure in the entire sample.

Table 3. Descriptive statistical analysis of the full sample variables.

	Minimum	Maximum	Average Value	Standard Deviation
ROA	−0.2122	0.3344	0.0883	0.0741
CDLI	25.0000	100.0000	80.8546	14.1486
NPM	−0.8475	0.4107	0.1102	0.1162
LEV	0.1578	0.9747	0.6472	0.2057
SIZE	22.4130	28.6619	25.1815	1.4681
GR	−0.3216	2.9728	0.0913	0.2342

4.1.2. Descriptive Statistical Analysis of Sub-Samples

Table 4 presents the minimum, maximum, average, and standard deviation of all variables in the carbon-intensive industries. The number of samples in this group is 320. The minimum value of return on assets (ROA) is −0.2122 and the maximum value is 0.3344. The minimum value of carbon disclosure (CDLI) is 34, the maximum value is 99, the average value is 79.1833, and the standard deviation is 14.3732, indicating that the quality of carbon disclosure varies among companies.

Table 4. Descriptive statistical analysis of variables (IND = 1).

	Minimum	Maximum	Average Value	Standard Deviation
ROA	−0.2122	0.3344	0.0938	0.0698
CDLI	34.0000	99.0000	79.1833	14.3732
NPM	−0.8475	0.4107	0.0841	0.1443
LEV	0.2654	0.9549	0.5741	0.1404
SIZE	23.1833	27.2987	24.7454	0.8002
GR	−0.3216	2.9728	0.1224	0.3147

Table 5 presents the minimum, maximum, average, and standard deviation of all variables in non-carbon-intensive industries. The number of samples in this group is 432. The minimum return on assets (ROA) is −0.0928 and the maximum is 0.3220. The minimum carbon disclosure (CDLI) is 25, the maximum is 100, the average is 82.0926, and the standard deviation is 13.8948, indicating that the quality of carbon disclosure also varies among non-carbon-intensive industries.

Table 5. Descriptive statistical analysis of variables (IND = 0).

	Minimum	Maximum	Average Value	Standard Deviation
ROA	−0.0928	0.3220	0.0843	0.0771
CDLI	25.0000	100.0000	82.0926	13.8948
NPM	−0.1051	0.3551	0.1295	0.0855
LEV	0.1578	0.9747	0.7013	0.2287
SIZE	22.4130	28.6619	25.5046	1.7441
GR	−0.2643	1.0015	0.0683	0.1459

4.1.3. Mean Difference

Comparing the statistical values of the main variables in carbon-intensive industries and carbon-non-intensive industries, it can be seen from Table 6 that the average return on assets (ROA) are 0.0938 and 0.0843, with the standard deviations of 0.0698 and 0.0771, respectively. And the value of return on assets in the two groups of industries do not differ significantly. The mean values of carbon disclosure (CDLI) are 79.1833 and 82.0926, respectively, indicating that the average quality of carbon disclosure in non-carbon-intensive industries is higher, with the standard deviations of 14.37332 13.8948, respectively. To further test whether the main variables are significantly different among different groups are, this paper adopts mean difference analysis to identify.

Table 6. Descriptive statistics for grouping of main variables.

	IND	N	Average	Standard Deviation	Standard Error of Mean
ROA	1	320	0.0938	0.0698	0.0064
	0	432	0.0843	0.0771	0.0061
CDLI	1	320	79.1833	14.3732	1.3121
	0	432	82.0926	13.8948	1.0917

As can be seen from Table 7 that the significance probability (Sig.) of the return on assets (ROA) is 0.2800, that is, there is no significant difference in ROA between the two sub-samples. The carbon disclosure (CDLI) has the significance probability (Sig.) of is 0.0880, which means there is a significant difference in CDLI between the two sub-samples. Therefore, based on the above analysis, it is theoretical and scientific to group carbon-intensive industries (IND = 1) and non-carbon-intensive industries (IND = 0) for discussion.

Table 7. Comparison of differences in means between sub-samples.

	IND = 1	IND = 0	Difference in Mean	t	Sig.
ROA	0.0938	0.0843	0.0095	1.08	0.2800
CDLI	79.1833	82.0926	−2.9093	−1.71	0.0880

Note: *t*-test using independent samples.

4.2. Correlation Analysis between Carbon Disclosure and Current Financial Performance

Tables 8 and 9 respectively present the correlation coefficients between the variables of carbon-intensive industries and carbon-non-intensive industries. Comparing the correlation coefficients in the two tables, it is found that the positive and negative signs of the correlation coefficients are consistent and do not differ significantly. In addition, the correlation coefficients between the control variables of debt to asset ratio (LEV) and enterprise scale (SIZE) are close to 0.5 in Table 8 and exceed 0.5 in Table 9. However, the multicollinearity results show that the VIF values of LEV and SIZE in Table 8 are 1.14 and 1.22, respectively, and the VIF values of LEV and SIZE in Table 9 are 2.33 and 2.28, respectively, which are far less than 10. This indicates that there is no problem of multicollinearity in Model 1.

Table 8. Correlation analysis between variables (IND = 1).

	ROA	CD	NPM	LEV	SIZE	GR
ROA	1					
CDLI	−0.1563	1				
NPM	0.6409 **	−0.1401	1			
LEV	−0.3000 **	0.2527 **	−0.0694	1		
SIZE	−0.1645	0.0295	−0.0282	0.3438 **	1	
GR	0.2158 *	−0.2912 **	0.1346	−0.0175	0.0131	1

Note: * indicates that the two variables are significantly correlated at the 5% level, ** indicates that the two variables are significantly correlated at the 1% level.

Table 9. Correlation analysis between variables (IND = 0).

	ROA	CD	NPM	LEV	SIZE	GR
ROA	1					
CDLI	−0.1889 *	1				
NPM	0.4797 **	−0.2718 **	1			
LEV	−0.5098 **	0.2329 **	−0.2757 **	1		
SIZE	−0.7362 **	0.3399 **	−0.2357 **	0.7357 **	1	
GR	0.1338	−0.0356	0.1049	−0.2178 **	−0.1723 *	1

Note: * indicates that the two variables are significantly correlated at the 5% level, ** indicates that the two variables are significantly correlated at the 1% level. Numbered lists can be added as follows.

4.3. Regression Analysis

4.3.1. Sub-Sample Regression Analysis

(1) The Impact of Carbon Disclosure on Current Financial Performance

The results of the regression analysis with carbon disclosure (CDLI) as the explanatory variable are presented in Tables 10 and 11 for the carbon-intensive and non-carbon-intensive industries, respectively.

Tables 10 and 11 respectively show the results of the regression analysis of carbon-intensive industries and non-carbon-intensive industries with carbon disclosure (CDLI) as the explanatory variable. The adjusted R^2 values are 0.4761 and 0.6600, respectively, indicating that the degree of explanation of the return on assets (ROA) of all independent variables in the two samples is 47.61% and 66.00%, respectively. The *p*-value of Model 1 in both sets of samples was 0.0000, indicating that Model 1 passed the significance test. In the group with carbon-intensive industries, the coefficient of carbon disclosure (CDLI) is 0.0157, which is positive but does not pass the significance test. Therefore, the hypothesis H1 has not been verified. However, in the group of non-carbon-intensive industries, the coefficient of carbon disclosure (CDLI) is 0.0009 with a *p*-value of 0.0010, indicating a significant correlation at the 1% level. Therefore, the hypothesis of H2 is verified.

Table 10. Regression analysis with CDLI as explanatory variable (IND = 1).

Variable	Coefficient	t-Value	*p*-Value
CDLI	0.0157	0.48	0.6330
NPM	0.2940 ***	7.83	0.0000
LEV	−0.1194 ***	−2.68	0.0090
SIZE	−0.00589	−0.86	0.3900
GR	0.03107 ***	3.26	0.0010
Constant term	0.26702	1.54	0.1270
Adjusted R^2		0.4761	
F−statistic of the model		15.18	
Sig.		0.0000	
Number of samples		320	

Note: *** indicate that the two variables are significantly correlated at the 1% level.

Table 11. Regression analysis with CDLI as explanatory variable (IND = 0).

Variable	Coefficient	t−Value	*p*−Value
CDLI	0.0009 ***	3.39	0.0010
NPM	0.3362 ***	6.50	0.0000
LEV	0.0529 *	1.75	0.0830
SIZE	−0.0361 ***	−8.08	0.0000
GR	−0.0034	−0.13	0.8990
Constant term	0.8556 ***	9.47	0.0000
Adjusted R^2		0.6600	
F−statistic of the model		58.77	
Sig.		0.0000	
Number of samples		432	

Note: *, *** indicate that the two variables are significantly correlated at the 10% and 1% level.

(2) The Intertemporal Impact of Carbon Disclosure on Financial Performance

It can be seen from the above that H2 has been verified, that is, in non−carbon−intensive industries, carbon disclosure has a significant positive impact on current period financial performance. Therefore, this paper will further investigate the intertemporal effects of various explanatory variables on next period financial performance based on this hypothesis.

As can be seen from Table 12 that the coefficient of Carbon Disclosure (CDLI) is 0.0008 (slightly smaller than the coefficient of 0.0009 in Table 11), but its *p*−value is 0.0030, which is significant at the 1% level, the same level of significance as in Table 11. Therefore, the positive impact of carbon disclosure on financial performance can be extended to the next period, and hypothesis H3 is thus verified.

Table 12. The intertemporal impact of carbon disclosure on financial performance (IND = 0).

Variable	Coefficient	t-Value	*p*-Value
CDLI	0.0008 ***	2.99	0.0030
NPM	0.2644 ***	5.39	0.0000
LEV	0.0615 *	1.97	0.0510
SIZE	−0.0367 ***	−7.84	0.0000
GR	−0.0229	−0.71	0.4810
Constant term	0.8768 ***	9.19	0.0000
Adjusted R^2		0.6034	
F-statistic of the model		51.88	
Sig.		0.0000	
Number of samples		432	

Note: *, *** indicate that the two variables are significantly correlated at the 10% and 1% level.

4.3.2. Full Sample Regression Analysis

In order to compare with the regression results of the sub−samples after grouping, the regression analysis for the full sample is also conducted and the result is briefly summarized in Table 13. From which it can be seen that, unlike the sub−sample companies, carbon disclosure and financial performance are not significantly correlated for the full sample companies, thus further verifying the necessity of group research as described above.

Table 13. Regression analysis for the full sample.

Model	Number of Samples	Sig.	Adjusted R^2	Variable	Coefficient	t-Value	*p*-Value
Model 1	752	0.0000	0.5537	CDLI	0.0003	1.40	0.1630

4.4. Robustness Test

In this paper, the return on equity (ROE) is chosen as a substitute variable for financial performance to do the robustness test. According to the classic DuPont analysis system, it is known that ROE = ROA*1/(1−debt ratio), so the return on equity (ROE) and return on assets (ROA) is relatively close, and the regression results are expected to be consistent when the two are used as explanatory variables. The results of robustness test are shown in Table 14.

Table 14. Robustness test results.

Variable	Model 1		Model 2
	IND = 1	IND = 0	IND = 0
CDLI	0.0005 (0.83)	−0.0008 (−0.23)	0.0023 ** (2.42)
NPM	0.5679 *** (3.97)	0.8411 *** (4.62)	0.7596 *** (5.16)
LEV	0.5662 *** (2.84)	0.8613 *** (5.17)	0.8378 *** (5.50)
SIZE	−0.0287 ** (−2.05)	−0.1236 *** (−6.99)	−0.1387 *** (−6.68)
GR	0.0346 ** (2.11)	−0.0154 (−0.22)	−0.1250 (−1.47)
Constant term	0.4468 (1.37)	2.7141 *** (6.63)	2.8638 *** (6.88)
Adjusted R2	0.3764	0.2958	0.4827
F-statistic of the model	4.85	17.77	18.78
Sig.	0.0005	0.0000	0.0000
Number of samples	320	432	432

Note: **, *** indicate that the two variables are significantly correlated at the 5% and 1% level.

For carbon−intensive industries, when ROE is used as a surrogate variable for financial performance, the sign and significance level of the regression coefficient are almost the same as when the return on assets (ROA) is used as the explained variable, that is, although the regression coefficient of carbon disclosure (CDLI) is positive, it has not passed the significance test, which is completely consistent with the regression analysis above.

In non−carbon−intensive industries, when ROE is used as a substitute variable for financial performance, the regression coefficients of key variables are consistent with the above regression results, but the significance level is slightly different. The regression coefficient of carbon disclosure (CDLI) is still positive, but at a lower level of significance compared to using return on assets (ROA) as a financial performance.

In addition, when ROEt + 1 is used as an alternative for the financial performance in the model, the positive impact of carbon disclosure on financial performance will have a significant contribution to financial performance in the next period and is significant at the 5% level.

In conclusion, when the return on equity (ROE) is used as a substitute variable for financial performance, the analysis results of all the main variables of the model in this article are still valid and pass the robustness test.

5. Discussion and Conclusions

Based on previous research and theoretical foundations, this paper combines the average and distribution characteristics of carbon disclosure and divides the sample enterprises into two groups, namely carbon−intensive enterprises, and non−carbon−intensive enterprises, to investigate the impact of carbon disclosure on the current financial performance and whether the significant impact on the current financial performance will be deferred to the next period in these two groups of companies. As this article selects the Fortune 500 companies, which are of great concern to the public. Beside this, these companies have made great contributions to environmental issues and can be regarded as leaders in carbon disclosure. And they have a relatively comprehensive understanding of carbon management, which they have implemented and integrated into their corporate culture, providing practical experience for the implementation of carbon emission reduction to the world. Based on the combination of theory and previous empirical research, this article draws the following research conclusions.

In carbon−intensive industries, although carbon disclosure by the company can promote the improvement of financial performance in the current period, the improvement is small, and it does not pass the significance test. This indicates that the improvement in the quality of carbon disclosure of carbon−intensive companies has a relatively small impact on financial performance. The company has not been greatly rewarded for its high−quality carbon disclosure behavior, and the impact of the carbon disclosure on their current financial performance is still limited despite the high evaluation of their carbon management from the public.

In non−carbon−intensive industries, the carbon disclosure of company can significantly contribute to the improvement of financial performance in the current period. The higher the quality of carbon disclosure, the better the financial performance of the company, and the impact of carbon disclosure on current financial performance can be extended to the next period. Through the analysis of carbon disclosure data, it can be seen that many non−carbon−intensive companies actively respond to the development of low−carbon economy. Taking the financial industry as an example, as a leader in carbon emission reduction in this industry, they vigorously implement carbon strategies and advocate green development of enterprises. Based on the conclusions, non−carbon−intensive industries can get better financial performance with carbon disclosure and this impact will last to the next period. However, in carbon−intensive industries, the improvement of financial performance by discoursing carbon information is small. This can also how that despite the disclosure of carbon information, based on the carbon−intensive characteristics, these companies do not necessarily get better financial performance from disclosing carbon information, which means that carbon−intensive industries also need to achieve the goal of attracting stakeholders by developing low−carbon awareness and reducing carbon emissions. For carbon−non−intensive industries, in order to obtain better financial performance, they should continue to maintain the behavior of disclosing carbon information.

The paper has deepened the understanding of carbon disclosure by enterprises from the theoretical level, which has certain significance for the practice of carbon disclosure of Chinese enterprises, and also promotes the development of empirical research related to carbon disclosure. At the same time, the empirical research in this paper also makes up for the deficiencies of existing research as few scholars have studied whether the significant impact of carbon disclosure on financial performance is deferred to the next period and few scholars have conducted comparative analysis of for carbon−intensive industries and carbon−non−intensive industries. Therefore, this research investigates the impact of carbon disclosure on current financial performance for carbon−intensive industries and carbon−non−intensive industries, and on this basis, the inter−temporal impact of carbon disclosure on financial performance.

Since this paper takes the world's top 500 enterprises as the research object, these enterprises have already had a fairly high awareness of carbon management in terms of carbon information disclosure, the conclusions obtained are not applicable to all enterprises though. However, under the global trend of promoting low carbon development, the study has certain guiding significance, and the Top 500 enterprises have played an exemplary role, which is worth learning from for other countries' enterprises such as China. To this end, this paper proposes the following policy recommendations based on the above findings.

(1) Increase publicity and raise low carbon awareness.

Take China as an example, from the situation of the world's top 500 and China's top 100 enterprises for CDP response, China's top 100 enterprises are less conscious of carbon information disclosure, which shows that for other Chinese enterprises, low carbon awareness is quite weak and will seriously hinder China's vision to achieve carbon emission reduction. Low carbon awareness is the first thing that needs to be advocated, because only by forming low carbon awareness can we take action to reduce carbon emissions and carry out carbon management, achieve sustainable development of enterprises, jointly create a green development atmosphere, avoid the risks brought about by climate change, seize the opportunities of low carbon development, and be brave enough to take on the challenges to achieve a virtuous cycle of low carbon economy.

(2) Improve laws and regulations and regulate carbon disclosure channels.

Laws and regulations are the basis for enterprises to regulate their behavior. In the absence of sound laws and regulations, enterprises often pursue short−term interests at the expense of the environment, which will cause serious damage to the environment in the long run. Although public opinion, media attention, etc. will, to a certain extent, prompt enterprises to disclose carbon information, but their roles are limited. Laws and regulations are the boundaries that enterprises cannot cross, mandatory application of carbon emission reduction targets or disclosure of carbon information to guide the behavior of enterprises are more effective. In addition, the government should also increase the rewards for enterprises that independently disclose carbon information and achieve the emission reduction target or even exceed the emission reduction target, and formulate a standardized incentive policy, while those that violate the relevant laws and regulations should be strictly punished, so as to strengthen enterprises' awareness of the initiative to disclose carbon emissions and cultivate their sense of social responsibility in the legal system.

(3) Establish a unified carbon information disclosure system.

For example, at present, there is no unified carbon information disclosure system in China. Although some enterprises are conscious of reducing their carbon emissions, the accounting methods among enterprises are different and not comparable, and the evaluation standards among enterprises are also different. The complex assessment procedures and the lack of comparability, even with information lacking authenticity, will not only confuse the judgment of corporate investors and mislead corporate investments, but also discourage corporate compliance with the principles of low−carbon development. For example, companies may reduce their carbon emission statistics by changing their carbon emission accounting methods. Therefore, the government should gradually establish a unified carbon information disclosure system and develop a standard and unified accounting method, so that every enterprise can have evidence to follow and evidence to rely on, and investors and other relevant stakeholders can also benefit from it, so that every enterprise can pay close attention to its own carbon emissions and lay the foundation for the national carbon emission reduction target.

(4) Standardize the way of carbon information disclosure in CDP reports.

Because carbon information disclosure has shifted from a voluntary to a mandatory requirement in many jurisdictions, the format and content of CDP reports could be considered to be formulated as a formal GHG statement. Based on our research analysis, we observe that there is room for improvement in the current version of the CDP report. For example, we believe that there should be industry−specific disclosure guidelines and that there should be more information disclosure at the project level. In addition, companies need to upgrade their accounting systems to match the current needs of the low carbon economy.

As the paper mentioned before, although previous studies have been conducted to find the relationship between carbon disclosure and financial performance, and the number of literatures on financial performance research from the perspective of environmental information disclosure is relatively large, scholars still cannot reach a consistent research conclusion. One of the reasons for this situation can be attributed to the lack of a unified carbon information disclosure index. If the above−mentioned suggestions are adopted, the establishment of a unified carbon information disclosure system and standardization of the carbon information disclosure method in the CDP report can effectively resolve this problem. Coupled with the guidance and regulation of laws and regulations, companies with higher low−carbon ceremonies will consciously and proactively disclose carbon information in accordance with the normative model, which can also provide better indicator choices for relevant research. Besides, on the basis of being able to obtain uniform and standardized carbon information disclosure measurement indicators, scholars can choose more different research subjects to study the relationship between carbon disclosure and financial performance and. While at the same time, because of the consistency of indicators measurement methods, the results of all these studies will become more comparable to provide the impact of carbon disclosure on financial performance among different industries in the future.

Author Contributions: Conceptualization, N.Z. and J.Z.; methodology, W.L. and N.Z.; software, J.Z. and W.L.; validation, W.L. and J.Z.; formal analysis, J.Z. and W.L.; investigation, W.L.; resources, W.L.; data curation, W.L. and J.Z.; writing—original draft preparation, W.L. and J.Z.; writing—review and editing, W.L.; visualization, W.L. and J.Z.; supervision, N.Z.; project administration, N.Z.; funding acquisition, N.Z. All authors have read and agreed to the published version of the manuscript.

Funding: This research received no external funding.

Institutional Review Board Statement: Not applicable.

Informed Consent Statement: Not applicable.

Data Availability Statement: The carbon disclosure data used in this article are all sourced from the Carbon Disclosure Project (CDP), available at https://www.cdp.net/ (accessed on 15 June 2021). The financial data used come from Wind Information, available at https://www.wind.com.cn/ (accessed on 15 June 2021).

Acknowledgments: This work was supported by the National Social Science Fund of China (Grant No. 20BGL099).

Conflicts of Interest: The authors declare no conflict of interest.

References

1. Liu, Y.; Zhou, Z.; Zhang, X.; Xu, X.; Chen, H.; Xiong, Z. Net global warming potential and greenhouse gas intensity from the double rice system with integrated soil–crop system management: A three-year field study. *Atmos. Environ.* **2015**, *116*, 92–101. [CrossRef]
2. Sands, F. The United Nations framework convention on climate change. *Rev. Eur. Community Int. Environ. Law* **1992**, *1*, 270. [CrossRef]
3. Stocker, T.F.; Qin, D.; Plattner, G.-K.; Tignor, M.; Allen, S.K.; Boschung, J.; Nauels, A.; Xia, Y.; Bex, V.; Midgley, P.M. IPCC, 2013: Climate Change 2013: The Physical Science Basis. Contribution of Working Group I to the Fifth Assessment Report of the Intergovernmental Panel on Climate Change. Available online: https://www.ipcc.ch/report/ar5/wg1/ (accessed on 15 June 2021).
4. Global Warming of 1.5 °C. Available online: https://www.ipcc.ch/sr15/ (accessed on 15 June 2021).
5. The Global Risks Report. 2020. Available online: https://www.weforum.org/reports/the-global-risks-report$-$2020 (accessed on 15 June 2021).
6. Luo, L. The influence of institutional contexts on the relationship between voluntary carbon disclosure and carbon emission performance. *Account. Financ.* **2019**, *59*, 1235–1264. [CrossRef]
7. Li, D.; Huang, M.; Ren, S.; Chen, X.; Ning, L. Environmental legitimacy, green innovation, and corporate carbon disclosure: Evidence from CDP China 100. *J. Bus. Ethics* **2018**, *150*, 1089–1104. [CrossRef]
8. Meng, X.H.; Zeng, S.X.; Shi, J.J.; Qi, G.Y.; Zhang, Z.B. The relationship between corporate environmental performance and environmental disclosure: An empirical study in China. *J. Environ. Manag.* **2014**, *145*, 357–367. [CrossRef] [PubMed]
9. Zhu, N.; Qian, L.; Jiang, D.; Mbroh, N. A simulation study of China's imposing carbon tax against American carbon tariffs. *J. Clean. Prod.* **2020**, *243*, 118467. [CrossRef]
10. Liu, L. A Study on Corporate Carbon Emission Information Disclosure Model. *Commun. Financ. Account.* **2015**, *16*, 23–26.
11. Hahn, R.; Reimsbach, D.; Schiemann, F. Organizations, climate change, and transparency: Reviewing the literature on carbon disclosure. *Organ. Environ.* **2015**, *28*, 80–102. [CrossRef]
12. Uyar, A.; Karaman, A.S.; Kilic, M. Is corporate social responsibility reporting a tool of signaling or greenwashing? Evidence from the worldwide logistics sector. *J. Clean. Prod.* **2020**, *253*, 119997. [CrossRef]
13. Mobley, S.C. The Challenges of Socio-Economic Accounting. *Account. Rev.* **1970**, *45*, 762–768.
14. Gray, R.; Owen, D.; Maunders, K. Corporate Social Reporting: Emerging Trends in Accountability and the Social Contract. *Account. Audit. Account. J.* **1988**, *1*, 6–20. [CrossRef]

15. Friedman, M. The Social Responsibility of Business is to Increase Its Profit. *The New York Times Magazine*, 13 September 1970; 122–126.
16. Deegan, C.; Gordon, B. A Study of the Environmental Disclosure Practices of Australian Corporations. *Account. Bus. Res.* **1996**, *26*, 187–199. [CrossRef]
17. Karim, K.E.; Lacina, M.J.; Rutledge, R.W. The Association between Firm Characteristics and the Level of Environmental Disclosure in Financial Statement Footnotes. *Adv. Environ. Account. Manag.* **2006**, *3*, 77–109.
18. Dhaliwal, D.; Radhakrishnan, S.; Tsang, A.; Yong, G.Y. Nonfinancial Disclosure and Analyst Forecast Accuracy: International Evidence on Corporate Social Responsibility Disclosure. *Account. Rev.* **2012**, *87*, 723–759. [CrossRef]
19. Lys, T.; Naughton, J.P.; Wang, C. Signaling Through Corporate Accountability Reporting. *SSRN Electron. J.* **2015**, *60*, 56–72. [CrossRef]
20. Hoi, C.K.; Zhang, H. Is Corporate Social Responsibility (CSR) Associated with Tax Avoidance? Evidence from Irresponsible CSR activities. *Account. Rev.* **2013**, *88*, 2025–2059. [CrossRef]
21. Hassel, L.; Nilsson, H.; Nyquist, S. The Value Relevance of Environmental Performance. *Eur. Account. Rev.* **2005**, *14*, 41–61. [CrossRef]
22. Chapple, L.; Clarkson, P.M.; Gold, D.L. The Cost of Carbon: Capital Market Effects of the Proposed Emission Trading Scheme (ETS). *Abacus* **2013**, *49*, 1–33. [CrossRef]
23. Griffin, P.A.; Lont, D.H.; Sun, E.Y. The Relevance to Investors of Greenhouse Gas Emission Disclosures. *Contemp. Account. Res.* **2017**, *34*, 1265–1297. [CrossRef]
24. Saka, C.; Oshika, T. Disclosure effects, carbon emissions and corporate value. *Sustain. Account. Manag. Policy J.* **2014**, *5*, 22–45. [CrossRef]
25. Velte, P.; Stawinoga, M.; Lueg, R. Carbon performance and disclosure: A systematic review of governance-related determinants and financial consequences. *J. Clean. Prod.* **2020**, *254*, 120063. [CrossRef]
26. Lemma, T.T.; Feedman, M.; Mlilo, M.; Park, J.D. Corporate carbon risk, voluntary disclosure, and cost of capital: South African evidence. *Bus. Strategy Environ.* **2019**, *28*, 111–126. [CrossRef]
27. Lueg, K.; Krastev, B.; Lueg, R. Bidirectional effects between organizational sustainability disclosure and risk. *J. Clean. Prod.* **2019**, *229*, 268–277. [CrossRef]
28. Wegener, M. The Carbon Disclosure Project, an Evolution in International Environmental Corporate Governance: Motivations and Determinants of Market Response to Voluntary Disclosures. Master's Thesis, Brock University, St. Catharines, ON, Canada, 2010.
29. Ziegler, A.; Busch, T.; Hoffmann, V.H. Disclosed corporate responses to climate change and stock performance: An international empirical analysis. *Energy Econ.* **2011**, *33*, 1283–1294. [CrossRef]
30. Schiager, H.; Haukvik, G.D. The Effect of Voluntary Environmental Disclosure on Firm Value: A Study of Nordic Listed Firms. Master's Thesis, Norges Handelshøyskole School, Bergen, Norway, 2012.
31. Ganda, F. The influence of carbon emissions disclosure on company financial value in an emerging economy. *Environ. Dev. Sustain.* **2018**, *20*, 1723–1738. [CrossRef]
32. Iskandar, D.; Fran, E. The Effect of Carbon Emissions Disclosure and Corporate Social Responsibility on the Firm Value with Environmental Performance as Variable Control. *Res. J. Financ. Account.* **2016**, *7*, 9.
33. Siddique, A.; Akhtaruzzaman, M.; Rashid, A.; Hammami, H. Carbon disclosure, carbon performance and financial performance: International evidence. *Int. Rev. Financ. Anal.* **2021**, *75*, 101734. [CrossRef]
34. Borghei, Z.; Leung, P.; Guthrie, J. Voluntary greenhouse gas emission disclosure impacts on accounting-based performance: Australian evidence. *Australas. J. Environ. Manag.* **2018**, *25*, 321–338. [CrossRef]
35. Brouwers, R.; Schoubben, F.; Cynthia Van Hulle, C.V. The influence of carbon cost pass through on the link between carbon emission and corporate financial performance in the context of the European Union Emission Trading Scheme. *Bus. Strategy Environ.* **2018**, *27*, 1422–1436. [CrossRef]
36. Freedman, M.; Jaggi, B. An analysis of the association between pollution disclosure and economic performance. *Account. Audit. Account. J.* **1988**, *1*, 43–58. [CrossRef]
37. Murray, A.; Sinclair, D.; Power, D.; Gray, R. Do Financial Markets Care About Social and Environmental Disclosure? *Account. Audit. Account. J.* **2006**, *19*, 228–255. [CrossRef]
38. Anderson, J.; Frankle, A. Voluntary Social Reporting an Iso-Beta Portfolio Analysis. *Account. Rev.* **1980**, *19*, 467–479.
39. Stanwick, S.D.; Stanwick, P.A. The relationship between environmental disclosures and financial performance: An empirical study of US firms. *Corp. Soc. Responsib. Environ. Manag.* **2000**, *7*, 155–164. [CrossRef]
40. Nor, N.M.; Bahari, N.A.; Adnan, N.A.; Kamal, S.M.; Ali, I.M. The Effects of Environmental Disclosure on Financial Performance in Malaysia. *Procedia Econ. Financ.* **2016**, *35*, 117–126. [CrossRef]
41. Luo, L.; Lan, Y.C.; Tang, Q. Corporate incentives to disclose carbon information: Evidence from the CDP Global 500 report. *J. Int. Financ. Manag. Account.* **2012**, *23*, 93–120. [CrossRef]
42. Luo, L.; Tang, Q.; Lan, Y.C. Comparison of propensity for carbon disclosure between developing and developed countries: A resource constraint perspective. *Account. Res. J.* **2013**, *26*, 6–34. [CrossRef]
43. Matsumura, E.M.; Prakash, R.; Vera-Muñoz, S.C. Firm-value effects of carbon emissions and carbon disclosures. *Account. Rev.* **2013**, *89*, 695–724. [CrossRef]

44. Zhao, X.M.; Li, Y.Y. The Company Performance and the Information Disclosure Quality of Carbon Emission- From the Evidence of China's Listed Enterprises. *J. Xi'an Shiyou Univ.* **2013**, *22*, 22–27.
45. Zhao, X.; Yan, G. Business Performance and Carbon Disclosure Level in Heavy Polluting Industries: An Empirical Study Based on CDP China Report 2008–2011. *Commun. Financ. Account.* **2014**, *18*, 68–70.
46. Li, X.; Shi, Y. The Impact of the Quality of Carbon Information Disclosure on Corporate Financial Performance in the Concept of Green Development. *Econ. Manag.* **2016**, *7*, 119–132.
47. Zhu, N.; Bu, Y.; Jin, M.; Mbroh, N. Green financial behavior and green development strategy of Chinese power companies in the context of carbon tax. *J. Clean. Prod.* **2020**, *245*, 118908. [CrossRef]
48. Piesiewicz, M.; Ciechan-Kujawa, M.; Kufel, P. Differences in Disclosure of Integrated Reports at Energy and Non-Energy Companies. *Energies* **2021**, *14*, 1253. [CrossRef]
49. Frost, G.R.; Wilmshurst, T.D. Corporate environmental reporting: A test of legitimacy theory. *Account. Audit. Account. J.* **2000**, *13*, 10–26.
50. Clarkson, P.M.; Li, Y.; Richardson, G.D.; Vasvari, F.P. Revisiting the relation between environmental performance and environmental disclosure: An empirical analysis. *Account. Organ. Soc.* **2008**, *33*, 303–327. [CrossRef]
51. Al-Tuwaijri, S.A.; Christensen, T.E.; Hughes, K.I. The relations among environmental disclosure, environmental performance, and economic performance: A simultaneous equations approach. *Soc. Sci. Electron. Publ.* **2004**, *29*, 447–471. [CrossRef]
52. Jenkins, H.; Yakovleva, N. Corporate Social Responsibility in the Mining Industry: Exploring Trends in Social and Environmental Disclosure. *J. Clean. Prod.* **2004**, *14*, 274–284. [CrossRef]
53. Freedman, M.; Jaggi, B. Pollution disclosures, pollution performance and economic performance. *Omega* **1982**, *10*, 167–176. [CrossRef]
54. Richardson, A.J.; Welker, M. Social disclosure, financial disclosure and the cost of equity capital. *Account. Organ. Soc.* **2001**, *26*, 597–616. [CrossRef]
55. Johnson, H.H. Does it pay to be good? Social responsibility and financial performance. *Bus. Horiz.* **2003**, *46*, 34–40. [CrossRef]
56. Stanny, E.; Ely, K. Corporate Environmental Disclosures about the Effects of Climate Change. *Corp. Soc. Responsib. Environ. Manag.* **2008**, *15*, 338–348. [CrossRef]
57. Hsu, A.W.; Wang, T. Does the market value corporate response to climate change? *Omega* **2013**, *41*, 195–206. [CrossRef]
58. Plumlee, M.; Marshall, S.; Brown, D. Voluntary Environmental Disclosure Quality and Firm Value. Roles of Venue and Industry Type. *Soc. Sci. Electron. Publ.* **2009**, *34*, 336–361.
59. Dhaliwal, D.S.; Li, O.Z.; Tsang, A.; Yang, Y.G. Voluntary nonfinancial disclosure and the cost of equity capital: The initiation of corporate social responsibility reporting. *Account. Rev.* **2011**, *86*, 59–100. [CrossRef]
60. Clarkson, P.M.; Fang, X.; Li, Y.; Richardson, G. The relevance of environmental disclosures: Are such disclosures incrementally informative? *J. Account. Public Policy* **2013**, *32*, 410–431. [CrossRef]
61. De Villiers, C.; Marques, A. Corporate social responsibility, country-level predispositions, and the consequences of choosing a level of disclosure. *Account. Bus. Res.* **2016**, *46*, 167–195. [CrossRef]
62. He, Y.; Tian, Q.; Wang, K. Carbon Disclosure, Carbon Performance, and Cost of Capital. *Account. Res.* **2014**, *1*, 79–86, 95. [CrossRef]

energies

Article

Technical and Economic Analysis of the Supercritical Combined Gas-Steam Cycle

Marcin Jamróz [1], Marian Piwowarski [1,*], Paweł Ziemiański [2] and Gabriel Pawlak [3]

1 Faculty of Mechanical Engineering, Gdansk University of Technology, ul. Gabriela Narutowicza 11/12, 80-233 Gdansk, Poland; marcinjamroz@poczta.fm
2 Faculty of Management and Economics, Gdansk University of Technology, ul. Gabriela Narutowicza 11/12, 80-233 Gdansk, Poland; pawel.ziemianski@pg.edu.pl
3 Economica Consulting, 80-215 Gdansk, Poland; biuro@economica.com.pl
* Correspondence: marian.piwowarski@pg.edu.pl; Tel.: +48-58-347-1429

Abstract: Combined cycle power plants are characterized by high efficiency, now exceeding 60%. The record-breaking power plant listed in the Guinness Book of World Records is the Nishi-Nagoya power plant commissioned in March 2018, located in Japan, and reaching the gross efficiency of 63.08%. Research and development centers, energy companies, and scientific institutions are taking various actions to increase this efficiency. Both the gas turbine and the steam turbine of the combined cycle are modified. The main objective of this paper is to improve the gas-steam cycle efficiency and to reach the efficiency that is higher than in the record-breaking Nishi-Nagoya power plant. To do so, a number of numerical calculations were performed for the cycle design similar to the one used in the Nishi-Nagoya power plant. The paper assumes the use of the same gas turbines as in the reference power plant. The process of recovering heat from exhaust gases had to be organized so that the highest capacity and efficiency were achieved. The analyses focused on the selection of parameters and the modification of the cycle design in the steam part area in order to increase overall efficiency. As part of the calculations, the appropriate selection of the most favorable thermodynamic parameters of the steam at the inlet to the high-pressure (HP) part of the turbine (supercritical pressure) allowed the authors to obtain the efficiency and the capacity of 64.45% and about 1.214 GW respectively compared to the reference values of 63.08% and 1.19 GW. The authors believe that efficiency can be improved further. One of the methods to do so is to continue increasing the high-pressure steam temperature because it is the first part of the generator into which exhaust gases enter. The economic analysis revealed that the difference between the annual revenue from the sale of electricity and the annual fuel cost is considerably higher for power plants set to supercritical parameters, reaching approx. USD 14 million per annum. It is proposed that investments in adapting components of the steam part to supercritical parameters may be balanced out by a higher profit.

Keywords: combined gas-steam cycles; efficiency; heat exchange in Heat Recovery Steam Generators (HRSG); economic analysis; cost management; managerial decisions

Citation: Jamróz, M.; Piwowarski, M.; Ziemiański, P.; Pawlak, G. Technical and Economic Analysis of the Supercritical Combined Gas-Steam Cycle. *Energies* **2021**, *14*, 2985. https://doi.org/10.3390/en14112985

Academic Editor: Panagiotis Fragkos

Received: 18 April 2021
Accepted: 19 May 2021
Published: 21 May 2021

1. Introduction

In recent years, electricity generation and conversion issues have been in the spotlight of many scientific institutions, research and development centers, energy companies, and even state governments. In the European Union as well as outside it, particular emphasis is placed on improving the efficiency of electricity generation [1,2]. Currently, high-efficiency steam turbines set to supercritical and ultra-supercritical parameters are preferred in conventional energy systems [3,4]; unfortunately, this is a coal technology.

Another marked tendency is to build combined gas-steam cycles, which allow very high efficiencies, reaching 60%, to be achieved [5]. In addition to commercial activities, various research on gas turbines with an external combustion chamber is being conducted, also allowing the electricity generation efficiency to be improved [6,7]. On the other hand, another strong and applied tendency is distributed energy [8,9], which uses renewable energy sources [10,11]. Electricity generation using more gaseous fuels from biomass gasification or biomass itself is also under analysis [12,13]. Power plants that use gas microturbines and closed cycles [14], and polygeneration [15] are developing fast. It seems that the use of the microturbines technology in distributed energy will also be developed [16,17]. This technology has seen considerable improvements in terms of efficiency and applicability due to the use of generators based on rare-earth magnets [18,19]. Research on improvements to efficiency and durability and lower building costs of energy systems based on fuel cells [20,21] or PV cells [22,23] is underway. The construction of power plants and heat and power plants using organic Rankine cycles (ORCs) [24] is a separate direction for energy development. Due to their very low temperature, ORC plants have a low efficiency of approx. 10%, only exceptionally being able to reach approx. 20% [25]. It is possible to improve the cycle efficiency by increasing the upper temperature and modifying the plant design [26,27]. Despite very considerable developments in distributed energy and alternative energy sources, it should be concluded that high-efficiency power plants with steam turbines set to supercritical parameters, combined gas-steam cycle power plants, and nuclear power plants [28] will be developed in terms of high installed capacities.

The review above showed that researchers put a lot of effort into improving the efficiency of electricity generation, whereas the aim of this paper is to improve the efficiency of the combined gas-steam cycle—and to perform an economic evaluation of such improvement. The concept of the combined cycle power plant is to use hot exhaust gases leaving the gas turbine in order to generate steam in a heat recovery steam generator, which will be used to drive the steam turbine. Plants of this type use two thermodynamic cycles. The Brayton cycle is used in the gas turbine, while the Rankine cycle in the steam part. The gas turbine cycle is referred to as a topping cycle, while the steam part cycle as a bottoming cycle [29]. The names stem from the fact that high-temperature heat is supplied in the topping cycle, whereas heat at a much lower temperature leaves the bottoming cycle. A heat recovery steam generator (HRSG) links these two cycles. In recent years, combined cycles have moved from the concept of improving the efficiency of electricity generation systems to the most preferred fossil fuel power plants [30]. One distinctive feature of combined cycles is that power is generated from both the gas turbine and the steam turbine when burning fuel. The comparison of the volume of CO_2 emitted to the amount of energy generated is much more favorable for combined cycles than for a typical coal-fired condensing power plant [31]. Gas-steam cycles are becoming increasingly popular due to a number of advantages such as high efficiency-now exceeding 60% (the current record, 63.08%, was set in Japan in the Nishi-Nagoya power plant) [32], high operational reliability, capacity reaching 1 GW, high flexibility translating into short startup times to reach full load, process automation possibilities, and low greenhouse gas emissions [33,34].

As already mentioned, the heat recovery steam generator, based on a countercurrent heat exchanger, is what links the gas and steam cycles. The generator design includes three zones. The first one is the economizer zone in which water is heated to the saturation temperature. Then it is transferred to the evaporator zone in order to generate steam. The last zone is the superheater zone, where the temperature is increased on purpose in order to improve both the steam turbine capacity and rate of heat recovery from exhaust gases. A point at which the water evaporation process starts is crucial in the entire heat recovery process. It is the point where the temperature difference between the heat transfer exhaust gas and the receiving water must be the lowest. This difference is referred to as a pinch point. The smaller it is, the higher the efficiency of the gas-steam cycle efficiency is, but it is achieved thanks to the larger heat exchange surface—which affects the cost of the entire installation [35]. Forced circulation heat recovery steam generators are mainly used; however, there are also plants that use assisted or natural circulation heat recovery steam generators. The heat recovery process in a single-pressure HRSG is given in Figure 1, while an example of a single-pressure steam-gas combined cycle power plant is presented in Figure 2 [36]. Heat recovery steam generators must face many challenges, which include a high rate of heat recovery from exhaust gases, allowing for small pressure losses at the steam side, as well as corrosion resistance and low pressure losses at the exhaust gas side. To ensure high heat recovery efficiency and low exergy losses, the difference between the heat transfer medium and the receiving medium should be as low as possible [37]. This means that a large heat exchange surface area is required. Heat recovery steam generators that are currently designed have very low pressure losses from 25 to 30 mbar at the exhaust gas side [38]. Designs where practically any fuel is burnt in the generator are also analyzed [39,40].

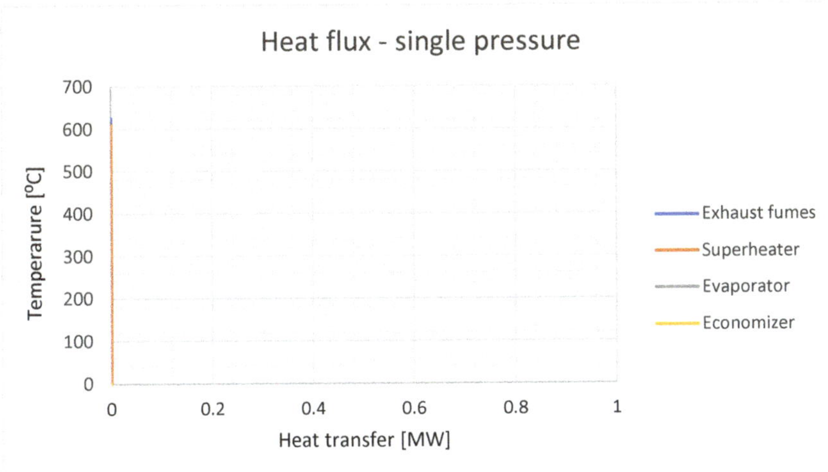

Figure 1. Heat exchange process in an HRSG (interpretations in temperature-heat flux).

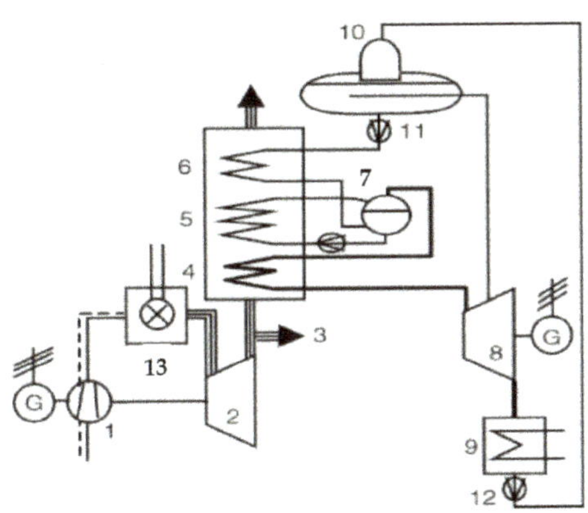

Figure 2. Schematic diagram of a single-pressure gas-steam cycle, where 1—Compressor, 2—Gas turbine, 3—Bypass to the stack, 4—Superheater zone, 5—Vaporiser zone, 6—Economiser zone, 7—Drum, 8—Steam turbine, 9—Condenser, 10—Deaerator, 11—Main pump, 12—Circulation pump, 13—Combustion chamber.

A gas turbine with the highest possible efficiency should be used in order to obtain a high-efficiency gas-steam cycle. This means that new materials able to withstand higher temperatures must be used and developed. Apart from the main components such as the steam turbine, gas turbine, and the heat recovery steam generator, there have been significant developments in all systems and installations linked directly with the cycle.

They all serve the same purpose—to minimize losses. At present, the following trends can be observed in order to achieve the aforementioned goals [35,39,41]:

- Higher combustion temperatures to increase gas turbine efficiency and steam cycle efficiency thanks to higher steam parameters;
- New designs for gas turbines with higher efficiencies;
- Higher gas turbine capacity to benefit from the economies of scale;
- Lower operating costs due to the use of remote control;
- Lower emissions, in particular NO_2 emissions, to reduce environmental impact;
- Better cycle loading capabilities to control partial load and frequency;
- Development of hydrogen-fueled gas turbines.

For new gas-steam cycles, an existing gas turbine model is selected and then the steam part is designed and optimized according to given requirements [42]. There are also models designed from the ground up for this type of systems. Contemporary high-capacity combined cycles use steam at a pressure up to 17 MPa, with the steam temperature at the steam turbine inlet reaching 580 °C. For steam turbines of over 100 MW, superheating is now used as standard. The trend can be expected to continue in the near future. Large power stations will operate at higher parameters, namely pressure and temperature. Before increasing steam parameters, the following issues must be analyzed [38]:

- A higher operating temperature requires the use of much more expensive alloys in the heat recovery steam generator, steam turbine, and pipelines. A higher investment cost must be justified by higher capacity and efficiency;
- A higher pressure leads to higher thickness of walls in all components, reducing thermal flexibility and increasing costs;
- A higher pressure combined with superheating reduces the main steam flow. This leads to issues with the design of high-pressure turbines.

A higher steam pressure does not necessarily mean that the efficiency of the entire combined cycle is higher. Optimization is achieved not only due to the steam cycle efficiency but also due to the extent to which exhaust gas heat is used to generate steam. The cycle efficiency will go up as the pressure increases but up to a certain point. At the same time, overall efficiency and total capacity will decrease. That is why the selection of the fresh steam pressure depends on the steam turbine efficiency, the steam cycle efficiency, and the efficiency at which heat from exhausts gases is recovered in the heat recovery steam generator.

In combined cycles, the use of multi-pressure systems (two and three-pressure heat recovery steam generators) allows to reduce the temperature difference between fluids, and thus to increase the efficiency of the entire cycle. The key task is to select the values of these pressures and divide the flow rates to obtain the highest efficiency.

As the temperature of gas turbine exhaust gases increases, the selection of an appropriate cycle configuration becomes crucial. Most frequently, a three-pressure cycle is the optimum solution for a system in which exhaust gases at a temperature of approx. 600 °C are available. A trend towards single-pressure cycles will be observed when exhaust gases at temperatures of approx. 750–800 °C start to be used.

Current combined cycle power units achieve a net efficiency of approx. 58.5% thanks to a number of improvements [38,43]. Power units currently in use are considered here. Research and tests aimed at reaching the efficiency of more than 60% are also underway. In terms of improving the efficiency of combined gas-steam cycles, three main development directions are taken: selection of optimum thermodynamic parameters in the steam part; selection of the optimum type of heat recovery steam generator; and the selection of the most favorable distribution of heated areas in the heat recovery steam generator-all with the highest cost-effectiveness of the solution in mind.

In this paper, the authors proposed using a supercritical steam pressure to improve the efficiency (Figure 3). Further on in the paper, design calculations for steam turbines were made and an economic analysis was carried out, which may form a basis for making appropriate investment and managerial decisions in terms of selecting the most optimal solution.

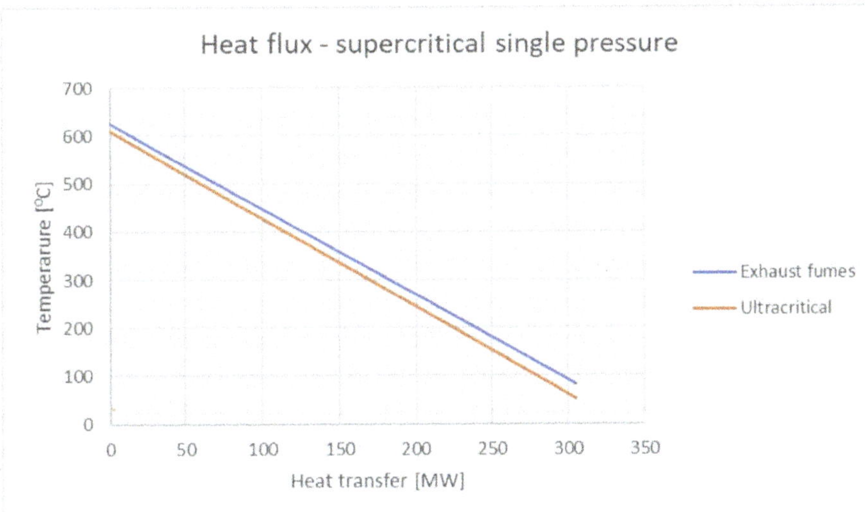

Figure 3. Example of the heat exchange process in a single-pressure heat recovery steam generator with supercritical steam pressure (interpretations in temperature-heat flux).

2. Modelling

A combined cycle power plant with the highest efficiency—the Nishi-Nagoya power plant located in Japan—was chosen as a reference power plant. Commissioned in March 2018, this power plant reaches a gross efficiency of 63.08% [32,43]. It is configured as 3 × 3 × 1. This means that the power unit includes 3 gas turbines whose exhaust gases move to 3 heat recovery steam generators. Heat is recovered from exhaust gases and steam is generated there. Then steam moves to the steam turbine. The objective of this paper is to analyze the improvement of the efficiency of this combined cycle by applying supercritical parameters in the HP part of the steam turbine. A three-pressure combined cycle power plant with superheating set to subcritical parameters (Figure 4a) and supercritical parameters (Figure 4b) was analyzed. A number of assumptions were made for the calculations and their values are presented in Table 1 [29,33,35,44].

Assuming that the pressure in the heat recovery steam generator is constant, the heat flux in exhaust gases leaving the gas turbine may be given as

$$\dot{Q}_{GTexh} = \dot{m}_{exh} * c_{pexh} * \Delta t = \dot{m}_{exh} * c_{pexh} * \left(t_4^{TG} - t_1^{TG}\right) = \dot{Q}_{max}, \quad (1)$$

where: $\dot{Q}_{max}$—heat flux in exhaust gases, $\dot{m}_{exh}$—exhaust gas mass flow rate, c_{pexh}—specific heat of exhaust gases, t_4^{TG}—exhaust gas temperature at gas turbine outlet, t_1^{TG}—temperature of exhaust gases leaving the heat recovery steam generator.

The entire steam generation process is divided into 3 stages. Heat fluxes can also be given for each of these 3 stages.

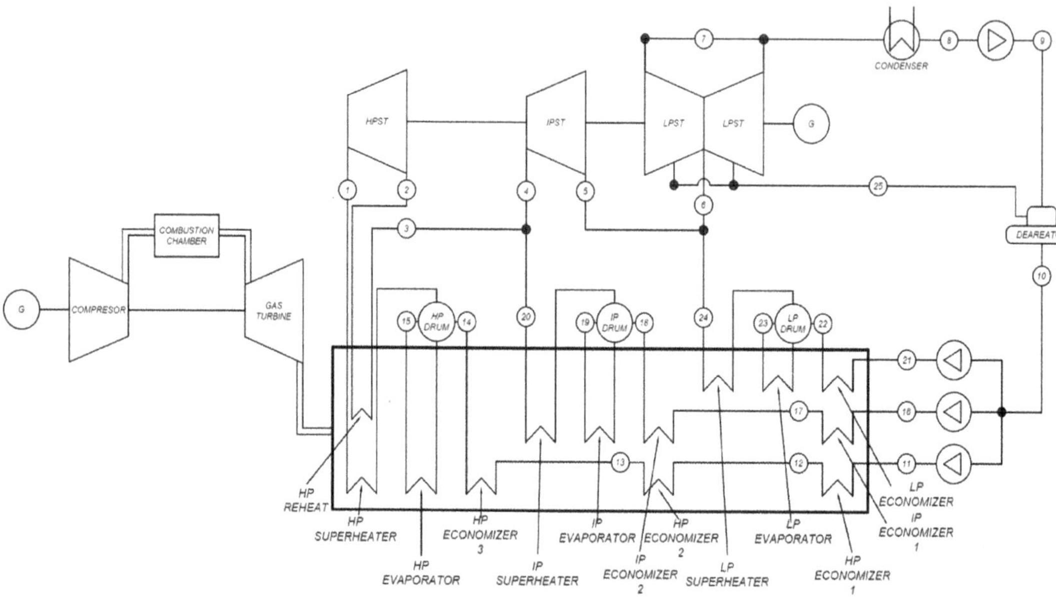

(a)

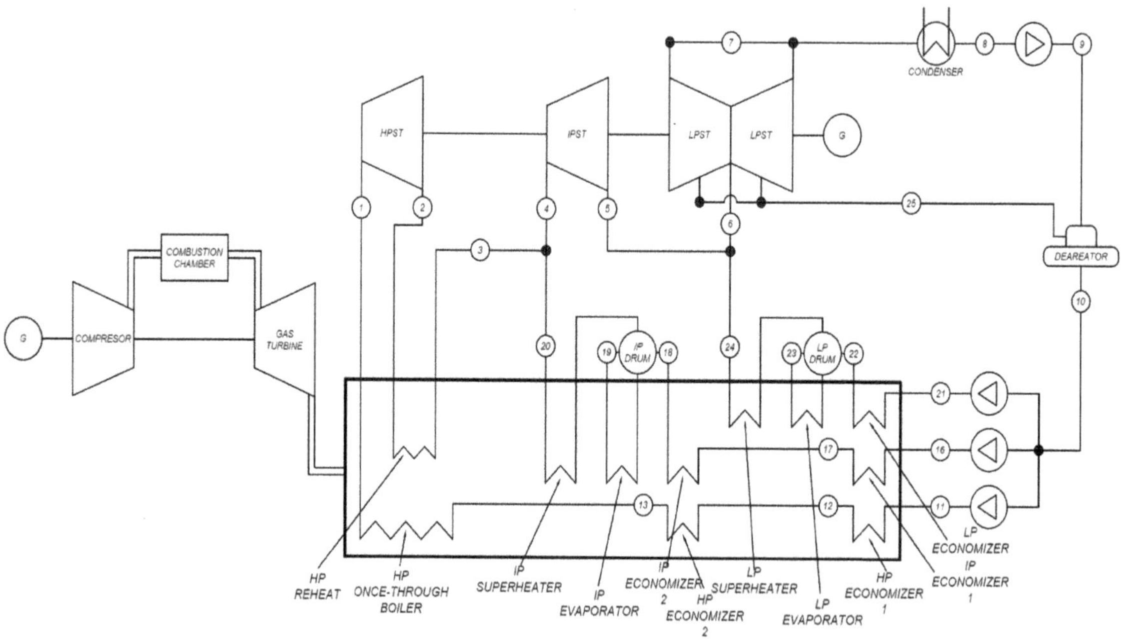

(b)

Figure 4. Three-pressure cycle with superheating (**a**) and HP turbine set to supercritical parameters (**b**).

Table 1. Assumed values of the efficiencies of particular cycle elements.

Description	Symbol	Value	Unit
Gas turbine capacity	N_{TG}	280	MW
Flow rate of exhaust gases	$\dot{m}_{exh}$	510	kg/s
Outlet temperature of exhaust gases	t_G	626	°C
Superheat temperature	t_G	610	°C
Mean specific heat of exhaust gases	c_{pexh}	1.1	kJ/kg K
Pressure drop in the heat exchanger	p_i/p_{i-1}	0.01	–
Temperature increase in the deaerator	Δt_D	30	°C
Pinch point in the HRSG	Δt_{HRSG}	3	°C
HP turbine efficiency	η_{HP}	0.93	–
IP turbine efficiency	η_{IP}	0.94	–
LP turbine efficiency	η_{LP}	0.93	–
Pump efficiency	η_P	0.85	–
Mechanical efficiency	η_m	0.99	–
Leakage efficiency (losses)	η_n	0.985	–
Generator efficiency	η_G	0.98	–

- Water heating

$$\dot{Q}_{eco} = \dot{m}_{med} * c_{pwater} * \Delta t = \dot{m}_{med} * c_{pwater} * \ (t_3 - t_2) \ = \dot{Q}_{max}, \tag{2}$$

where: $\dot{m}_{med}$—mass flow rate in the steam cycle, c_{pwater}—specific heat of water, t_3—water temperature at economizer outlet, t_2—water temperature at economizer inlet,

- Evaporation

$$\dot{Q}_{evo} = \dot{m}_{med} * \Delta h = \dot{m}_{TP} * \ (h_4 - h_3), \tag{3}$$

where: h_4—specific enthalpy at the end of evaporation process, h_3—specific enthalpy in saturation point.

- Superheating

$$\dot{Q}_{sup} = \dot{m}_{med} * \Delta h = \dot{m}_{med} * \ (h_5 - h_4), \tag{4}$$

where: h_5—specific enthalpy at the end of superheating process.

The total heat flux used to generate steam is a sum of the three heat fluxes given above:

$$\dot{Q}_{TP} = \dot{Q}_{eco} + \dot{Q}_{evo} + \dot{Q}_{sup}. \tag{5}$$

Minimum temperature differences between the media that give off and receive heat are observed in two places:

- Between exhaust gases and water ($\Delta t_{exhaust-water}$);
- Between exhaust gases and steam ($\Delta t_{exhaust-steam}$).

The gas-steam cycle efficiency can be defined as [38]

$$\eta_{GS} = \frac{N_{TG} + N_{Tp}}{\dot{Q}_D} = \eta_{TG}\left(1 + \frac{N_{TG}}{N_{Tp}}\right), \tag{6}$$

where: η_{GS}—gas-steam cycle efficiency, η_{TG}—gas turbine efficiency.

Three-pressure cycle with superheating

High pressure mass flow:

$$\dot{m}_{HP} = \frac{\dot{m}_{exh} * c_{pexh} * \ \left(t_{exh}^{TG} - \left(t_{sat}^{HP} + \Delta t\right)\right)}{(h_1 - h_{14}) + (h_3 - h_2)}. \tag{7}$$

Intermediate pressure mass flow:

$$\dot{m}_{IP} = \frac{\dot{m}_{ehx} * c_{pspal} * \left(\left(t_{sat}^{HP} + \Delta t \right) - \left(t_{sat}^{IP} + \Delta t \right) \right) - \dot{m}_{HP} * (h_{14} - h_{13})}{(h_{20} - h_{18})}. \tag{8}$$

Low pressure mass flow:

$$\dot{m}_{LP} = \frac{\dot{m}_{exh} * c_{pexh} * \left(\left(t_{nas}^{IP} + \Delta t \right) - \left(t_{nas}^{LP} + \Delta t \right) \right) - \dot{m}_{HP} * (h_{13} - h_{12}) - \dot{m}_{IP} * (h_{18} - h_{17})}{(h_{24} - h_{22})}. \tag{9}$$

Outlet temperature of gas turbine fumes leaving HRSG:

$$t_{out}^{TG} = \left(t_{sat}^{LP} + \Delta t \right) - \frac{\dot{m}_{HP} * (h_{12} - h_{11}) + \dot{m}_{IP} * (h_{17} - h_{16}) + \dot{m}_{LP} * (h_{22} - h_{21})}{\left(\dot{m}_{spal} * Cp_{spal} \right)}. \tag{10}$$

Three-pressure cycle with superheating and set to supercritical steam parameters
High-pressure steam mass flow rate:

$$\dot{m}_{HP} = \frac{\dot{m}_{exh} * c_{pexh} * \left(t_{exh}^{TG} - (t_{13} + \Delta t) \right)}{(h_1 - h_{13}) + (h_3 - h_2)}. \tag{11}$$

Intermediate-pressure steam mass flow rate:

$$\dot{m}_{IP} = \frac{\dot{m}_{exh} * c_{pexh} * \left((t_{13} + \Delta t) - \left(t_{sat}^{IP} + \Delta t \right) \right) - \dot{m}_{HP} * (h_{13} - h_{12})}{(h_{20} - h_{18})}. \tag{12}$$

Low-pressure steam mass flow rate:

$$\dot{m}_{LP} = \frac{\dot{m}_{exh} * c_{pexh} * \left(\left(t_{sat}^{IP} + \Delta t \right) - \left(t_{sat}^{LP} + \Delta t \right) \right) - \dot{m}_{IP} * (h_{18} - h_{17})}{(h_{24} - h_{22})}. \tag{13}$$

Outlet temperature of exhaust gases:

$$t_{out}^{TG} = \left(t_{sat}^{LP} + \Delta t \right) - \frac{\dot{m}_{HP} * (h_{12} - h_{11}) + \dot{m}_{IP} * (h_{17} - h_{16}) + \dot{m}_{LP} * (h_{22} - h_{21})}{\dot{m}_{exh} * c_{pexh}} \tag{14}$$

where:

$\dot{m}_{exh}$ mass flow rate for exhaust gases (kg/s),
$\dot{m}_{HP}$ mass flow rate for high-pressure steam (kg/s),
$\dot{m}_{IP}$ mass flow rate for intermediate-pressure steam (kg/s),
$\dot{m}_{LP}$ mass flow rate for low-pressure steam (kg/s),
h specific enthalpy (kJ/kg),
c_{pexh} specific heat of exhaust gases (kJ/kg K),
t_{sat}^{HP} high-pressure steam saturation temperature (°C),
t_{sat}^{IP} intermediate-pressure steam saturation temperature (°C),
t_{sat}^{LP} ow-pressure steam saturation temperature (°C),
t_{exh}^{TG} exhaust gases temperature leaving gas turbine (°C),
Δt pinch point (°C),
t_{out}^{TG} temperature of exhaust gases flowing from the heat recovery steam generator (°C).

Knowing mass flow rates, it was possible to develop an initial design of steam turbine flow channels. For one-dimensional calculations of the axial turbine stage, required data include pressure at stage inlet, temperature at stage inlet, mass flow rate, isentropic enthalpy drop in the stage. These values are obtained through cycle calculations. On the other hand, thermodynamic parameters at the turbine inlet (index 0) and parameters after the isentropic transformation (index 2s) are determined based on tables containing medium properties (using REFPROP-Reference Fluid Thermodynamic and Transport Properties [45]). Other input parameters for stage calculations are assumed and then optimized if necessary (they include reactivity, velocity ratio, velocity coefficients, flow coefficients, angle at which absolute velocity leaves the guide vane lattice, rotational speed, number of stages, etc.). Further on in the paper, steps taken to make detailed thermodynamic and transport calculations for the axial turbine according to the following algorithm that follows the one-dimensional axial turbine stage were presented. The assumptions for the detailed calculations of the turbine stages are presented in Table 2, the given values were selected from the presented ranges depending on the turbine stage. [44]

Table 2. Assumptions for calculation of steam turbine stages.

Description	Symbol	Value	Unit
Isentropic enthalpy drop in the stage	H_s	36–190	kJ/kg
Degree of reaction	ρ	0.3–0.65	–
Velocity ratio	v	0.5–0.65	–
Velocity loss coefficient in the nozzle	ϕ	0.94–0.98	–
Velocity loss coefficient in the rotor	Ψ	0.93–0.96	–
Flow coefficient in the nozzle	μ_2	0.92–0.95	–
Flow coefficient in the rotor	μ_2	0.92–0.95	–
Nozzle output angle	α_1	9–24	°
Angular velocity	ω	314.16	1/s
Flow rate	$\dot{m}$	52.9–241	kg/s

The isentropic enthalpy drop in the rotor was given by

$$H_{sb} = \rho * H_s. \tag{15}$$

The isentropic enthalpy drop in the guide vanes was given by

$$H_{sn} = H_s \; * \; (1 - \rho) \, . \tag{16}$$

The peripheral velocity was given by

$$u = \nu * \sqrt{2 * H_s} \, . \tag{17}$$

The average diameter was given by

$$D_{av} = \frac{2 * u}{\omega} \, . \tag{18}$$

The nozzle blade height was given by

$$l_n = \frac{\dot{m} * v_{1s}}{\mu_1 * \Pi * D_{av} * c_{1s} * \sin(\alpha_1)}. \tag{19}$$

The rotor blade height was given by

$$l_b = \frac{\dot{m} * v_{2s}}{\mu_2 * \Pi * D_{av} * w_{2s} * \sin(\beta_2)}. \tag{20}$$

The isentropic stator outlet velocity was given by

$$c_{1s} = \sqrt{2 * H_{sn} + c_0^2}. \tag{21}$$

The stator outlet velocity was given by

$$c_1 = c_{1s} * \phi. \tag{22}$$

The specific enthalpy in point 1 was given by

$$h_1 = h_0 + \frac{C_0^2}{2000} - \frac{C_1^2}{2000}. \tag{23}$$

The specific enthalpy in point 2 s (without losses of specific entropy) was given by

$$h_{2s} = h_1 - H_{sb}. \tag{24}$$

The relative flow angle at rotor inlet was given by

$$\beta_1 = arctg\left(\frac{C_1 * \sin\alpha_1}{C_1 * \cos\alpha_1 - u}\right). \tag{25}$$

The rotor inlet relative velocity was given by

$$w_1 = arctg\left(\frac{C_1 * \sin\alpha_1}{\sin\beta_1}\right). \tag{26}$$

The rotor isentropic exit velocity was given by

$$w_{2s} = \sqrt{2 * H_{sb} + w_1^2}. \tag{27}$$

The rotor outlet velocity was given by

$$w_2 = w_{2s} * \Psi. \tag{28}$$

The rotor exit relative angle was given by

$$\beta_2 = \frac{\dot{m} \ * v_{2s}}{l * \mu_2 * \Pi * D_{av} * w_{2s}}. \tag{29}$$

The rotor exit angle was given by

$$\alpha_2 = arctg\left(\frac{w_2 * \sin\beta_2}{w_2 * \cos\beta_2 - u}\right). \tag{30}$$

The rotor exit velocity was given by

$$c_2 = \frac{w_2 * \sin\beta_2}{\sin\alpha_2}. \tag{31}$$

The losses in stator was given by

$$\Delta h_n = h_1 - h_{1s}. \tag{32}$$

The losses in rotor was given by

$$\Delta h_b = \left(\frac{w_{2s}^2}{2000} - \frac{w_2^2}{2000}\right). \tag{33}$$

The exit losses was given by

$$\Delta h_{ex} = \frac{C_2^2}{2000}. \tag{34}$$

The summary losses were given by

$$\Delta h = \Delta h_n + \Delta h_b + \Delta h_{ex}. \tag{35}$$

The specific enthalpy in point 2 was given by

$$h_2 = h_1 + \left(\frac{w_1^2}{2000} - \frac{w_2^2}{2000}\right). \tag{36}$$

The peripheral work was given by

$$lu = Hs + \frac{c_0^2}{2000} - \Delta h. \tag{37}$$

The peripheral capacity was given by

$$Nu = \dot{m} * l_u. \tag{38}$$

The peripheral efficiency was given by

$$\eta_u = \frac{l_u}{H_s + \frac{(C_0^2 - C_2^2)}{2000}}. \tag{39}$$

The internal capacity was given by

$$Ni = Nu * \eta i, \tag{40}$$

where:

H_b isentropic enthalpy drop in the rotor (kJ/kg),
H_n isentropic enthalpy drop in the nozzle (kJ/kg),
H_s isentropic enthalpy drop in the stage (kJ/kg),
ρ degree of reaction (–),
u peripheral velocity (m/s),
v velocity ratio (–),
D_{av} average diameter (m),
ω angular velocity (1/s),
c_0 nozzle inlet velocity (m/s),
c_{1s} nozzle outlet velocity, without losses (m/s),
c_1 nozzle outlet velocity (m/s),
c_2 rotor exit velocity (m/s),
$\dot{m}$ flow rate (kg/s),
v_{1s} specific volume at nozzle outlet, without losses (m^3/kg),
μ_1 flow coefficient in the guide vanes (–)
α_1 nozzle output angle (°),
α_2 rotor exit angle (°),
l_b rotor blade height (m),
l_n nozzle blade height (m),
β_1 relative flow angle at rotor inlet (°),
β_2 relative flow angle at rotor exit (°),
w_{2s} relative velocity at rotor outlet, without losses (m/s),
w_2 relative velocity at rotor exit (m/s),
Ψ velocity loss coefficient in the rotor (–),
ϕ velocity loss coefficient in the nozzle (–
v_{2s} specific volume at rotor exit, without losses (m^3/kg),
μ_2 flow coefficient in the rotor (–)
Δh_n losses in the nozzle (kJ/kg),
Δh_b losses in the rotor (kJ/kg),
Δh_{ex} exit losses (kJ/kg),
Δh summary losses in the stage (kJ/kg),
l_u peripheral work (kJ/kg),

N_i internal power (kW),
N_u peripheral power (kW),
η_i internal efficiency (–),
Π Greek letter PI-denotes the ratio of circumference to diameter (–).

3. Results and Discussion

The calculations were made for a three-pressure cycle with superheating set to subcritical and supercritical steam parameters at the inlet to the HP part. The analysis sought to find the most favorable thermodynamic parameters for the cycle, which would lead to achieving the highest efficiency. For the subcritical cycle, the pressures were selected from a range of 10 MPa to 21 MPa for the HP part, 1 MPa to 4 MPa for the IP part, and 0.2 MPa to 0.7 MPa for the LP part. For the supercritical cycle, the pressure ranges were 22 MPa to 35 MPa for the HP part, 1 MPa to 4 MPa for the IP part, and 0.2 MPa to 0.7 MPa for the LP part. Examples of diagrams showing the effect of the pressure in individual steam turbine parts on the capacity are given in Figures 5 and 6 for the subcritical cycle and the supercritical cycle, respectively. The highest subcritical cycle efficiency was achieved at pressures of 21 MPa, 4 MPa, and from 0.3 MPa for the HP, IP, and LP part, respectively. In the case of the supercritical cycle efficiency, the pressures are 35 MPa, 1 MPa, and from 0.2 MPa for the HP, IP, and LP part, respectively. The most important and beneficial results are given in Table 3. The highest efficiency 64.45% was obtained for the three-pressure cycle with superheating set to supercritical steam parameters, which is higher than the reference power plant efficiency by about 1.25 percentage points (63.08%).

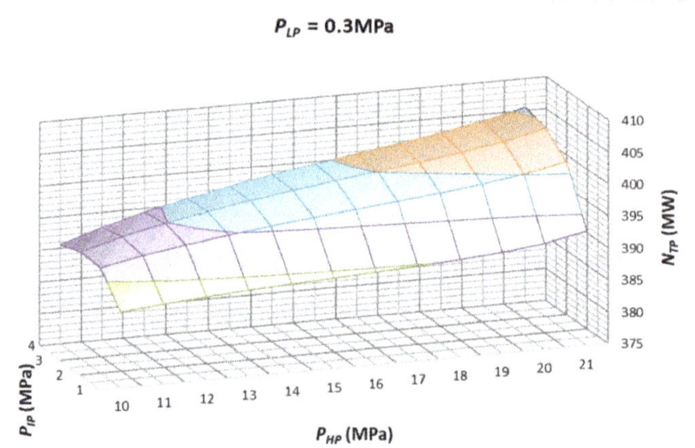

Figure 5. Example of a diagram showing the effect of the pressure in individual steam turbine parts on the capacity for the subcritical cycle.

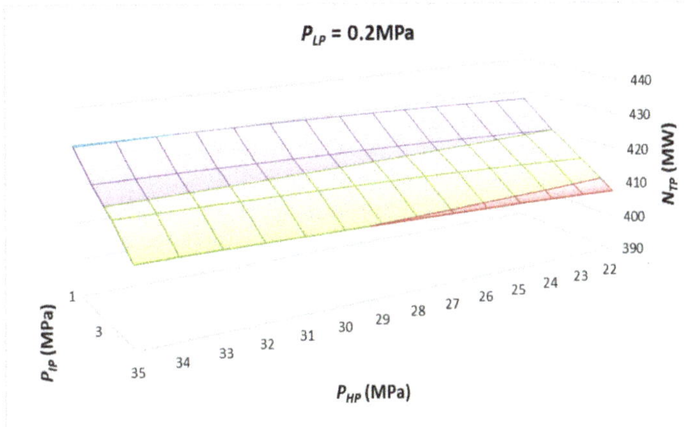

Figure 6. Example of a diagram showing the effect of the pressure in individual steam turbine parts on the capacity for the supercritical cycle.

Table 3. Summary of the results of the cycle calculations.

Description	Symbol	Subcritical Value	Supercritical Value	Unit
Electric power of steam turbine	N_{TP}	350	374	MW
Electric power of gas turbines	N_{TG}	840	840	MW
Electric power of power unit	N_{PG}	1.19	1.214	GW
Steam pressure at HP inlet	p_{HP}	21	35	MPa
Steam temperature at HP inlet	t_{HP}	600	600	°C
Steam pressure at IP inlet	p_{IP}	4	1	MPa
Steam temperature at IP inlet	t_{IP}	581.7	587.4	°C
Steam pressure at LP inlet	p_{LP}	0.3	0.2	MPa
Steam temperature at LP inlet	t_{LP}	221.6	315.2	°C
Pressure after LP expansion	p_C	3	3	kPa
Enthalpy drop in HP	H_{HP}	484.35	804.27	kJ/kg
Enthalpy drop in IP	H_{IP}	709.41	509.94	kJ/kg
Enthalpy drop in LP	H_{LP}	663.69	692.78	kJ/kg
HP steam flow rate (from one generator)	$\dot{m}_{HP1}$	62.06	64.11	kg/s
IP steam flow rate (from one generator)	$\dot{m}_{IP1}$	8.23	3.95	kg/s
LP steam flow rate (from one generator)	$\dot{m}_{LP1}$	12.94	14.44	kg/s
Total HP steam flow rate	$\dot{m}_{HP}$	187.5	197.7	kg/s
Total IP steam flow rate	$\dot{m}_{IP}$	29.41	8.66	kg/s
Total LP steam flow rate	$\dot{m}_{LP}$	24.02	34.67	kg/s
Flow rate of the LP steam to the deaerator	$\dot{m}_D$	4.48	4.15	kg/s
Temperature at which exhaust gases leave the HRSG	t_{exh}	86.16	51.9	°C
Gas-steam cycle efficiency	$\eta_{G\text{-}S}$	63.08	64.45	%

Detailed thermodynamic and transfer calculations were made for the three-pressure cycle with superheating set to subcritical and supercritical steam parameters. The steam part includes a high-pressure (HP) turbine, an intermediate-pressure (IP) turbine, and a low-pressure (LP) turbine. The HP part has 19 stages when using supercritical parameters and 14 stages when using subcritical parameters. This means that there is a difference of only five stages between these casings. For IP and LP casings, both the subcritical and supercritical turbine have the same number of stages, i.e., 15 stages for the IP part and nine stages for the LP part. The LP part was divided into two casings, each with two outlets. The reason behind this was to provide a considerable reduction in the amount of steam flowing through the casings. As a result, the final stages had vanes of acceptable lengths.

Selected results from the detailed calculations of the steam turbine stages are presented in Table 4. The given values in the ranges show the results depending on the position of the stage within the steam turbines.

Table 4. Selected results from detailed calculations of steam turbine stages.

Description	Symbol	Value	Unit
Internal work	l_i	32–180	kJ/kg
Internal power	N_i	3–13	MW
Internal efficiency	η_i	0.9–0.95	–
Peripheral velocity	u	128–350	m/s
Average diameter	D_{av}	0.8–2.4	m
Rotor blade height	l_b	0.025–0.8	m
Summary losses in the stage	Δh	4–20	kJ/kg
Relative flow angle at rotor inlet	β_1	22–80	°
Relative flow angle at rotor exit	β_2	20–47	°
Rotor exit angle	α_2	60–88	°

4. Economic Analysis

In the liberalized energy market, the profitability of investments in new power plants is considered just like any other investment. The investor usually determines the rate of return from investment. It may include the internal rate of return (*IRR*) or the return on equity (*ROE*). The expected return on investment depends on the risk of the project. The *ROE* depends on electricity prices and generation costs, which include interest and depreciation costs, fuel costs, as well as operating and running costs. Apart from the fuel cost, the electricity price is an enigma. In liberal economies, this price may fluctuate greatly. Typically, private investors or independent electricity generators strive to find electricity consumers to whom they could sell electricity under multi-year contracts. Such contracts may also include a provision that the electricity price may be associated with the cost of the fuel that is burnt at the power plant. This significantly reduces investment risk. With this investment structure, the risk is low, allowing the investor to receive a loan on very favorable terms. The remaining risks relate directly to the power plant and include costs, efficiency, and reliability. In fossil fuel power plants, both the electricity market and the fuel market must be monitored and analyzed on an ongoing basis. There is a correlation between these two markets. If, for example, a majority of generators use natural gas as a fuel in the market of a given country, the electricity price will go up if the natural gas price will increase, and vice versa. Interest and depreciation costs have a direct impact on the cost of electricity produced. The main factors include the debt-to-equity ratio and loan terms such as interest and depreciation. The debt-to-equity ratio mainly depends on the risk of the venture. Investment projects of independent electricity generators with a low market risk may be highly leveraged, meaning that a very large portion of the investment project can be financed through a loan. For well-planned projects, 80% of the costs may be covered by a loan, with the remaining 20% being covered by generator's own resources. For risky projects, up to 50% of the costs may be covered by a loan [39]. Another important aspect includes the type of electricity to be generated because its price varies throughout the day. The on-peak electricity price is higher than the off-peak price. Power plants for baseload generation sell electricity at a lower price than peaking power plants. The following types of power plants can be distinguished [38]:

- Base load power plants (>5000 h/a);
- Load following power plants (2000–5000 h/a);
- Peaking power plants (<2000 h/a).

Power plants that require low capital expenditures and use an expensive fuel are suitable for peaking generation. On the other hand, power plants that require high capital expenditures and use a cheap fuel are suitable for baseload generation.

The economic analysis for a power plant operating at subcritical parameters will be presented further on in this paper. The analysis aims to verify the extent to which the investment may prove profitable. The analysis will also be the starting point for presenting savings that can be achieved when using supercritical parameters. The operation of a power plant that uses supercritical parameters involves higher electricity generation and thus higher revenue and lower fuel costs resulting from higher efficiency.

However, the construction of the supercritical plant is associated with higher costs [46]. In the case of the plant described in the current paper the costs of the following components are higher:

- More expensive alloys used in the HRSG supercritical part, HP part turbine, supercritical pump, and supercritical pipelines;
- Greater wall thickness in all cycle elements operating at supercritical pressures (HRSG, HP turbine, pump, and pipelines)-which will make the installation more expensive;
- More expensive pump for supercritical parameters;
- More stages in the HP supercritical turbine (19 stages) compared to the HP subcritical turbine (14 stages), which increases the investment costs.

A higher initial construction cost results in a higher cost of financing when financial resources need to be obtained externally. The difference between the electricity price and the fuel price will be used as a clear indication of how much higher the construction cost of a power plant in the supercritical variant can be to find the investment economically justifiable as compared to the subcritical variant. It can be a basis for making appropriate data-driven decisions that lead to better results [47,48].

In the following part of the article, the currently available data is used to approximate costs and revenue in the case of the subcritical variant. Further analysis pertains to the supercritical variant. The authors are aware that the above-mentioned differences in construction and financing costs are unavoidable. Therefore any advantage of the supercritical plant in terms of the calculated revenue should be regarded as a cautious and approximated indicator of its potential profitability. This advantage needs to cover the entire higher cost.

A number of assumptions had to be made for the economic analysis. They are given in Table 5. The power plant was assumed to be used for baseload generation, with its operating time being 7500 h per annum. Knowing the installed capacity, it was necessary to determine the capital expenditures required to build the power plant. The average cost of building 1 kW of installed capacity for combined gas-steam cycles was obtained from statistical data available at the U.S. Energy Information Administration website [49] and articles [50,51]. The planned operating life of the investment project was 25 years, with the construction time being two years. The investment project was financed through both investor's own resources (25%) and a loan (75%). The loan repayment method was based on a fixed annual amount, which means the sum of instalments and interest is the same every year. The following parameters were also assumed to be constant throughout the analyzed time period: electricity price, inflation rate, discount rate, WIBOR (Warsaw InterBank Offered Rate), fuel price, rate of excise duty, interest rate on the loan, interest rate on investor's own resources. The following parameters were calculated: *CF* (cash flow), *NPV* (net present value), annual income from the sale of electricity, gross profit, net profit, generation costs, income tax, excise duty, instalments, and interest on the loan. The calculation results are presented in the Table 5. Diagrams showing the change in the *CF* and *NPV* over the entire duration of the investment project are also presented.

Table 5. Parameters used for the economic analysis of the subcritical-type power plant.

Parameter	Unit	Value
Installed capacity of the entire power unit	kW	1,190,260
Capacity of gas turbines	MW	840
Average cost of building 1 kW of installed capacity	$	950
Average electricity price	USD/MWh	64.5
Fuel price	USD/MWh	13.15
Annual operating time	h/a	7500
Construction time	years	2
Operating life	years	25
Financing through investor's own resources	%	25
Financing through a loan	%	75
Inflation	%	2
Loan interest rate	%	5
Interest rate on investor's own resources	%	4
Income tax rate	-	0.19
Bank margin for granting a loan	%	5
WIBOR rate	%	0.27
Excise duty	USD/MWh	1.32

Power plant construction cost:

$$Cc = C_u * N_e \,. \tag{41}$$

c_u—unit cost of building 1 kW of installed capacity [$/kW], N_e—installed capacity [kW].
Annual electricity generation:

$$W = N_e * \tau \,. \tag{42}$$

τ Annual power plant operating time [h/a].
Annual revenue from the sale of electricity:

$$S = W * P_u \,. \tag{43}$$

P_u—electricity unit price [$/MWh].
Annual fuel cost:

$$Cfc = \frac{N_{TG} * \tau * C_{fu}}{\eta_{TG}}. \tag{44}$$

N_{TG}—installed capacity of gas turbines (MW), C_{fu}—unit fuel cost ($/MWh), η_{TG}—gas turbine efficiency.

Actual loan interest rate:

$$r_a = \frac{r_l - i}{1 + i} + m. \tag{45}$$

r_l—WIBOR rate (%), i–inflation (%), m—bank margin (%).
Annual interest on the loan:

$$F = (l_s * E - (n-1) \ * \ L) \ * \ r_a \,. \tag{46}$$

E—total capital expenditures ($), l_s—share of the loan in financing the investment project (%), L—total loan instalments for the year ($).

Discount rate:

$$r = l_s * r_a * \ (1 - t) \ + \ r_{or} * \ (1 - l_s) \,. \tag{47}$$

t—income tax rate (%), r_{or}, r_a—interest rate on investor's own resources, bank loan (%).
Generation costs:

$$Cgen = C_f + C_s + C_o. \tag{48}$$

C_f—fuel cost ($), C_s—staff costs ($), C_o—costs such as overhaul costs, environmental charges, insurance fees. They were assumed to be 4% of the capital expenditures ($).

Excise duty:

$$E = W * e\,. \tag{49}$$

e—excise duty rate ($/MWh).

Gross profit:

$$P = S \; - \; Cgen. \tag{50}$$

Income tax:

$$T = P * t. \tag{51}$$

t—income tax ($) rate.

Net profit:

$$Pnet = P - T - E\,. \tag{52}$$

Cash flow:

$$CF = S \; - \; T \; - \; Cgen. \tag{53}$$

NPV (Net Present Value):

$$NPV = \sum_{b}^{t} \frac{CF}{(1+r)^t}. \tag{54}$$

NPV—net present value in a given year ($), t—investment project year under consideration ($), b—investment project starting point ($), r—discount rate (%).

During the calculations, the following additional assumptions were made for the two variants:

Depreciation values for the planned investment project were calculated linearly for the entire power plant operating life, i.e., 4% per annum;

The entire bank margin 5% of the loan value-was recognized in expenses in the first loan repayment year;

The loan repayment method was the equal total payment method in which the proportions of the value of instalments and interest are variable over the entire loan period;

The values used for the cost–benefit analysis are given in Table 6. Financial values are given in USD.

Table 6. *NPV* over the years for the subcritical-type power plant in thousands USD.

			Year of Analysis					
No.	**Specification**	**Investment Project Value**	**1**	**2**	**3**	...	**24**	**25**
1	Net cash flow (*NCF*)	−1,130,747	226,677	268,909	268,729	...	261,736	261,177
2	Discount rate	0	6.20%	6.20%	6.20%	...	6.20%	6.20%
3	Discounted cash flow	−1,130,747	212,620	236,591	221,770	...	56,307	52,702
4	Cumulative cash flow	−1,130,747	−918,127	−681,536	−459,766	...	1,998,872	2,051,574

The net cash flow (*NCF*) presented in row one of Table 6 shows the nominal values of free cash flow in each subsequent year of analysis. The successive discounted cash flow values according to Equation (38) (i.e., discounted by discount rates that successively increase) give discounted (actual) values for each subsequent year as shown in row three of Table 6.

Table 7 shows the accumulated results of the cost–benefit analysis for the construction of the subcritical-type power plant (assuming that the cost of building 1 kW is USD 950/kWh).

Table 7. Economic analysis results for the subcritical-type power plant.

Discount Rate	6.2%
NPV	Thou. USD 2,051,574
IRR	15.26%
Payback period of the investment project in years	5.3

The sum of initial capital expenditures (the negative investment project value in Table 6) and positive discounted cash flow values yields the net present value (*NPV*) of the investment project. If positive, it means that the investment project will be profitable. The higher it is, the more profitable the investment project. The calculations give the *NPV* of thousands USD 2,051,574 with initial capital expenditures of thousands USD 1,130,747.

The profitability of the planned investment project is also confirmed by the internal rate of return (*IRR*) given in Table 7, which is much higher than the assumed discount rate. The internal rate of return gives the discount rate at which the cumulated discounted annual net incomes equal the initial investment value, resulting in the situation when the net present value (*NPV*) equals 0 (zero). In other words, the *IRR* refers to the discount rate at which the actual cash flow covers the planned capital expenditures in full. The calculations show that the rate of return on the investment project is considerably higher (15.26%) than the minimum rate of return accepted by the investor and expressed by the discount rate (6.2%).

The calculations also demonstrate that investing in a subcritical-type power plant is a potentially profitable venture with a high possible rate of return. It should be noted that the analysis carried out in this paper does not consider all elements (such as staff costs) and thus should be considered an approximate simulation. Still, the high internal rate of return and a relatively short payback period of the investment project point to a high probability of the economic profitability of the project implemented with the use of the parameters assumed in the paper.

The annual electricity generation, annual revenue from the sale of electricity, and fuel cost for the subcritical and supercritical variants are given in Table 8. They are the basis for determining the amount by which the cost of the supercritical variant may be higher than the subcritical variant so that the investment in the former is economically justified.

Table 8. Annual electricity generation, annual revenue from the sale of electricity, and fuel cost for subcritical and supercritical variants.

Equation	Equation No.	Value for the Subcritical Variant	Value for the Supercritical Variant
Annual electricity generation in MWh	42	8,926,950	9,104,550
Annual revenue from the sale of electricity in thousands of USD	43	575,553	587,004
Annual fuel cost in thousands of USD	44	131,163	128,619

For the subcritical-type power plant, the difference between the annual revenue from the sale of electricity and the annual fuel cost is thousands USD 444,391. For the supercritical-type power plant, the difference is thousands USD 458,385, i.e., it is higher by thousands USD 13,994 per annum. Assuming that this difference is constant for 25 years, the accumulated difference between the values obtained for the subcritical and supercritical variants is USD 349,860,209. This is the amount of possible savings over 25 years if a supercritical-type power plant is erected. If the construction cost of this type of power plant as well as costs of financial services are lower than these savings, investing into the supercritical-type power plant might be economically justifiable. Rational investment and managerial decisions related to the construction of a more profitable power plant variant may be made once the aforementioned difference is known.

It is important to emphasize once again that the obtained difference should be regarded as a cautious indication of the possible higher profitability of the supercritical variant as it needs to cover its higher cost. It is, however, worth mentioning that a potential increase in energy prices in the future should positively affect the profitability of the supercritical variant in comparison to the subcritical one.

Investments in supercritical-type power plants would also entail the introduction of innovation as described in the Schumpeterian innovation theory [52]. Schumpeter defined entrepreneurs as people who primarily bring innovations, mainly through the introduction of new technologies or improvements to the existing ones. In line with this theory, innovations are actions that have potential benefits not only for the investor but also for the economy as a whole.

5. Conclusions

Current gas-steam power plants boast a relatively high efficiency of approx. 60%, with some already exceeding 60%. In terms of improving the efficiency of combined gas-steam cycles, three main development directions are taken: selection of optimum thermodynamic parameters in the steam part; selection of the optimum type of heat recovery steam generator; and the selection of the most favorable distribution of heated areas in the heat recovery steam generator—all with the highest cost-effectiveness of the solution in mind. In this paper, the authors proposed using a supercritical steam pressure in the turbine HP part to improve the efficiency. The paper includes the most advantageous parameters of pressure and temperature at inlets to HP, IP, and LP parts of steam turbines, while the development of an initial design of flow channels of these parts was possible through the thermodynamic and transfer calculations. The efficiency of 64.45% and the capacity of about 1.214 GW were obtained against the reference values of 63.08% and 1.19 GW. The authors believe that the efficiency can be improved further. Research on this is and will continue to be conducted. One of the methods to do so is to continue increasing the high-pressure steam temperature because it is the first part of the generator into which exhaust gases enter. The economic analysis revealed that the difference between the annual revenue from the sale of electricity and the annual fuel cost is considerably higher for power plants set to supercritical parameters, reaching approx. USD 14 million per annum. Therefore, it seems that any investments in adapting some components of the steam part to supercritical parameters might be balanced out by the profit. These issues must also be taken into account when making decisions on the selection of the more favorable type of power plant.

Author Contributions: Conceptualization, M.P., M.J., P.Z., and G.P.; methodology, M.P.; software, M.J.; validation, M.P., M.J., and P.Z.; formal analysis, M.P., M.J., P.Z., and G.P.; investigation, M.P., M.J., P.Z., and G.P; resources, M.P., M.J., and P.Z.; data curation, M.P., M.J., P.Z., and G.P.; writing—original draft preparation, M.P., M.J., P.Z., and G.P.; writing—review and editing, M.P., M.J., and P.Z.; visualization, M.P., M.J., and P.Z.; supervision, M.P.; project administration, M.P.; funding acquisition, M.P., M.J., and P.Z. All authors have read and agreed to the published version of the manuscript.

Funding: This research received no external funding.

Institutional Review Board Statement: Not applicable.

Informed Consent Statement: Not applicable.

Data Availability Statement: Not applicable.

Conflicts of Interest: The authors declare no conflict of interest. The funders had no role in the design of the study; in the collection, analyses, or interpretation of data; in the writing of the manuscript, and in the decision to publish the results.

References

1. Herran, D.S.; Tachiiri, K.; Matsumoto, K. Global energy system transformations in mitigation scenarios considering climate uncertainties. *Appl. Energy* **2019**, *243*, 119–131. [CrossRef]
2. Veum, K.; Bauknecht, D. How to reach the EU renewables target by 2030? An analysis of the governance framework. *Energy Policy* **2019**, *127*, 299–307. [CrossRef]
3. Piwowarski, M. Optimization of steam cycles with respect to supercritical parameters. *Pol. Marit. Res.* **2009**, *16*, 45–51. [CrossRef]
4. Tumanovskii, A.G.; Shvarts, A.L.; Somova, E.V.; Verbovetskii, E.K.; Avrutskii, G.D.; Ermakova, S.V.; Kalugin, R.N.; Lazarev, M.V. Review of the coal-fired, over-supercritical and ultra-supercritical steam power plants, Pleiades Publishing. *Therm. Eng.* **2017**, *64*, 83–96. [CrossRef]
5. Maheshwari, M.; Singh, O. Thermodynamic study of different configurations of gas-steam combined cycles employing intercooling and different means of cooling in topping cycle. *Appl. Therm. Eng.* **2019**, *162*, 114249. [CrossRef]
6. Kosowski, K.; Tucki, K.; Piwowarski, M.; Stępień, R.; Orynycz, O.; Włodarski, W.; Bączyk, A. Thermodynamic Cycle Concepts for High-Efficiency Power Plans. Part A: Public Power Plants 60+. *Sustainability* **2019**, *11*, 554. [CrossRef]
7. Mikielewicz, D.; Kosowski, K.; Tucki, K.; Piwowarski, M.; Stępień, R.; Orynycz, O.; Włodarski, W. Gas Turbine Cycle with External Combustion Chamber for Prosumer and Distributed Energy Systems. *Energies* **2019**, *12*, 3501. [CrossRef]
8. Capros, P.; Kannavou, M.; Evangelopoulou, S.; Petropoulos, A.; Siskos, P.; Tasios, N.; Zazias, G.; de Vita, A. Outlook of the EU energy system up to 2050: The case of scenarios prepared for European Commission's "clean energy for all Europeans" package using the PRIMES model. *Energy Strat. Rev.* **2018**, *22*, 255–263. [CrossRef]
9. Renn, O.; Marshall, J.P. Coal, nuclear and renewable energy policies in Germany: From the 1950s to the "Energiewende". *Energy Policy* **2016**, *99*, 224–232. [CrossRef]
10. Brown, T.; Schlachtberger, D.; Kies, A.; Schramm, S.; Greiner, M. Synergies of sector coupling and transmission reinforcement in a cost-optimised, highly renewable European energy system. *Energy* **2018**, *160*, 720–739. [CrossRef]
11. Mikielewicz, D.; Kosowski, K.; Tucki, K.; Piwowarski, M.; Stępień, R.; Orynycz, O.; Włodarski, W. Influence of Different Biofuels on the Efficiency of Gas Turbine Cycles for Prosumer and Distributed Energy Power Plants. *Energies* **2019**, *12*, 3173. [CrossRef]
12. Kosowski, K.; Tucki, K.; Piwowarski, M.; Stępień, R.; Orynycz, O.; Włodarski, W. Thermodynamic Cycle Concepts for High-Efficiency Power Plants. Part B: Prosumer and Distributed Power Industry. *Sustainability* **2019**, *11*, 2647. [CrossRef]
13. Corrêa, P.S.P., Jr.; Zhang, J.; Lora, E.E.S.; Andrade, R.V.; Pinto, L.R.D.M.E.; Ratner, A. Experimental study on applying biomass-derived syngas in a microturbine. *Appl. Therm. Eng.* **2019**, *146*, 328–337. [CrossRef]
14. Kosowski, K.; Piwowarski, M. Design Analysis of Micro Gas Turbines in Closed Cycles. *Energies* **2020**, *13*, 5790. [CrossRef]
15. Thu, K.; Saha, B.B.; Chua, K.J.; Bui, D.T. Thermodynamic analysis on the part-load performance of a microturbine system for micro/mini-CHP applications. *Appl. Energy* **2016**, *178*, 600–608. [CrossRef]
16. Kosowski, K.; Stępień, R.; Włodarski, W.; Piwowarski, M.; Hirt, Ł. Partial admission stages of high efficiency for a microturbine. *J. Vib. Eng. Technol.* **2014**, *2*, 441–448.
17. Soares, C. *Microturbines: Applications for Distributed Energy Systems*, 1st ed.; Butterworth-Heinemann: Amsterdam, The Netherlands, 2007; ISBN 978-0750684699.
18. Włodarski, W. Control of a vapour microturbine set in cogeneration applications. *ISA Trans.* **2019**, *94*, 276–293. [CrossRef]
19. Włodarski, W. A model development and experimental verification for a vapour microturbine with a permanent magnet synchronous generator. *Appl. Energy* **2019**, *252*, 113430. [CrossRef]
20. Saadabadi, S.A.; Thattai, A.T.; Fan, L.; Lindeboom, R.E.; Spanjers, H.; Aravind, P. Solid Oxide Fuel Cells fuelled with biogas: Potential and constraints. *Renew. Energy* **2019**, *134*, 194–214. [CrossRef]
21. Kwaśniewski, T.; Piwowarski, M. Design Analysis of Hybrid Gas Turbine-Fuel Cell Power Plant in Stationary and Marine Applications. *Pol. Marit. Res.* **2020**, *27*, 107–119. [CrossRef]
22. Herrando, M.; Pantaleo, A.M.; Wang, K.; Markides, C.N. Solar combined cooling, heating and power systems based on hybrid PVT, PV or solar-thermal collectors for building applications. *Renew. Energy* **2019**, *143*, 637–647. [CrossRef]
23. Hsu, P.-C.; Huang, B.-J.; Wu, P.-H.; Wu, W.-H.; Lee, M.-J.; Yeh, J.-F.; Wang, Y.-H.; Tsai, J.-H.; Li, K.; Lee, K.-Y. Long-term Energy Generation Efficiency of Solar PV System for Self-consumption. *Energy Procedia* **2017**, *141*, 91–95. [CrossRef]
24. Quoilin, S.; Declaye, S.; Tchanche, B.F.; Lemort, V. Thermo-economic optimization of waste heat recovery Organic Rankine Cycles. *Appl. Therm. Eng.* **2011**, *31*, 2885–2893. [CrossRef]
25. Tchanche, B.F.; Lambrinos, G.; Frangoudakis, A.; Papadakis, G. Low-grade heat conversion into power using organic Rankine cycles—A review of various applications. *Renew. Sustain. Energy Rev.* **2011**, *15*, 3963–3979. [CrossRef]
26. Vescovo, R.; Spagnoli, E. High Temperature ORC Systems. *Energy Procedia* **2017**, *129*, 82–89. [CrossRef]
27. Piwowarski, M.; Kosowski, K. Advanced Turbine Cycles with Organic Media. *Energies* **2020**, *13*, 1327. [CrossRef]
28. Zohuri, B.; McDaniel, P. *Combined Cycle Driven Efficiency for Next Generation Nuclear Power Plants*, 2nd ed.; Springer International Publishing AG: Cham, Switzerland, 2018.
29. Boyce, P. *Handbook for Cogeneration and Combined Cycle Power Plants*; ASME Press: New York, NY, USA, 2010; ISBN 978-0-7918-5953-7.
30. Gülen, C.S. *Gas Turbine Combined Cycle Power Plants*; CRC Press: Boca Raton, FL, USA, 2020; ISBN 9780367199579.
31. Ziółkowski, P.; Kowalczyk, T.; Lemański, M.; Badur, J. On energy, exergy, and environmental aspects of a combined gas-steam cycle for heat and power generation undergoing a process of retrofitting by steam injection. *Energy Convers. Manag.* **2019**, *192*, 374–384. [CrossRef]

32. Nishi-Nagoya Combined-Cycle Power Plant. Available online: https://www.power-technology.com/pro-jects/nishi-nagoya-combined-cycle-power-plant/ (accessed on 7 September 2020).
33. Jeffs, E. Generating power at high efficiency. In *Combined Cycle Technology for Sustainable Energy Production*; Woodhead Publishing Limited: Cambridge, UK, 2008; ISBN 978-1-84569-454-8.
34. Alobaid, F.; Ströhle, J.; Epple, B.; Kim, H.-G. Dynamic simulation of a supercritical once-through heat recovery steam generator during load changes and start-up procedures. *Appl. Energy* **2009**, *86*, 1274–1282. [CrossRef]
35. Kotowicz, J.; Job, M.; Brzęczek, M. Maximisation of Combined Cycle Power Plant Efficiency. *Acta Energetica* **2015**, *4*, 42–48. [CrossRef]
36. Zahoransky, R. *Energy Technology. Systems for Conventional and Renewable Energy Conversion. Compact Knowledge for Studies and Work*, 8th ed.; Springer Publishing House: Wiesbaden, Germany, 2019; (In German). [CrossRef]
37. Zhang, G.; Zheng, J.; Yang, Y.; Liu, W. Thermodynamic performance simulation and concise formulas for triple-pressure reheat HRSG of gas–steam combined cycle under off-design condition. *Energy Convers. Manag.* **2016**, *122*, 372–385. [CrossRef]
38. Kehlhofer, R.; Rukes, B.; Hannemann, F.; Stirnimann, F. *Combined-Cycle Gas and Steam Turbine Power Plants*, 3rd ed.; PennWell Corporation: Oklahoma City, OK, USA, 2009.
39. Bartnik, R. *Gas and Steam Power Plants and Combined Heat and Power Plants. Energy and Economic Efficiency*; WNT Publishing House: Warsaw, Poland, 2017; ISBN 978-83-01-19311-9. (In Polish)
40. Ibrahim, T.K.; Rahman, M. Effects of isentropic efficiencies on the performance of combined cycle power plants. *Int. J. Automot. Mech. Eng.* **2015**, *12*, 2914–2928. [CrossRef]
41. Polyzakis, A.; Koroneos, C.; Xydis, G. Optimum gas turbine cycle for combined cycle power plant. *Energy Convers. Manag.* **2008**, *49*, 551–563. [CrossRef]
42. Ziółkowski, P.; Badur, J.; Ziółkowski, P.J. An energetic analysis of a gas turbine with regenerative heating using turbine extraction at intermediate pressure—Brayton cycle advanced according to Szewalski's idea. *Energy* **2019**, *185*, 763–786. [CrossRef]
43. Hattori, Y.; Hyomori, K. State-of-the-Art Technologies for High-Efficiency Combined-Cycle Power Generation Systems. *Toshiba Rev. Glob. Ed.* **2016**, *2*, 1.
44. Kosowski, K.; Domachowski, Z.; Próchnicki, W.; Kosowski, A.; Stępień, R.; Piwowarski, M.; Włodarski, W.; Ghaemi, M.; Tucki, K.; Gardzilewicz, A.; et al. *Steam and Gas Turbines with the Examples of Alstom Technology*; Alstom: Saint-Ouen, France, 2007; ISBN 978-83-925959-3-9.
45. Lemmon, E.; Huber, M.; McLinden, M.O. *NIST Standard Reference Database 23: NIST Reference Fluid Thermodynamic and Transport Properties-REFPROP Version 9.1*; National Institute of Standards and Technology, NIST NSRDS: Gaithersburg, MD, USA, 2013.
46. Hospers, G.-J. Joseph Schumpeter and his legacy in innovation studies. *Knowl. Technol. Policy* **2005**, *18*, 20–37. [CrossRef]
47. Chelst, K.; Canbolat, Y.B. *Value-Added Decision Making for Managers*; CRC Press: Boca Raton, FL, USA, 2011.
48. Hammond, J.S.; Keeney, R.L.; Raiffa, H. *Smart Choices: A Practical Guide to Making Better Decisions*; Harvard Business Review Press: Boston, MA, USA, 2015.
49. U.S. Energy Information Administartion. Available online: https://www.eia.gov/electricity/generatorcosts/ (accessed on 7 September 2020).
50. Pauschert, D. *Study of Equipment Prices in the Power Sector, Energy Sector Management Assistance Program*; Technical Paper 122/09; The International Bank for Reconstruction and Development: Washington, DC, USA, 2010.
51. Bassily, A. Enhancing the efficiency and power of the triple-pressure reheat combined cycle by means of gas reheat, gas recuperation, and reduction of the irreversibility in the heat recovery steam generator. *Appl. Energy* **2008**, *85*, 1141–1162. [CrossRef]
52. de la Calle, A.; Bayon, A.; Pye, J. Techno-economic assessment of a high-efficiency, low-cost solar-thermal power system with sodium receiver, phase-change material storage, and supercritical CO_2 recompression Brayton cycle. *Sol. Energy* **2020**, *199*, 885–900. [CrossRef]

energies

Article

Simulating the Evolution of Business Models for Electricity Recharging Infrastructure Development by 2030: A Case Study for Greece

Stergios Statharas [1,2], Yannis Moysoglou [2], Pelopidas Siskos [1,2,*] and Pantelis Capros [1]

1 School of Electrical and Computer Engineering, E3MLab, National Technical University of Athens, 9 Iroon Polytechniou Street, Zografou, 15773 Athens, Greece; statharas@e3modelling.com (S.S.); kapros@central.ntua.gr (P.C.)
2 E3-Modelling SA, Panormou 70-72, 11524 Athens, Greece; moysoglou@e3modelling.com
* Correspondence: psiskos@e3modelling.com; Tel.: +30-2106775696

Abstract: It is widely accepted that the market uptake of electric vehicles is essential for the decarbonisation of transport. However, scaling up the roll out of electric vehicles (EV) is challenging considering the lack of charging infrastructure. The latter is, currently, developing in an uneven way across the EU countries. A charging infrastructure with wide coverage addresses range limitations but requires high investment with uncertain returns during the early years of deployment. The aim of this paper is to assess how different policy options affect EV penetration and the involvement of private sector in infrastructure deployment. We propose a mathematical programming model of the decision problem and the interaction between the actors of EV charging ecosystem and apply it to the case of Greece from the time period until 2030. Greece represents a typical example of a country with ambitious targets for EV penetration by 2030 (10% of the total stock) with limited effort made until now. The results indicate that it is challenging to engage private investors in the early years, even using subsidies; thus, publicly financed infrastructure deployment is important for the first years. In the mid-term, subsidization on the costs of charging points is necessary to positively influence the uptake of private investments. These are mainly attracted from 2025 onwards, after a critical mass of EVs and infrastructure has been deployed.

Keywords: electric vehicles; electricity recharging infrastructure; business models; equilibrium programming; Greek EV mobility 2030; private investments in infrastructure

Citation: Statharas, S.; Moysoglou, Y.; Siskos, P.; Capros, P. Simulating the Evolution of Business Models for Electricity Recharging Infrastructure Development by 2030: A Case Study for Greece. *Energies* **2021**, *14*, 2345. https://doi.org/10.3390/en14092345

Academic Editor: Hongwen He

Received: 13 March 2021
Accepted: 19 April 2021
Published: 21 April 2021

Publisher's Note: MDPI stays neutral with regard to jurisdictional claims in published maps and institutional affiliations.

1. Introduction

The mitigation of the GHG emissions from transport constitutes a strategic objective within the energy system decarbonisation strategy. Electrification of private transport is considered as one of the most promising options for transport decarbonisation. This is supported by several long-term scenarios which foresee a massive uptake of electric vehicles (EV), at least in the period after 2030, as a cost-effective option for decarbonisation in the EU [1–5]. The low-carbon transition scenarios envisage electrification of private transport modes associated with substantial reductions in battery costs driven by economies of scale and development of a large network for recharging EVs.

Even though the private transport electrification strategy for the long-term seems clear enough, the existing market barriers, such as the high battery costs and lack of charging infrastructure [6–9], still need to be alleviated. There is a growing consensus that battery costs will continue to decrease over the next decade, continuing the trend that has been observed over the last 5 years [10]. However, even though EVs are close to reaching cost parity on a lifetime costing basis, deployment is still slow due to a number of barriers, including psychological factors, affecting consumers' decisions [11]. Implementing complementary policies focused on the demand side, such as subsidies [12], access to bus lanes, parking privileges [13], or on the supply side, such as zero emission vehicle quotas on the total annual sales imposed on car manufacturers [14,15] can alleviate these barriers of EVs by 2030.

However, why do we consider the charging infrastructure as a critical market barrier for a wider EV uptake in the EU? The answer is that electricity recharging points have developed in an uneven way among the EU countries. Currently, the number of charging points developed in France, Netherlands, and Germany represent the vast majority (more than 65%) of the total number of charging points in the EU. This would mean that an EV user would have issues aiming to travel outside the three abovementioned countries. While home charging may prove sufficient for the majority of small EVs during the first stages of EV deployment, the development of a robust charging network which enables the supply of fast charging services at key traffic nodes is necessary to achieve high EV penetration [16,17]. Several studies confirm a positive correlation between the existence of a public charging infrastructure system and EV adoption rates, although there is uncertainty regarding the direction of causality [7,18,19]. Nevertheless, there is evidence that infrastructure is crucial to support the electrification of a greater share of vehicle miles travelled [20]; new BEV purchasers show high willingness to pay (WTP) for paying the use of recharging through public infrastructure [21], yet, there are diminishing returns of such charging infrastructure placement [22]. Moreover, publicly accessible recharging infrastructure is critical if EVs are to be adopted by population segments that do not have off-street parking option [23,24]. As the decade 2021–2030 will set the ground for the massive transition towards transport electrification [25], investments in charging infrastructure have to start as soon as possible.

This "chicken–egg" problem [26,27], which characterises the problematic interdependence between EV penetration and infrastructure development, needs to be solved. The development of viable business models for charging infrastructure is critical in order to turn charging infrastructure deployment into an appealing activity for private investors while ensuring an affordable charging price, as well as an adequate network of charging points for EV users. With the term business model, we refer to the ways businesses generate cash flow by offering a product or service (brief overview of the main related business models and their advantages and disadvantages is presented in Section 2.2). The prevailing choice for the success of EVs may not be limited to a single model but can be the result of the simultaneous presence of public and private investments assumed by different models, effectively adapting to the evolving charging needs and aspirations of market actors.

The novelty of the present work is twofold: (i) first, we provide a modelling framework that simulates the interactions and the decision making of the actors involved in EV charging; (ii) then, apply the model to the specific case of the Greece and assess different pathways of EV fleet and charging deployment by 2030. We propose a mathematical program based on a game-theoretic analysis of the actions of the actors of the electromobility system. Through the interactions of the decision makers, the modelling simulates the evolution of the business model for the development of the recharging infrastructure. We carry out scenario analysis to assess the development of the different business models and the uptake of private investments under alternative assumed trajectories on the future costs of EVs and charging points. The quantitative analysis also assesses the impacts of the subsidization of private investors for developing infrastructure to trigger a transition from public to private business model and accelerate the uptake of EVs. The analysis applies specifically to the case of Greece and focuses on the time period between 2021 and 2030. Greece represents a typical example of a country with ambitious targets for EV penetration by 2030, while at the same time limited or no effort has been made up to now. We take on board the Greek national objective to reach an EV penetration of approximately 10% of the total fleet of cars in 2030, as part of the Greek 2030 national energy and climate targets. This is an ambitious target, considering that the total number of charging points in Greece was about 300 (EAFO statistics) and significantly lower compared to many other EU countries.

The approach proposed aims to advance empirical modelling on this field. Previous research has compared and assessed the viability of selected business models assuming a specific business model each time. However, evaluating the impacts of the business models in a static manner does not allow grasping dynamic effects related with the evolution of EV costs and charging infrastructure. In addition, static approaches do not allow a proper simulation of potential impacts from the demand side (i.e., consumers purchasing EVs). Hence, the developments related to the emergence and the dynamic evolution of electricity charging business models until 2030 as a result of the interactions between the system's agents are yet to be explored. In Section 2.1 we present the literature review on this topic.

The present paper is structured as follows: Section 2 provides a literature review and presents, briefly, the main characteristics of infrastructure business models; Section 3 describes the modelling methodology developed to answer the research questions posed; Section 4 presents the results on different scenarios applied to the case of Greece; Section 5 concludes the paper.

2. Background

2.1. Literature Review and Motivation for Research

The lack of public charging infrastructure is identified in the literature as a major barrier in the development of the EV market. The analysis in [7] shows that the charging infrastructure coverage can guarantee high EV adoption rates and suggests that the installation of charging stations may be more effective than financial measures. In [6], the review of the factors which influence the economics of public infrastructure, concludes that psychological factors, such as range anxiety, are not yet well understood. Nevertheless, the existence of public charging infrastructure is considered as a key factor for mitigating the range anxiety of EV drivers and encouraging EV usage. Ref. [19] examines the relation between daily driving distance and the power of charging service offered, based on data of EVs and infrastructure usage in the UK and the US. The authors claim that fast chargers enable the usage of EVs for trips of distance above their single-charge range and suggest that the existence of fast charging infrastructure can help overcome range limitations of EVs.

Yet, the actual charging behaviour of EV users needs to be better understood in order to assess the financial viability (from the perspective of the private investor revenues) of public charging stations. Several scientific papers analyse data on charging behaviour to estimate the actual utilization of different charging types. Ref. [16] analyses the charging patterns of electric vehicles using data from the Western Australian Electric Vehicle Trial and the EV Charging Research Network in Perth. Their analysis confirms that most of the EV charging takes place at business location and at home. The authors conclude that slow and semi-fast public charging stations will not be properly utilized when the number of EVs is low and that a fast-DC charging network should be favoured as it will satisfy the segment of charging demand associated with the need for quick full recharge.

In a similar spirit, [17] analyses the consumers' charging behaviour using data on infrastructure usage in Ireland. For fast charging stations in particular, the authors suggest that car park locations recorded the highest usage frequencies, which, in turn, indicates that public fast charging infrastructure can be financially viable in the short- to mid-term. The authors, also, claim that the development of a highly connected network of strategically placed fast charging stations should receive priority. Less favourable findings regarding the utilization of fast public infrastructure are presented in [28]. Choosing Austria as a case study, the authors estimate that less than 2% of charging events involve the utilization of public charging infrastructure implying profitability issues at the present state. The authors further suggest that fast charging located at highways is likely to be profitable since the willingness-to-pay (WTP) for a fast recharge is expected to be significant. Similar are the findings from a case study based on California's charging network [21], especially related with fast chargers used in both intercity routes and intraregional travels.

A few studies have proposed methodologies to assess the viability of public charging infrastructure assuming different underlying business models. Ref. [29] explores the economic feasibility of different charging infrastructure possibilities, in terms of charging power, ownership of the charging station, and accessibility. The authors conclude that home charging is preferred by users who have this capability while the success of a business model for fast charging depends heavily on the utilization of the infrastructure. In [30], authors study the pricing for using charging infrastructure in the context of public–private partnerships and propose a charging pricing model based on a system dynamics technique. Their results indicate that the charging price is heavily influenced by the operating cost, electricity price, and charging volume. In a subsequent work [31], they propose a game theoretic modelling to decide the EV public charging pricing, considering the interests of government, consumers and charging facility operators. The model is applied in different regions of China and the results suggest that governmental subsidies to charging station operators positively affect the uptake of EVs.

While previous works have explored the conditions of viability of selected cases of business models assuming a fixed business model each time, the emergence and evolution of electricity recharging business models, over time, as a result of the interactions between the system's agents have yet to be explored.

2.2. Defining Business Models for EV Recharging Infrastructure

A business model can generally be defined as the way in which the members of an economic community generate and share value. According to [32], a business model defines the relationships between different actors (i.e., the direction and the type of the value offered and the flow of payments among the actors). Scaling this abstract definition down to the level of EV charging businesses, the value is associated with extending the effective range of EVs and mitigating range anxiety.

A successful charging infrastructure system generates value indirectly for car manufacturers, parking operators, and retailers via increased EV sales, the attraction of customers who own EVs and the increased customer dwelling time in the business premises. In addition, value is generated indirectly for society overall regardless of their participation in electro-mobility; this value is associated with indirect positive externalities stemming from the reduction in GHG and pollutants emissions. The success of an infrastructure business model depends also on its ability to capture this indirect value.

In the case of the charging infrastructure ecosystem, cash flow is generated from EV consumers who pay the charging providers for their services. The charging providers, in return, pay the electric utilities for the electricity. Alternative schemes foresee the possibility for recovering the capital costs of the charging infrastructure from the electricity bills of all consumers and not only the ones using the infrastructure. Private investors then see the recovery of capital costs as less risky compared to a case where only the EV users pay for the service.

Different business models for EV charging emerge when market actors are assigned different roles. The literature has identified a set of roles that are common among different EV charging business models [29,33]. These roles include:

- the electro-mobility service provider (EMSP), who offers electro-mobility services to the end users. The offered services may include, apart from vehicle charging, navigating services;
- the charging station operator (CSO), who is involved in the management, monitoring, and maintenance of a charging station and offers charging to the EMSP based on a business-to-business (B2B) relationship (directly or through a third party);
- the Distribution System Operator (DSO), the owner and operator of the distribution network. The DSO is the entity that maintains and operates the distribution network and provides a platform that allows the connection between the charging station and the electrical utilities.

In accordance with the above, the case of home charging, for example, can be seen as a trivial model in which the EV user acts as EMSP and CSO by satisfying their charging needs using a low power (3.3 KW) home charger. In the case of infrastructure deployment as a competitive market, the market actors (e.g., private investor or private operator) develop, own, and operate publicly accessible high-power (typically around 50 KW) charging stations. This private model meets the evolving needs of EV charging and enjoys the benefits of a competitive market, i.e., optimal cost solutions and high utilization of infrastructure [34]. However, there is the risk of developing charging infrastructure in areas where investors face the lowest investment risks. In addition, the participation of the private sector is subject to the "chicken–egg" dilemma. The UK and Germany are two examples of applications of such a free market approach. In the UK, public–private partnerships, and revenue-sharing arrangements for the rollout of charging infrastructure are gaining increasing popularity in the last few years [35]. In Germany, major utilities in cooperation with car manufacturers are planning the development of charging infrastructure [36], effectively implementing a private market model which captures the value generated for car manufacturers and utilities via increased sales of EVs and increased consumption of electric energy, respectively. Other approaches propose optimising the energy demand of EVs by maximizing owners' profits [37,38] or focusing on the EV charging profiles.

Contrastingly, a prominent example of a public model that assumes a central planning approach for the development of the charging infrastructure is the so-called DSO model; the DSO ensures the deployment of public infrastructure and the roles of EMSP and CSO may either be taken by the DSO or by private actors. Such a model ensures the uniform development of the infrastructure even in regions with lower utilization rates. However, DSO models are not easy to adapt to changing charging needs as infrastructure may be developed based on grid adequacy and may not respond sufficiently to the penetration of EVs; thus, these models are more appropriate for the short-term [39,40]. Typical examples can be found in literature. For instance, in Italy, a DSO model was initially deployed but was later replaced by a free-market model when the regulatory authority perceived investing interest from market actors [39]. In Austria, Ireland, and Luxembourg, a DSO model is adopted where the DSOs own and operate the infrastructure as an extension of their regulated roles—however, the commercial operation of charging stations can be assigned to an external party [41].

Hybrid models may emerge from combining elements of private and public approaches. For example, the charging network may be centrally planned, and the development and operation of charging infrastructure may be assigned to private agents via public tenders. Norway presents an example of such a model; the location of the charging stations is planned along the road network and the charging infrastructure is owned and operated by charging operators who compete for public funding [42]. This model combines the advantage of even infrastructure deployment that is common to central planning approaches with the advantages of decreased costs and higher infrastructure utilization. Nonetheless, this hybrid approach may introduce delays due to the tendering procedures and the danger of binding deployment of charging stations regardless of changes to the EV ecosystem. Table 1 summarizes the characteristics of the different business model approaches.

To resolve the "chicken–egg" problem of the development of the charging infrastructure and the promotion of EVs, the EC put forward a proposal (COM (2016) 864) which aims to harness the advantages of the two approaches. In this respect, Member States may allow the DSOs to be engaged in the development and operation of the recharging infrastructure in case no private investors have expressed interest to invest. After all, government and industry need innovative business models to attract investments [43]. This kind of policy making: (i) supports the development of recharging infrastructure at the early years (if no private investors are interested to be engaged in this business), in order to avoid delaying the uptake of electric vehicles; and (ii) allows the transition towards free market conditions when conditions are mature and a critical mass of charging points and EVs are in place.

Table 1. Characteristics of private, public, and hybrid business models for infrastructure.

Business Model Category	Infrastructure Deployment	Advantages	Disadvantages	Application Examples
Private free market model	Private	Competition ensures optimal cost and utilization of infrastructure	Uneven spatial deployment, chicken-egg dilemma	Germany, UK, Italy
Public DSO-type model	Public	Deployment of infrastructure even in remote areas, resolution of chicken egg dilemma	Problematic adaptation to changing charging needs	Austria, Ireland, Italy, Luxembourg
Public tenders hybrid model	Private (central planning)	Even infrastructure deployment, merits of competition	Delays due to tendering procedures, binding infrastructure deployment	Norway

3. Methodology

3.1. Modelling the Agents of Electro-Mobility

In this section, we provide an overview of the proposed methodology, focusing on the system modelling and, specifically, the interactions between the agents. In Section 3.2, we present a brief game theoretic analysis of the system and in Section 3.3 we present in detail the relevant mathematical formulation.

We consider a set of decision makers which includes the EV consumers, the private sector investors, the policy-maker and the DSO. Our modelling does not aim to simulate the agents' decision-making in full detail; the methodology is focused on the decision-making of agents relevant to be studied, effectively "projecting" the agents' decision making onto the "space" of electro-mobility. This way we circumvent the difficulty of providing an unnecessarily complex modelling of the agents' behaviour as we reduce the latter to a set of conditions that can be efficiently formulated as an equilibrium problem with equilibrium constraints. In what follows, we elaborate on how we model the behaviour of each agent (Table 2):

- The private investors generally allocate their capital in investing options that maximize the overall returns while minimizing the investment risks and considering the opportunity costs. To model their behaviour (i.e., investing in charging infrastructure), we utilize the internal rate of return (IRR) criterion to model whether the private investor will engage in the investment or not. We assume a decision threshold R which represents an estimation of the opportunity cost incurred to the investor by the former choice, i.e., it represents a rate of return that the investor could achieve by investing in other choices. If the IRR of charging infrastructure investments is below that threshold, the investor has no incentive to invest. The modelling considers only the case of fast charging points and assumes exogenously that a specific part of the electricity needs is provided by other types of charging points. The price of charging services comprises of the electricity price and the tariff for recovering the capital cost of the investment. The former is exogenous to the modelling and is provided by the PRIMES energy systems model [44]. The latter is endogenously calculated based on the utilisation of the charging points assuming that the tariff is calculated using the levelized cost approach. The overall price of the charging service is capped to an upper limit and, thus, considered being regulated. We carry out sensitivity analysis around this hypothesis;
- The policy maker promotes the decarbonisation of the energy system aiming to mitigate GHG emissions in the most efficient way. Decarbonisation entails setting a target on EV penetration. The EV penetration targets are exogenous to our modelling. If the electro-mobility system fails to achieve the target without the policy maker's intervention, the latter chooses to either subsidize private investors, respecting an

assumed subsidy budget per charging station, or to allow the DSO to deploy the EV infrastructure for some part of the modelled period (see Section 3.2). Further, we assume that the policy maker prioritizes the deployment of infrastructure by private agents (free market model) against employing a DSO model. This assumption effectively implements the EC proposal (COM (2016) 864);

- The DSO is modelled as an agent with a trivial behaviour that is activated by the policy maker and is influenced by the participation of private investors. Hence, the DSO model may not be activated in cases of high participation of private investors. Alternatively, the DSO model takes place in case of low interest from private investors that would hamper EV deployment and risk not meeting the penetration target. In this case, the DSO develops charging infrastructure that allows reaching the EV penetration targets, respecting an annual budget for charging infrastructure expenditures. It is assumed that the DSO follows a central planning approach in building charging stations and may not exceed an assumed budget for public infrastructure expenditures. The pricing of the charging service provided by the DSO is assumed to be regulated; the infrastructure costs are socially recovered via increases in electricity price;
- Consumers decide whether to purchase a conventional vehicle or an EV. They make their choice considering total cost of vehicle ownership (depending on capital, maintenance, fuel costs, and mileage), as well as perceived cost and, in particular, the lack of charging infrastructure (range anxiety). In our modelling the evolution of the capital and maintenance costs is exogenous. The rest of the cost components are endogenous. Fuel costs are calculated by adding the charging price (which are endogenous) and the electricity provided by the charging station. Range anxiety is an endogenous feature of the model as it relates to the availability of infrastructure which is the result of the choices of private investors and DSO model activation. The modelling of range anxiety draws from the PRIMES-TREMOVE model [45] and for the purposes of the present paper follows a reduced form approach.

Table 2. Taxonomy of the decision makers of the electro-mobility system considered in modelling.

Decision Makers	Objectives and Behaviour	Notes
Private investor	Seeks to allocate capital in profitable investing options. If charging businesses display a certain level of profitability, the private investor expresses interest to invest in charging infrastructure.	• Private investors are risk avert and require policy insurance regarding the recovery of their costs.
Consumers	Modelled to select purchasing either an EV or a conventional vehicle depending on the total cost of ownership of the options and the density of charging infrastructure.	• Consumer choice is modelled to account for range anxiety • Consumers consider charging prices as set by the operator of the charging stations
Policy maker	Sets concrete targets regarding the envisaged penetration of EVs. Needs to ensure the availability of recharging points to promote uptake of EVs.	• The policy maker is assumed to prioritize private deployment and operation of infrastructure over public. • The policy maker either employs a DSO model or subsidizes private investors inciting them to deploy charging infrastructure.
DSO	DSO's role in infrastructure deployment is activated via a public DSO-type model only if the private sector does not express interest in infrastructure investments.	• The DSO develops the charging infrastructure up to the required level, also respecting an annual budget on investments

3.2. A Game Theoretic View on the Interaction among the Agents

We first consider the interaction between the consumers and the private investors. The actors can be seen as participating in a game which rewards cooperation between actors; value is generated by a successfully deployed electro-mobility system and shared with the actors only if both groups choose to participate, (i.e., the consumers purchase EVs and the investors develop and operate the charging infrastructure). Otherwise, if only one of the parties chooses to participate, this party suffers a loss: a consumer buying an EV would not be able to utilize his vehicle; an investor who chooses to invest while the EV penetration is at low levels would not be able to recover the investment costs. Both parties can ensure a state-of-business payoff by choosing not to get involved in the electro-mobility system. For both parties, the payoff of successful electro-mobility deployment is assumed to be greater than their state-of-business payoff. The described game is essentially the archetypal strategic setting of "stag hunt" (We note that the assumption of a regulated charging price is essential for the strategic setting we present. Should the level of charging price be part of the actors' strategies, then the resulting game would have been a bargaining game, with the charging price deciding how the value generated by a successful electromobility system is distributed among the two parties.) The two pure Nash equilibria are the following: both groups opting for electro-mobility (the payoff dominant equilibrium) and both groups averting from it (the risk dominant equilibrium or the "chicken–egg" situation).

The actors' strategic sets in our actual modelling are closed intervals instead of discrete sets. As we aim to model the decision making of a large number of consumers (i.e., Greek consumers) with idiosyncratic behaviours, for a given level of infrastructure deployment and vehicle costs, the market share of EVs purchased can be anywhere between 0 and 100% of the total new vehicle registrations. Each individual consumer has a distinct turning point on his or her decision to purchase EV with respect to infrastructure coverage, assuming given EV and ICE conventional vehicle costs. Under standard assumptions on the distribution of consumers' turning points, the total consumers' EV charging demand D with respect to the level i of infrastructure deployment can be approximated by discrete choice-based functional forms $c = D(i)$.

Likewise, the strategic set for private investors comprises of all the possible amounts of capital they invest in charging infrastructure. If the estimated IRR of charging infrastructure is below the threshold R, the investors choose not to invest. If the conditions are favourable enough that the estimated IRR is greater than the threshold R, the investors engage in the business. Once investments in recharging infrastructure start accumulating, their utilization rate may decrease, depending on the extent of the uptake of EVs. Consequently, the investors decelerate infrastructure investments down to the point where the estimated IRR is equal to R.

More formally, the set of strategies for consumers is defined as $C = [0, V]$ and the set of strategies for investors as $I = [0, U]$. Let $c_1, c_2 \in C$ and $i_1, i_2 \in I$. Let $\mathrm{IRR}(c, i)$ denote the internal rate of return achieved when the EV demand is c and the total infrastructure investments is i. Then, the preference relations over strategic profiles of the actors have the following properties:

For consumers' preference $\geq^C$: For $c_1 = D(i_1)$ we have $(c_1, i_1) \geq^C (c_2, i_1)$ for any $c_2 \neq c_1$, where $D(i)$ is the EV demand function with respect to level i of infrastructure investments (most preferred technology choice with respect to given infrastructure coverage). A demand greater than c_1 means that some consumers buy EVS although they do not perceived EVs to be the best choice. Similarly, a demand less than c_1 means that some consumers who prefer EVs buy conventional vehicles.

For investors' preference $\geq^I$: If $\mathrm{IRR}(c_1, i_1) \geq R$ and $i_1 \geq i_2$ then $(c_1, i_1) \geq^I (c_1, i_2)$, otherwise, if $\mathrm{IRR}(c_1, i_1) < R$, then $(c_1, i_2) \geq^I (c_1, i_1)$ (prefers to increase investments as long as EV business are profitable enough).

The introduction of the third player, the policy maker, completes the strategic setting of our modelling. His or her strategy set includes the following: take no action, subsidize the deployment of private charging stations up to a budget B, or enable the DSO to develop the charging infrastructure, making infrastructure investments within a budget K. This actor strictly prefers any outcome in which electro-mobility is successfully deployed (with respect to the specified penetration target) to any outcome in which it does not. Among outcomes of electro-mobility's success, the policy maker prefers not taking action to subsidizing the private infrastructure deployment and prefers the latter to allowing DSO to deploy the charging infrastructure.

In accordance to the above, the strategy set for the policy maker is defined as $P \subseteq [0, B] \times [0, K]$. For a strategy $(s, d) \in P$, s denotes the subsidy per charging station given to private investors and d denotes the infrastructure investments deployed by the DSO.

Extending the definition to account for the policy maker, a strategic profile is now defined as a triplet $(c, i, (s, d))$ with $c \in C, i \in I, (s, d) \in P$. The EV demand function now depends on both the DSO-deployed and the privately deployed infrastructure, i.e., $c = D(i, d)$. Likewise, the investors' estimated IRR is a function of the consumer demand, total infrastructure investments and the subsidy, i.e., $IRR = IRR(c, i, s, d)$.

Lastly, we note that the Nash equilibrium of the defined strategic setting can be shown to be unique and, depending on the exogenous parameters, can be any of the following: (i) the EV penetration target is achieved by the cooperated actions of consumers and private investors alone, (ii) the EV target is achieved via subsidizing privately deployed infrastructure, (iii) the EV target is achieved via a DSO-deployed infrastructure, (iv) the EV target is achieved by combined private and public investments, and (v) the EV target is not achieved. The latter situation may occur when the private sector was not incited and the budget restrictions on public investments did not allow DSO to develop the infrastructure to the required level.

3.3. Mathematical Formulation of the Problem

In this section, we formulate the strategic setting of Section 3.2 as a mixed complementarity problem (mcp). For simplicity, the formulation is given for a fixed year. Table 3 summarizes the set of variables and parameters of the formulation.

Table 3. The variables and parameters of the formulation.

Name	Type	Description
EV_1	endogenous	The number of EVs incited by deployed infrastructure when the DSO is not involved
EV_2	endogenous	The number of EVs incited by infrastructure, deployed either by private investors or DSO
EV_{ex}	endogenous	Existing stock of EVs inherited from previous periods.
T	exogenous	EV penetration target
B	exogenous	Upper bound of annual subsidy given per charging station ('000 euros)
K	exogenous	Annual DSO's budget for infrastructure investments ('000 euros)
sf	exogenous	Spatial factor denoting the geographical coverage a single charging station satisfies
SSf	exogenous	Self-Supply factor: fraction of a single EV's demand that can be satisfied by means of charging at home/work
$SSupp$	endogenous	Amount of charging demand that is supplied by chargers at home/work
C_{infra}	exogenous	Capital cost of a charging station
C_{EV}	exogenous	total cost of ownership for EVs
C_{conv}	exogenous	total cost of ownership for the typical ICE conventional vehicle
EV_{demand}	exogenous	the annual charging demand of a representative EV in kWh/year
$subs$	endogenous	The annual subsidy per charging station in '000 euros/year
d_1	endogenous	Total annual charging demand considering only the private investors involvement in GWh

Table 3. *Cont.*

Name	Type	Description
d_2	endogenous	Total charging demand incurred by the deployment of infrastructure by private investors and DSO
$Infr_1$	endogenous	New privately deployed infrastructure for the current period (number of charging stations)
$Infr_2$	endogenous	Total new infrastructure deployed in the current period
$Infr_{ex}$	exogenous	Existing infrastructure inherited from previous periods.
$infrMax$	exogenous	Maximum possible number of charging stations
Inv_{PA}	endogenous	Private agent's infrastructure investments in '000 euros
Inv_{DSO}	endogenous	DSO infrastructure investments in '000 euros
$Cann$	endogenous	Charging station's annual cash flow
IRR	endogenous	The internal rate of return for charging infrastructure investments
R	exogenous	Decision threshold on the value of IRR
u	endogenous	Annual demand satisfied by a single charging station
Pr	endogenous	Charging price in Euros/kWh

The formulation can be, intuitively, seen as having two phases. However, the model being formulated as a mixed complementarity problem is solved on a single shot. The use of MCP allows to use a dual variable of a constraint on the formulation of subsequent constraints. In the first phase, the model attempts to satisfy the EV penetration target allowing only private deployment of infrastructure, deciding the amount of subsidy if required. Variables with subscript 1 correspond to this phase. In the second phase, if the target is not achieved by private investments, the DSO is allowed to deploy the necessary infrastructure; the respecting variables have the subscript 2.

Let EV_1 be the number of electric vehicles and T the EV penetration target. The following constraint implements the component of policy maker's strategic choice that refers to subsidizing private investors. The complementary variable $subs$ represents the subsidy per charging station, in the form of annual payments, given to private investors to incite the deployment of infrastructure. The subsidy may not exceed an exogenous budget B.

$$EV_1 \; - T \; \geq \; 0 \bot subs \; \in \; [0, B] \tag{1}$$

Note that, since the complementary variable has an upper bound, it might be the case that in a feasible solution the inequality is not respected ($EV_1 < T$ and $subs = B$), meaning failure to achieve the target via a free market business model. This outcome is interpreted as the policy maker offering subsidy B and, nonetheless, the private investors not being interested to invest.

The next constraint (abstractly) implements the EV demand function considering the new private investments $Infr_1$ and the existing infrastructure $Infr_{ex}$. The number of new EVs is a function of the number of charging points, charging price Pr and vehicle costs C_{EV}, C_{conv}, with C_{EV}, C_{conv} being exogenous. D represents demand for EV and is based on discrete choice functional form. This constraint implements the consumers' choice for the first phase.

$$EV_1 = D(Infr_1, Infr_{ex}, \; Pr, C_{EV}, C_{conv}) \bot EV_1 \in \mathbb{R} \tag{2}$$

The following constraint calculates the annual charging demand (kWh/year) of the EV fleet for the first phase. The parameter EV_{demand} denotes the annual charging demand of a representative EV (kWh/year).

$$d_1 = (EV_1 + EV_{ex}) \cdot EV_{demand} \; \bot \; d_1 \in \mathbb{R} \tag{3}$$

An exogenous fraction of the total demand, denoted as SSf, is satisfied by home chargers based on the off-street parking capability of EV users or by charging at work.

$$SSupp = d_1 \cdot SSf \perp SSupp \in \mathbb{R} \tag{4}$$

Constraint (5) calculates the utilization u (kWh/year) of charging station. Function U gives the share of the demand that is satisfied by a single charging station, accounting for the spatial limitations of charging stations.

$$u = U(Infr_1, Infr_{ex}, sf) \cdot (d_1 - SSupp) \perp u \in \mathbb{R} \tag{5}$$

The annual cash flow $NCann$ for a charging station is given below. The charging price is denoted by pr while vc and fc denote the variable (e.g., electricity price) and fixed costs, respectively. The annual subsidy $subs$ is included in the calculation. The use of MCP allows to use the dual variable $subs$ of constraint (1) on the formulation of the annual cash flow. We note that the charging price pr comprises of the electricity price and the tariff associated to the recovery of the capital cost. The latter is calculated endogenously, using the levelized cost approach and depending on the rate of utilisation u.

$$NCann = u \cdot (pr - vc) - fc + subs \perp NCann \in \mathbb{R} \tag{6}$$

The next constraint calculates the IRR estimated by private investors. C_{infra} denotes the capital cost of a station.

$$\sum_t \left(\frac{NCann}{(1 + IRR)^t} \right) - C_{infra} = 0 \perp IRR \in \mathbb{R} \tag{7}$$

The decision making of private investors is implemented by constraint (8). Whenever IRR exceeds the exogenous threshold R, the agent chooses to invest, i.e., $Inv_{PA} > 0$. Although the complementary variable Inv_{PA} has no upper bound, it always takes finite values in a feasible solution. This is because, as Inv_{PA} increases beyond a certain value, IRR decreases (a) due to competition (controlled by the utilisation of the stations U of constraint (5)) and (b) due to the fact that, beyond a level of infrastructure coverage, marginal demand for EVs decreases (function D of constraint (2)). Thus, Inv_{PA} either takes a positive value and the constraint is satisfied as equality, or is zero if $IRR < R$.

$$IRR - R \leq 0 \perp Inv_{PA} \geq 0 \tag{8}$$

The new infrastructure deployed by private investors for the present period is simply derived by dividing the related investment expenditures by the unit cost of the charging point:

$$Infr_1 = \frac{Inv_{PA}}{C_{infra}} \perp Infr_1 \in \mathbb{R} \tag{9}$$

Constraints (10)–(12) refer to the DSO involvement in EV infrastructure deployment. If the target is not achieved and the private sector has not expressed interest even though the offered subsidy is at its maximum value ($subs =$ B) then, and only then, the DSO may invest in infrastructure.

Constraint (10) checks whether the target is satisfied considering the total infrastructure deployment; variable EV_2 denotes the number of EVs whose purchase was incited by total infrastructure (also see constraint (12)). If the target is not achieved, the complementary variable Inv_{DSO}, denoting DSO's investments, takes a positive value. In a feasible solution, either Inv_{DSO} takes a sufficiently large value to achieve the EV target, or the target was achieved by private infrastructure and Inv_{DSO} is equal to 0.

$$EV_2 - \mathrm{T} \geq 0 \perp Inv_{DSO} \in [0, K] \tag{10}$$

The number of charging station $Infr_2$ of this phase is simply the total investments divided by the capital cost $infra\_CC$ of a station.

$$Infr_2 = \frac{Inv_{PA} + Inv_{DSO} + Infr_{ex}}{C_{infra}} \perp Infr_2 \in \mathbb{R} \tag{11}$$

The number EVs incited by total charging infrastructure in the second phase is derived by the following constraint:

$$EV_2 = D(Infr_2, Infr_{ex}, Pr, C_{EV}, C_{conv}) \perp EV_2 \in \mathbb{R} \tag{12}$$

In the actual model, the mathematical program is solved iteratively for each time period, assuming 1-year time steps, inheriting the state of the electro-mobility system that resulted from the solution of the previous years. The inherited state concerns the existing stock of EVs and the already developed charging infrastructure and affects the utilization and profitability of new infrastructure, and the market penetration of new EVs. Lastly, we assume annual targets on EV penetration—the cumulative EV stock should amount to the desired level of penetration at the end of the modelled period, assuming it was feasible to achieve each year's target.

4. Scenarios and Results: The Evolution of Electro-Mobility in Greece for the 2021–2030 Period

4.1. Background

This section outlines the proposed methodology, presented in Section 3. We display the functionality of the proposed system modelling by applying it to the Greek case. The 2021–2030 decade is critical for ensuring transport embarks on a decarbonisation pathway by 2050. Recent plans by the Greek government [46] envisage reaching a deployment of approximately 10% of EVs in the total fleet of cars by 2030 (i.e., around 500,000 EVs), in view of the 2030 Energy and Climate target. The vast majority of the envisaged EV sales are expected to be small-sized cars. Yet, the current deployment of recharging points and the associated EV sales are negligible in Greece. To motivate EV sales and development of recharging points, Greece has adopted the EC proposal by retaining a DSO model as a short-term solution [34] and aiming at the adoption of a free market model for infrastructure deployment in the medium-term.

Given the above, the quantification of scenarios carried out in this research paper is built around the assumption that the EV deployment will need to reach approximately 10% of the total fleet in 2030. The scenarios explore the factors which influence the transition from a public DSO to a private free market model, considering alternative framework conditions (e.g., costs of EVs, costs of charging points), as well as potential subsidisation of the capital costs of the charging points.

In the following, we present the underlying assumptions and the description of the scenarios. Afterwards, we present the model results for scenarios and sensitivities we carry out.

4.2. Assumptions and Description of Scenarios

For our analysis, we consider three different cases for the evolution of EV purchase costs and infrastructure capital costs. The purchasing costs of EVs draw from literature and follow a central, an optimistic and a pessimistic trajectory, using assumptions from literature. We define a set of nine scenarios corresponding to the possible combinations of cost assumptions (see the scenario names and definitions in Table 4). The assumed costs are presented in Tables 5 and 6.

Table 4. Qualitative specifications of the nine scenarios.

Scenario Name	EV Purchase Cost	Charging Station Capital Cost
Low EV-Low Ch.Point	Low Cost (optimistic)	Low Cost (optimistic)
Low EV-Mid Ch.Point	Low Cost (optimistic)	Central Cost
Low EV-High Ch.Point	Low Cost (optimistic)	High Cost (pessimistic)
Mid EV-Low Ch.Point	Central Cost	Low Cost (optimistic)
Mid EV-Mid Ch.Point	Central Cost	Central Cost
Mid EV-High Ch.Point	Central Cost	High Cost (pessimistic)
High EV-Low Ch.Point	High Cost (pessimistic)	Low Cost (optimistic)
High EV-Mid Ch.Point	High Cost (pessimistic)	Central Cost
High EV-High Ch.Point	High Cost (pessimistic)	High Cost (pessimistic)

Table 5. Assumptions on the evolution of medium sized EVs purchasing cost (in euros).

Euros	2020	2025	2030
Low Cost		27,000	23,000
Moderate Cost	31,000	28,000	24,000
High Cost		29,000	27,000

Table 6. Assumptions on the capital cost in euros of L3 Charging Stations (in euros).

Euros	2020	2025	2030
Low Cost		36,500	28,000
Moderate Cost	44,000	40,000	35,000
High Cost		42,500	40,000

The rest of the assumptions are common and include the following:

- The price of electricity, which is the variable cost of a charging station and affects the fuel cost of EVs draws from the PRIMES model [44] and ranges from 0.163 euros/kWh in 2021 to 0.175 euros/kWh in 2030. The remuneration of the capital cost of the charging points is calculated endogenously in the model based on the levelized cost of infrastructure. The charging price is assumed to be capped at 0.32 euros/kWh to prevent overcharging of EV users;
- The share of demand that is self-supplied by means of home charging is assumed to range from around 75% in the start of the 2021–2030 decade to around 70% in 2030, drawing from [47];
- The maximum annual subsidy per charging station is assumed to be 4000 euros. Recall that, in our modelling, the actual annual subsidy is endogenously derived each year to ensure a certain level of profitability for private investors;
- The annual budget for public infrastructure investment is assumed to be 15 million euros. Public infrastructure investments are made whenever DSO model deploys. In this case, the amount of investments depends on the infrastructure coverage required to achieve the desired level of EV penetration, following a central planning approach;
- The IRR decision threshold for private investors to engage in the charging infrastructure development is assumed to be 5%. We carry out sensitivity analysis on this assumption;
- All the techno-economic assumptions on the competing vehicle technologies, apart from the purchase cost of EVs, are common among the scenarios. These assumptions include vehicle mileage, fuel consumption, maintenance and insurance costs, and vehicle economic lifetime. For the competing fuel technologies, we assume two representative vehicles: a medium-sized gasoline car and a medium sized EV;
- Lastly, for public infrastructure we consider L3 DC fast recharging stations assuming a typical charging power of 50 KW.

Table 5 presents the three assumed cases for the evolution of EV purchase costs and Table 6 presents the respecting cases for the evolution of charging station capital costs for the 2021–2030 period. For the year 2020, we assume the same costs for EVs and charging points across the scenarios. For the intermediate years (2021–2024 and 2026–2029), we have assumed a linear interpolation of the costs

4.3. Model Results

4.3.1. Penetration of EVs

The penetration of EVs in all scenarios in the first half of the 2021–2030 decade displays only marginal variation (see Table 7). In all scenarios, in 2021 the EV fleet consists of around 2–3000 vehicles with only small differences among the scenarios due to the common scenario assumption (i.e., costs of cars and charging points). The fleet of EVs reaches approximately 143,000 vehicles in 2025 with small variations among the scenarios (which is justified to a certain extent by the relatively small variations in the cost assumptions in 2025). This result is also driven by the fact that the DSO model is employed in the first years, as will be presented further on this section, which develops infrastructure based on planned coverage and applies common regulated charging prices.

Table 7. Total stock of EVs in Greece in '000 vehicles.

Scenario	2021	2025	2030
Low EV-Low Ch.Point	2.7	143	549
Low EV-Mid Ch.Point	2.6	143	535
Low EV-High Ch.Point	2.5	143	527
Mid EV-Low Ch.Point	2.6	143	530
Mid EV-Mid Ch.Point	2.5	143	518
Mid EV-High Ch.Point	2.5	142	511
High EV-Low Ch.Point	2.5	143	504
High EV-Mid Ch.Point	2.4	142	493
High EV-High Ch.Point	1.8	138	479

By contrast, the second half of the 2021–2030 decade displays a greater variation in EV penetration. The Low EV-Low Ch.Point scenario records the highest EV penetration (around 548,600 EVs) while the lowest EV penetration takes place in the High EV-High Ch.Point (479,000 EVs). The differences in the evolution of EV and charging station capital costs are enlarged towards the end of the studied period and, thus, have a greater impact on EV penetration compared to the first half of the decade. In addition, in the 2025–2030 period, the private sector becomes more active in infrastructure deployment (also see Section 4.3.2). As the infrastructure investments of the private sector react with charging demand in a feedback relationship, the private model causes rebound effects on the EV penetration in scenarios with more optimistic assumed costs. We note that, even in 2030, the EV fleet variation is within 10% of the 500,000 EVs target. This is the result of policy action taken to ensure the satisfaction of the target; in scenarios with less favourable cost assumptions: (i) the DSO model is employed for a longer period to help achieve a sufficient level of EV penetration and, thus, charging demand before the private agents start investing and (ii) the subsidization of private agents is stronger in the scenarios with less favourable infrastructure capital cost (see following section).

4.3.2. Deployment of Private Investments in Charging Infrastructure Development

The total number of available charging stations for 2021, 2025, and 2030 is presented in Table 8. It is generally observed that in all scenarios the number of charging stations follows the increasing trend of the EV fleet. Table 9 presents the deployment of the private investment (i.e., the share of publicly available charging stations deployed using private funds in the total number of available charging stations). All scenarios indicate an ever-growing involvement of the private investors in the EV recharging infrastructure business towards the end of the decade 2021–2030.

Table 8. Total available charging stations per time period.

Scenario	2021	2025	2030
Low EV-Low Ch.Point	79	1719	5245
Low EV-Mid Ch.Point	78	1708	4924
Low EV-High Ch.Point	78	1695	4746
Mid EV-Low Ch.Point	79	1765	5066
Mid EV-Mid Ch.Point	78	1757	4772
Mid EV-High Ch.Point	78	1751	4599
High EV-Low Ch.Point	79	1841	4822
High EV-Mid Ch.Point	78	1832	4541
High EV-High Ch.Point	73	1822	4312

Table 9. Share of charging stations deployed by the private sector in the total available charging stations.

Scenario	2021	2025	2030
Low EV-Low Ch.Point		34%	85%
Low EV-Mid Ch.Point		34%	84%
Low EV-High Ch.Point		33%	83%
Mid EV-Low Ch.Point		35%	84%
Mid EV-Mid Ch.Point	0%	35%	83%
Mid EV-High Ch.Point		31%	81%
High EV-Low Ch.Point		36%	82%
High EV-Mid Ch.Point		32%	80%
High EV-High Ch.Point		25%	76%

Interestingly, while in 2025 the scenarios with pessimistic EV cost assumptions (i.e., High) show a higher number of charging stations, the picture changes in 2030 when we observe a higher number of total available charging stations in the scenarios with the optimistic cost assumptions (i.e., Low). Such development is attributed to the fact that in the first half of the 2020–2030 decade, infrastructure is deployed for the most part by the DSO, driven mainly by the need to meet the policy target rather than for profitability reasons. In the second half of the decade, we notice that the private model turns out to be the prevailing model for infrastructure deployment. This is the result of the development of a critical mass of charging points (thanks to the DSO engagement in the early years), implying an adequate utilisation of the charging points and ensuring profitability. The assumed cost reductions in all scenarios are also a critical factor for such development. The consumers also tend to purchase more EVs as a result of the decreasing EV costs and the reduction in the range anxiety (due to the development of the critical mass of charging points). Especially, in the case of the optimistic cost assumptions, we observe higher EV penetration resulting in higher utilization of infrastructure, further inciting the private sector engagement.

The scenarios project a relatively similar success for the electro-mobility system and the private infrastructure model at the end of the 2021–2030 period. The policy intervention via the initial DSO-model deployment and the subsidization of private investments mitigates to some extent the higher infrastructure costs and the low EV penetration (due to the higher costs in the early years). Evidently, the level of intervention required differs significantly among the scenarios. The annual subsidy per private charging station from 2024 to 2028 is presented in Table 10. In all scenarios, before 2024, the low infrastructure utilization prevents the private investors from entering the infrastructure market. Until 2023 all the charging stations are publicly developed via the DSO model. In 2024, with the help of subsidization, the private investors are involved in infrastructure deployment for the first time. In 2024 the subsidy that private stations require is close to the maximum available budget per charging point of 4000 euros. The subsidization is gradually decreased in later years. In the Low EV-Low Ch.Point (most optimistic) scenario the private charging stations stop needing subsidy as until 2026. In contrast, in the High EV-High Ch.Point (most pessimistic) scenario the private stations still require some subsidization, up to 2028, to ensure an acceptable level of profitability.

Table 10. Annual subsidy given to private agents per charging station in euros.

Scenario	2024	2025	2026	2027	2028
Low EV-Low Ch.Point	3700	2600	1200	0	0
Low EV-Mid Ch.Point	3900	2800	1500	100	0
Low EV-High Ch.Point	4000	3000	1600	200	0
Mid EV-Low Ch.Point	3800	2700	1400	200	0
Mid EV-Mid Ch.Point	4000	3000	1700	400	0
Mid EV-High Ch.Point	4000	3200	1900	600	0
High EV-Low Ch.Point	3900	3000	1800	600	0
High EV-Mid Ch.Point	4000	3200	2000	900	0
High EV-High Ch.Point	4000	4000	2700	500	300

The subsidization of the privately deployed recharging points and the DSO employment is found to increase the government bill. Table 11 presents the cumulative government expenditures on the infrastructure investments made via the DSO model and the subsidies given to support infrastructure deployment by private agents. Naturally, the expenditure is greater in scenarios with higher infrastructure costs. This is not only because the costs of the charging points, per se, are higher, but also because the private investors are more reluctant to invest. In particular, the largest part of the total government expenditures is found to be necessary in the period up to 2025, when the DSO model is largely employed. For the period after 2025 government expenditures concern, for the most part, the subsidization of private investors which requires less capital compared to the public development of infrastructure. Towards the end of the period, the deployment of recharging infrastructure takes place without further policy support and does not require additional expenditures from the government.

Table 11. Cumulative government expenditures for infrastructure investments and subsidies in million.

Scenario	2021	2025	2030
Low EV-Low Ch.Point	3.5	52.7	53.7
Low EV-Mid Ch.Point	3.4	54.2	55.6
Low EV-High Ch.Point	3.4	55.9	57.6
Mid EV-Low Ch.Point	3.5	53.7	55.3
Mid EV-Mid Ch.Point	3.4	55.3	57.4
Mid EV-High Ch.Point	3.4	58.6	60.9
High EV-Low Ch.Point	3.5	55.5	58.1
High EV-Mid Ch.Point	3.4	59.6	62.7
High EV-High Ch.Point	3.2	71.6	75.8

Table 12 presents the endogenously calculated charging price for using the infrastructure; a part of the charging price includes the tariff set by the investors to recuperate their capital cost. As presented in the assumptions of the modelling implementation, the latter is calculated to be equal to the levelized cost of the use of the charging infrastructure. We acknowledge though that, sometimes, this practice may differ from reality, especially in the early years of EV deployment when the demand for charging might be low. In such cases, regulated prices might apply to avoid deterring private investors from entering the market and EV users be discouraged from high charging prices. We carry out a sensitivity analysis on the charging price assumptions, however, more research is needed on this topic. As confirmed by the results, the tariff for recovery of the capital cost of charging point is greater in scenarios with higher infrastructure capital costs. The levelized costs decrease in 2030 due to assumed reduction in infrastructure costs and higher utilisation rates of the infrastructure. The charging tariff, however, increases as it includes the price of electricity which, according to our assumptions, is approximately 0.16 euros/kWh and 0.175 euros/kWh in 2025 and 2030, respectively.

Table 12. Charging price in euro/kWh in 2025 and 2030.

Scenario	Tariff for Recovery of the Capital Cost of Charging Point		Total Charging Tariff (Incl. Electricity Price)	
	2025	**2030**	**2025**	**2030**
Low EV-Low Ch.Point	0.131	0.114	0.281	0.289
Low EV-Mid Ch.Point	0.137	0.126	0.287	0.301
Low EV-High Ch.Point	0.142	0.137	0.292	0.312
Mid EV-Low Ch.Point	0.133	0.116	0.283	0.291
Mid EV-Mid Ch.Point	0.141	0.131	0.291	0.306
Mid EV-High Ch.Point	0.147	0.142	0.297	0.317
High EV-Low Ch.Point	0.136	0.119	0.286	0.294
High EV-Mid Ch.Point	0.146	0.136	0.296	0.311
High EV-High Ch.Point	0.152	0.137	0.302	0.312

4.4. Sensitivity Analysis

The scenario analysis considered nine combinations of alternative trajectories for the costs of EVs and the charging points. To complement the scenario analysis, we present a sensitivity analysis on key elements which are expected to influence the model results. The aim of the sensitivity analysis is to quantify how the recharging infrastructure deployment is influenced by different thresholds of profitability criteria for private investors and policy support. In particular, we carry out sensitivity analysis with varying:

- charging prices of the private investors, by assuming that the capital cost recovery is based on pre-defined regulated prices;
- values of IRR decision threshold for private investors to engage in the recharging infrastructure development business;
- levels of the maximum available budget (per charging point) for subsidising private investors.

4.4.1. The Effects of Different Levels of Regulated Charging Tariffs on the Infrastructure Deployment

A sensitivity analysis was performed to assess the implementation of regulated charging prices. On the one hand, applying regulated prices may attract private investors to develop private charging points as it reduces the associated investment risk to a certain extent; on the other hand, regulated prices will also ensure affordable price for consumers to use charging services for their EVs. We use the central scenario Mid EV-Mid Ch.Point, as a starting point to quantify two new scenarios, namely: The Low Ch.Price and the High Ch.Price scenarios, which assume regulated charging prices of 0.32 and 0.22 euros/kWh, respectively. We note that in the Mid EV-Mid Ch.Point scenario the charging price is approximately 0.29 euros/kWh (see Table 12).

Table 13 compares the cumulative deployment of charging stations and the respective share of private investments in the total installations in 2030 of the Mid EV-Mid Ch.Point, Low Ch.Price, and High Ch.Price scenarios.

Table 13. Number of charging stations and respective share deployed by the private sector in 2030.

Scenario	Cumulative Number of Stations	Share of Stations Developed by Private Agents
Mid EV-Mid Ch.Point	4772	83%
High Ch.Price	4985	89%
Low Ch.Price	4547	77%

Both the number of stations and the degree of private sectors' involvement in the two sensitivity scenarios suggest that a higher charging price positively affects the deployment of infrastructure and the success of a private model for EV charging. The higher prices assumed relative to the central case are found to positively influence investors. These, together with the increased investments in charging points, are found to positively influence demand (despite the increase in the prices of the services) as the range anxiety factor is diminished. The increased profitability of charging businesses leads the private sector to deploy 89% of the 4985 stations in 2030 in scenario High Ch.Price. By contrast, 4547 fast charging stations are built by 2030 in scenario Low Ch.Price, lower than the 4772 stations of the Mid EV-Mid Ch.Point scenario and about 77% of the stations are deployed via private investments.

Similar results are also found when comparing the EV fleet evolution. The higher charging price also results in greater penetration of EVs in the High Ch.Price scenario (approximately 530,000 EVs in 2030) compared to the Mid EV-Mid Ch.Point, while in the Low Ch.Price scenarios approximately 505,000 vehicles are found to penetrate the market by 2030. The increased charging price for public charging only marginally affected the EV user variable costs, as a large part of charging demand is satisfied by home charging. The variable costs (i.e., charging costs) account for a relatively small fraction of the total cost of ownership. The larger infrastructure coverage of the High Ch.Price scenario mitigates the range anxiety to a significant extent, leading to a more favourable perceived cost of EVs compared to the rest scenarios.

4.4.2. Sensitivity Analysis on Private Investors' IRR

A second sensitivity analysis is performed to evaluate how the private investments in the recharging infrastructure business are influenced when investors expect higher returns on their investment. Higher investment returns are also associated with the opportunity cost of the businesses. The common assumption, presented in the scenarios above, is that investors engage when the expected IRR of their investment exceeds 5%. As part of the sensitivity analysis, we assume greater values for the IRR decision threshold of private investors, which translates to a greater reluctance of the private sector to participate in the deployment of infrastructure.

We define two additional sensitivity scenarios based, again, on scenario Mid EV-Mid Ch.Point: the scenario IRR_8, which assumes an 8% IRR decision threshold for private investors, and scenario IRR_12, which assumes a 12% decision threshold. The 8 and 12% IRR thresholds selected for the present sensitivity analysis were based on [48], on the financial viability of low-power charging stations in Greece, which suggests an IRR of around 10%. The 5% IRR decision threshold assumed in the 9 scenarios of Section 4.3 was selected based on the past decade's experience on solar panel investments in Greece. Table 14 shows the cumulative number of charging stations, the respective share developed by the private sector and the total EV fleet in 2030.

The increased decision threshold makes investors more reluctant to invest in infrastructure. In IRR_8 and IRR_12 scenarios the private sector involvement in infrastructure deployment in 2030 is less than that of scenario Mid EV-Mid Ch.Point; overall, 75 and 61% of total charging stations are privately deployed in IRR_8 and IRR_12, respectively, compared to the 83% share in the Mid EV-Mid Ch.Point scenario.

Table 14. Share of charging stations deployed by the private sector in 2030.

Scenario	Cumulative Number of Stations	Share of Stations Developed by Private Agents	EV Fleet
Mid EV-Mid Ch.Point	4772	83%	518.2
IRR_8	4186	75%	491.9
IRR_12	3408	61%	465.7

In addition, the greater IRR decision threshold of IRR_8 and IRR_12 negatively affect the total number of EVs and the charging stations built by 2030. We find that as the expectations for the investment returns increase, the deployment of private investments decreases. This is driven by the fact that to achieve an IRR of 12%, high utilisation of the charging points for long periods of time is needed. Interestingly, we observe that the share of stations developed by private investors reaches 61% of the total installations in 2030 in the scenario with an IRR of 12%, in contrast with the scenario with an IRR of 5% in which the equivalent share is 83%. Moreover, the total EV fleet in the IRR_12 scenario is well below 500,000 vehicles. This is the result of the failure to achieve the penetration target in some years of the modelled period; in IRR_12 the private actors have lesser engagement compared to the other scenarios and the public investments fail to achieve the needed infrastructure coverage due to budget limitations.

4.4.3. Sensitivity Analysis on the Available Budget

The third part of the sensitivity analysis aims to assess how the assumed subsidy budget affects the decision making of private investors and the success of a private model for infrastructure. The annual subsidy for the private deployment and operation of fast charging stations, as observed in the results of Section 4.3, acted primarily as a support for the transition period from a DSO model to a private model that occurs in the mid-term of the studied period. This part of the sensitivity analysis aims to answer whether a different subsidy budget can accelerate or delay the participation of private investors.

For this purpose, we define two new scenarios: the scenario High Sub, which assumes a 5000 budget for subsidies and the scenario Low Sub, which assumes a 2000 budget. Both scenarios are based, once again, on the assumptions of the Mid EV-Mid Ch.Point scenario. Keep in mind that the scenario Mid EV-Mid Ch.Point assumes an annual budget of 4000 euros. Table 15 presents the total charging stations built in 2021, 2025, and 2030 for the three scenarios and Table 16 shows the respective share of charging stations that is deployed by private investors.

Table 15. Total available charging stations per time period for the three scenarios.

Scenario	2021	2025	2030
Mid EV-Mid Ch.Point		1757	4772
High Sub	78	1803	4799
Low Sub		1690	4727

Table 16. Share of charging stations deployed by the private sector in the total available charging stations per time period.

Scenario	2021	2025	2030
Mid EV-Mid Ch.Point	0%	35%	83%
High Sub		52%	88%
Low Sub		0%	71%

Model results show that, while higher and lower subsidy budget leads to accelerated and delayed participation of the private sector, respectively, the total infrastructure deployed, especially towards the end of the period, is only marginally affected. This finding is consistent with the observation that subsidies are needed mainly in the mid-term and are irrelevant towards 2030. As the infrastructure is deployed via a DSO model, whenever the private investors are not incited to invest, in later years (i.e., 2027–2030) the investors find themselves in a similar state of the electro-mobility system and, thus, resort to similar decisions. This explains smaller variation in the share of private infrastructure in 2030 compared to 2025.

5. Conclusions

Electric vehicles present a key solution to decarbonize the passenger car segment; the EU strategy for the mitigation of GHG emissions in transport foresees a widespread adoption of EVs. A key barrier to the wider uptake of EVs is the lack of charging infrastructure, which poses mobility limitations to the EV users and causes range anxiety. In turn, the low uptake of electro-mobility prevents investors from engaging in the recharging infrastructure development business. This "chicken–egg" problem can be solved by strategically developing charging stations using public funds to achieve the deployment of a critical mass of infrastructure, which then will attract private investors.

The scientific contribution of the present work lies on the proposal and application of a model of the electro-mobility system which simulates the interplay between the penetration of EVs and the coverage of charging infrastructure, focusing on the transition from an initial public model for infrastructure deployment to a private model, which ensures the successful participation of the private sector. The aim of the work is to assess the impact of multiple factors on the evolution of EVs, including EV purchase costs, charging stations' capital costs, charging price, policy support via subsidies, and variations in investors behaviour.

The proposed methodology is applied on the case study of Greece, quantifying the impacts of the aforementioned factors on the penetration of EVs and the rollout of charging infrastructure. The national Greek objectives refer to an uptake of 10% EVs in the total fleet of cars by 2030, when currently the national EV sales are negligible. Greece represents a typical example of a country with ambitious EV targets for 2030, with limited development of recharging infrastructure up to now. This is why we think this work is policy relevant also for other countries with limited infrastructure development until today.

5.1. Policy Implications

The policy implications for Greece reveal that a DSO-type model needs to be employed early in the 2021–2030 period and be replaced by a private model for charging infrastructure with the help of subsidies in the mid-term. DSO deployment is based on public funds, which we acknowledge to be one form of governmental support to developing the charging infrastructure.

Another form of governmental support, which we consider in this paper, is via subsidizing private investors in entering the market. Price subsidization of the charging points deployed by private investors is necessary during a transition period from a DSO to a private business model. The transition period for Greece is found to take place around 2025. Providing higher subsidies to the private investors can accelerate the transition to private business model in Greece. A third form of governmental support is assuming regulated tariffs, which allow private investors to sufficiently recover the initial investment. In such cases, private investors find the recharging business more secure, thus attractive, and engage early in the 2020–2030 decade. We find that such development leads to an earlier accumulation of a critical mass of charging points, which quickly mitigates the consumers' range anxiety.

The analysis finds that the private model is deployed without the need for further policy help, when approaching 2030, leading to the successful rollout of EVs. The main reason for this is that a critical mass of charging points and EVs has been already developed, which means infrastructure is sufficiently used and range anxiety is dropping. We suggest that the complete transition to the private business model is the solution to the "chicken–egg" problem.

Undoubtedly, framework conditions such as the dropping costs of EV and of charging points can have a noticeable effect in the transition from public to private business models. Lower infrastructure costs result in earlier participation of private investors, increased profitability of charging businesses and less need for subsidization. Similarly, lower EV purchasing costs result in stronger EV penetration, which affects infrastructure deployment in a feedback relationship.

The key policy proposal is that the infrastructure should be deployed in the first years of the decade with the support of public funds to ensure the penetration of a critical mass of EVs. The increased demand for charging services and use of infrastructure, combined with sufficient amount of subsidizations, should encourage, in the mid-term, the participation of the private sector so as to let, eventually, the market forces lead the evolution of the system. Subsidizing the private investors to deploy charging points can be assessed after a critical mass of publicly-deployed charging points has been installed, since such an analysis needs to consider the opportunity costs of the public funds.

An interesting question arises as to whether the analysis can be applied to the whole of the EU. This is critical considering the uneven distribution of electricity recharging infrastructure among the EU countries, so far. In our view, the conclusions of this paper ought to be generalised to the whole EU and to individual EU countries in a careful way and considering a number of specificities. Statistics show varying EV deployment patterns across EU Member States; higher EV deployment has been recorded in countries with high household disposable income. The available household budget has been an important criterion for alleviating the cost of EVs barrier until now. In addition, government support (in the form of purchasing price subsidies) has also focused on reducing the cost barrier for EVs. So far, the government support for EV adoption is quite different among EU countries. The development of the charging infrastructure in the EU has been based on a mix of private and public funds. The paper concludes that public funding may play an important role in the initial deployment of EVs charging infrastructure; this conclusion seems to be more relevant in countries with low EV market uptake, low charging infrastructure development, and a relatively low household disposable income.

5.2. Limitations and Scope for Further Research

We acknowledge a number of limitations in the present study, which need to be addressed in future research.

A limitation of the study is that it considers the case of fast-charging points and not the possibility of semi-fast public charging points. The analysis considers only a single type of publicly available charging points (i.e., a fast DC recharging point of approximately 50 kW). This is because in Greece short driving distances mainly for commuting purposes may not require the use of the entire batter power, which the car owner can charge at their residence. Future research thus needs to assess the necessary investments in the development of other types of charging points based on their charging power (semi-fast and ultra-fast DC charging points).

The representation of the policy support can be refined in the future. In this paper, for the deployment of infrastructure we assumed policy support schemes separately for private investors, charging points developed by DSO and charging prices. A more refined approach could consider a total amount of available public funds to be optimally allocated to ensure the timely roll-out of EV charging infrastructure. Moreover, we considered the tariff set on charging points to be equal to the levelized cost of using the charging infrastructure. We acknowledge though that, sometimes, this practice may differ from reality, especially in the early years of EV deployment when demand for charging might be low.

In this study the car manufacturers are not represented as a separate entity, interacting with the other agents of electro-mobility ecosystem. Considering them in future work would allow further exploring different business models, such as in [49], where the decision over the cooperation between EV manufacturers and charging station operators leads to different overall profits, market shares, and vehicle attributes.

Author Contributions: Conceptualization, P.S. and P.C.; methodology, P.S. and P.C.; software, S.S. and Y.M.; validation, P.S. and P.C.; formal analysis, S.S. and Y.M.; investigation, S.S., P.S. and Y.M.; resources, P.S.; data curation, S.S. and Y.M.; writing—original draft preparation, S.S. and Y.M.; writing—review and editing, P.S.; visualization, Y.M..; supervision, P.S. and P.C.; project administration, P.S.; funding acquisition, P.C. All authors have read and agreed to the published version of the manuscript.

Funding: This research was funded by the European Union's Horizon 2020 research and innovation programme under grant agreement No 730403 "Innovation pathways, strategies and policies for the Low-Carbon Transition in Europe (INNOPATHS)".

Acknowledgments: This paper has received funding from the European Union's Horizon 2020 research and innovation programme under grant agreement No 730403 "Innovation pathways, strategies and policies for the Low-Carbon Transition in Europe (INNOPATHS)". The content of this paper does not reflect the official opinion of the European Union. Responsibility for the information and views expressed herein lies entirely with the author(s). The authors would like to thank the anonymous reviewers for their comments which greatly improved the manuscript.

Conflicts of Interest: The authors declare no conflict of interest.

References

1. EC. A Clean Planet for all A European Strategic Long-Term Vision for a Prosperous, Modern, Competitive and Climate Neutral Economy: In-Depth Analysis in Support of the Commission Communication COM(2018) 773. 2018. Available online: https://knowledge4policy.ec.europa.eu/publication/depth-analysis-support-com2018-773-clean-planet-all-european-strategic-long-term-vision_en (accessed on 21 February 2021).
2. Connolly, D.; Mathiesen, B.V.; Ridjan, I. A comparison between renewable transport fuels that can supplement or replace biofuels in a 100% renewable energy system. *Energy* **2014**, *73*, 110–125. [CrossRef]
3. McCollum, D.; Krey, V.; Kolp, P.; Nagai, Y.; Riahi, K. Transport electrification: A key element for energy system transformation and climate stabilization. *Clim. Chang.* **2014**, *123*, 651–664. [CrossRef]
4. Pietzcker, R.C.; Longden, T.; Chen, W.; Fu, S.; Kriegler, E.; Kyle, P.; Luderer, G. Long-term transport energy demand and climate policy: Alternative visions on transport decarbonization in energy-economy models. *Energy* **2014**, *64*, 95–108. [CrossRef]
5. Bosetti, V.; Longden, T. Light duty vehicle transportation and global climate policy: The importance of electric drive vehicles. *Energy Policy* **2013**, *58*, 209–219. [CrossRef]
6. Zhang, Q.; Li, H.; Zhu, L.; Campana, P.E.; Lu, H.; Wallin, F.; Sun, Q. Factors influencing the economics of public charging infrastructures for EV—A review. *Renew. Sustain. Energy Rev.* **2018**, *94*, 500–509. [CrossRef]
7. Sierzchula, W.; Bakker, S.; Maat, K.; Wee, B. The influence of financial incentives and other socio-economic factors on electric vehicle adoption. *Energy Policy* **2014**, *68*, 183–194. [CrossRef]
8. Statharas, S.; Moysoglou, Y.; Siskos, P.; Zazias, G.; Capros, P. Factors Influencing Electric Vehicle Penetration in the EU by 2030: A Model-Based Policy Assessment. *Energies* **2019**, *12*, 2739. [CrossRef]
9. Dong, J.; Lin, Z. Within-day recharge of plug-in hybrid electric vehicles: Energy impact of public charging infrastructure. *Trans. Res. Part D Transp. Environ.* **2012**, *17*, 405–412. [CrossRef]
10. Sioshansi, F.; Webb, J. Transitioning from conventional to electric vehicles: The effect of cost and environmental drivers on peak oil demand. *Econ. Anal. Policy* **2019**, *61*, 7–15. [CrossRef]
11. Wang, S.; Wang, J.; Li, J.; Wang, J.; Liang, L. Policy implications for promoting the adoption of electric vehicles: Do consumer's knowledge, perceived risk and financial incentive policy matter? *Trans. Res. Part A Policy Pract.* **2018**, *117*, 58–69. [CrossRef]
12. Myklebust, B. EVs in bus lanes—Controversial incentive. In Proceedings of the World Electric Vehicle Symposium and Exhibition, Barcelona, Spain, 17–20 November 2013.
13. Nie, Y.M.; Ghamami, M.; Zockaie, A.; Xiao, F. Optimization of incentive polices for plug-in electric vehicles. *Trans. Res. Part B Methodol.* **2016**, *84*, 103–123. [CrossRef]
14. Sykes, M.; Axsen, J. No free ride to zero-emissions: Simulating a region's need to implement its own zero-emissions vehicle (ZEV) mandate to achieve 2050 GHG targets. *Energy Policy* **2017**, *110*, 447–460. [CrossRef]
15. Siskos, P.; Moysoglou, Y. Assessing the impacts of setting CO_2 emission targets on truck manufacturers: A model implementation and application for the EU. *Trans. Res. Part A Policy Pract.* **2019**, *125*, 123–138. [CrossRef]
16. Speidel, S.; Bräunl, T. Driving and charging patterns of electric vehicles for energy usage. *Renew. Sustain. Energy Rev.* **2014**, *40*, 97–110. [CrossRef]

17. Morrissey, P.; Weldon, P.; Mahony, M.O. Future standard and fast charging infrastructure planning: An analysis of electric vehicle charging behaviour. *Energy Policy* **2016**, *89*, 257–270. [CrossRef]
18. Kihm, A.; Trommer, S. The new car market for electric vehicles and the potential for fuel substitution. *Energy Policy* **2014**, *73*, 147–157. [CrossRef]
19. Neaimeh, M.; Salisbury, S.D.; Hill, G.A.; Blythe, P.T.; Scoffield, D.R.; Francfort, J.E. Analysing the usage and evidencing the importance of fast chargers for the adoption of battery electric vehicles. *Energy Policy* **2017**, *108*, 474–486. [CrossRef]
20. Lin, Z.; Greene, D.L. Promoting the market for plug-in hybrid and battery electric vehicles: Role of recharge availability. *Trans. Res. Rec.* **2011**, *2252*, 49–56. [CrossRef]
21. Greene, D.L.; Kontou, E.; Borlaug, B.; Brooker, A.; Muratori, M. Public charging infrastructure for plug-in electric vehicles: What is it worth? *Trans. Res. Part D Trans. Environ.* **2020**, *78*, 102182. [CrossRef]
22. Kontou, E.; Liu, C.; Xie, F.; Wu, X.; Lin, Z. Understanding the linkage between electric vehicle charging network coverage and charging opportunity using GPS travel data. *Trans. Res. Part C Emerg. Technol.* **2019**, *98*, 1–13. [CrossRef]
23. Kley, F.; Lerch, C.; Dallinger, D. New business models for electric cars—A holistic approach. *Energy Policy* **2011**, *39*, 3392–3403. [CrossRef]
24. Patt, A.; Aplyn, D.; Weyrich, P.; Vliet, O. Availability of private charging infrastructure influences readiness to buy electric cars. *Trans. Res. Part A Policy Pract.* **2019**, *125*, 1–7. [CrossRef]
25. Siskos, P.; Zazias, G.; Petropoulos, A.; Evangelopoulou, S.; Capros, P. Implications of delaying transport decarbonisation in the EU: A systems analysis using the PRIMES model. *Energy Policy* **2018**, *121*, 48–60. [CrossRef]
26. Schroeder, A.; Traber, T. The economics of fast charging infrastructure for electric vehicles. *Energy Policy* **2012**, *43*, 136–144. [CrossRef]
27. Gnann, T.; Plötz, P.; Wietschel, M. How to address the chicken-egg-problem of electric vehicles? Introducing an interaction market diffusion model for EVs and charging infrastructure. In Proceedings of the ECEEE Summer Study, Toulon, France, 1–6 June 2015; pp. 873–884.
28. Baresch, M.; Moser, S. Allocation of e-car charging: Assessing the utilization of charging infrastructures by location. *Trans. Res. Part A Policy Pract.* **2019**, *124*, 388–395. [CrossRef]
29. Madina, C.; Zamora, I.; Zabala, E. Methodology for assessing electric vehicle charging infrastructure business models. *Energy Policy* **2016**, *89*, 284–293. [CrossRef]
30. Zhang, L.; Zhao, Z.; Xin, H.; Chai, J.; Wang, G. Charge pricing model for electric vehicle charging infrastructure public-private partnership projects in China: A system dynamics analysis. *J. Clean. Prod.* **2018**, *199*, 321–333. [CrossRef]
31. Zhang, L.; Yang, M.; Zhao, Z. Game analysis of charging service fee based on benefit of multi-party participants: A case study analysis in China. *Sustain. Cities Soc.* **2019**, *48*. [CrossRef]
32. Osterwalder, A.; Pigneur, Y. *Business Model Generation: A Handbook for Visionaries, Game Changers, and Challengers*; John Wiley & Sons: Hoboken, NJ, USA, 2010.
33. Adler, M.; Bagemihl, J.; Bernard, G.; Biser, T.; Caleno, F.; Sanchez, J.M.C.; Densley, D.; Exposito, E.D.; Flader, L.; Martin, J.G.; et al. Eurelectric. Deploying publicly accessible charging infrastructure for electric vehicles: How to organise the market? *Eurelectric Concept Pap.* **2013**. Dépôt légal: D/2013/12.105/35. Available online: https://www.eurelectric.org/media/1816/0702_emobility_market_model_final_ac-2013-030-0501-01-e.pdf (accessed on 21 February 2021).
34. Papathanasiou, S.; Schina, O. Suggestions for the function of electro-mobility market in Greece. *Rep. Regul. Auth. Energy* **2019**. (In Greek)
35. Energy Saving Trust. Procuring electric vehicle charging infrastructure as a local authority. *Rep. Energy Sav. Trust.* **2019**.
36. International Energy Agency. Germany Charging Infrastructure. 2019. Available online: http://www.ieahev.org/by-country/germany-charging-infrastructure/ (accessed on 20 April 2021).
37. Chen, D.; Jing, Z.; Tan, H. Optimal Bidding/Offering Strategy for EV Aggregators under a Novel Business Model. *Energies* **2019**, *12*, 1384. [CrossRef]
38. Guo, Y.; Liu, W.; Wen, F.; Salam, A.; Mao, J.; Li, L. Bidding Strategy for Aggregators of Electric Vehicles in Day-Ahead Electricity Markets. *Energies* **2017**, *10*, 144. [CrossRef]
39. Schiavo, L.; Bonafede, D.; Celaschi, S.; Colzi, F. Regulatory Issues in the Development of Electro-Mobility Services: Lessons Learned from the Italian Experience. In Proceedings of the 1st e-Mobility Power System Integration Symposium, Berlin, Germany, 23 October 2017.
40. Caleno, F.; Coppola, G. DSO business model for speeding up EVs mass market. In Proceedings of the 22nd International Conference on Electricity Distribution, Stockholm, Sweden, 10–13 June 2013.
41. Eurelectric. Charging Infrastructure for electric vehicles. *Eurelectric Position Pap.* **2016**.
42. Lorentzen, E.; Haugneland, P.; Bu, C.; Hauge, E. Charging infrastructure experiences in Norway—The worlds most advanced EV market. In Proceedings of the EVS30 International Battery, Hybrid and Fuel Cell Electric Vehicle Symposium, Stuttgart, Germany, 9–11 October 2017.
43. Burger, S.P.; Luke, M. Business models for distributed energy resources: A review and empirical analysis. *Energy Policy* **2017**, *109*, 230–248. [CrossRef]
44. Capros, P.; Zazias, G.; Evangelopoulou, S.; Kannavou, M.; Fotiou, T.; Siskos, P.; Sakellaris, K. Energy-system modelling of the EU strategy towards climate-neutrality. *Energy Policy* **2019**, *134*, 110960. [CrossRef]

45. Siskos, P.; Capros, P.; de Vita, A. CO_2 and energy efficiency car standards in the EU in the context of a decarbonisation strategy: A model-based policy assessment. *Energy Policy* **2015**, *84*, 22–34. [CrossRef]
46. Hellenic Republic Ministry of the Environment and Energy. National Energy and Climate Plan. December 2019. Available online: https://ec.europa.eu/energy/sites/ener/files/el_final_necp_main_en.pdf (accessed on 20 April 2021).
47. Nicholas, M.; Hall, D.; Lutsey, N. Quantifying the electric vehicle charging infrastructure gap across U.S. markets. *ICCT White Pap.* **2019**.
48. Vagropoulos, S.; Kleidaras, A.; Bakirtzis, A. Financial Viability of Investments on Electric Vehicle Charging Stations in Workplaces with Parking Lots under Flat Rate Retail Tariff Schemes. In Proceedings of the Universities Power Engineering Conference, Cluj-Napoca, Romania, 2–5 September 2014.
49. Kang, N.; Feinberg, F.M.; Papalambros, P.Y. Integrated decision making in electric vehicle and charging station location network design. *J. Mech. Des.* **2015**, *137*, 061402. [CrossRef]

energies

MDPI

Article

Reducing the Decarbonisation Cost Burden for EU Energy-Intensive Industries

Panagiotis Fragkos *, Kostas Fragkiadakis and Leonidas Paroussos

E3Modelling S.A., Panormou 70-72, PO 11523 Athens, Greece; fragkiadakis@e3modelling.com (K.F.); paroussos@e3modelling.com (L.P.)
* Correspondence: fragkos@e3modelling.com

Abstract: Carbon leakage features prominently in the climate policy debate in economies implementing climate policies, especially in the EU. The imposition of carbon pricing impacts negatively the competitiveness of energy-intensive industries, inducing their relocation to countries with weaker environmental regulation. Unilateral climate policy may complement domestic emissions pricing with border carbon adjustment to reduce leakage and protect the competitiveness of domestic manufacturing. Here, we use an enhanced version of GEM-E3-FIT model to assess the macro-economic impacts when the EU unilaterally implements the EU Green Deal goals, leading to a leakage of 25% over 2020–2050. The size and composition, in terms of GHG and energy intensities, of the countries undertaking emission reductions matter for carbon leakage, which is significantly reduced when China joins the mitigation effort, as a result of its large market size and the high carbon intensity of its production. Chemicals and metals face the stronger risks for relocation to non-abating countries. The Border Carbon Adjustment can largely reduce leakage and the negative activity impacts on energy-intensive and trade-exposed industries of regulating countries, by shifting the emission reduction to non-abating countries through implicit changes in product prices.

Keywords: unilateral climate policy; GEM-E3-FIT; carbon leakage; industrial relocation; border carbon adjustment

Citation: Fragkos, P.; Fragkiadakis, K.; Paroussos, L. Reducing the Decarbonisation Cost Burden for EU Energy-Intensive Industries. *Energies* **2021**, *14*, 236. https://doi.org/10.3390/en14010236

Received: 7 December 2020
Accepted: 31 December 2020
Published: 5 January 2021

1. Introduction

Climate change is challenging for public policymakers, as the global nature of greenhouse gas (GHG) emissions increases the difficulty of implementing unilateral emission reduction actions as the country taking climate action bears most of the mitigation costs. As a global agreement on ambitious emission reduction seems unlikely and current pledges are not sufficient to meet Paris goals [1], individual countries implement unilateral climate policies hoping that other regions will adopt similar policies [2]. Unilateral action forgoes cost savings from "where-flexibility" [3], as the cheapest mitigation options across regions and sectors are not the first to be explored. That the unilateral climate policies lead to differential carbon prices across countries indicates an unexploited potential for cost savings. The application of unilateral carbon pricing results in structural adjustment of domestic production and consumption but also affects its competitiveness in international markets through changes in relative costs and trade patterns between countries. In this context, the cost-efficiency of climate policies is negatively impacted carbon leakage, i.e., the relocation of emissions to regions with weaker environmental regulation. Leakage occurs via two interlinked channels: (i) the energy channel, through adjustments in international energy prices, and (ii) the industrial competitiveness channel [4]. In the energy channel, reduced fossil fuel consumption in carbon abating countries depresses global prices for oil, gas and coal, which triggers increased fossil fuel consumption and emissions in non-abating countries. The competitiveness channel is triggered by increased production costs of energy-intensive and trade-exposed (EITE) industries in regulating countries relative to international competitors, resulting in industrial relocation to non-regulating countries.

The socio-economic implications of unilateral climate policies are commonly analyzed with computable general equilibrium (CGE) models, which complement theoretical analysis as they allow researchers to conduct quantitative experiments in the form of scenarios exploring the impacts of specific policies [5]. National authorities and policymakers in the climate field place a high focus on domestic economy and employment effects, while the reaction of other countries is highly uncertain. These economic impacts are related to "loss of competitiveness" triggered by changes in output, export volumes and terms of trade, which may lead to delayed or weakened climate policies [5].

In the EU climate policy debate, industrial competitiveness is a key issue. Recently, the EU Green Deal [6] acknowledged that European ambitious climate policies may weaken the competitiveness of energy-intensive industries leading to relocation of their activities to countries with weaker climate ambition. In case of differentiated climate policy efforts across regions, the European Commission (EC) proposes a Border Carbon Adjustment (BCA) mechanism, for selected sectors, to ensure that the price of imported products reflects their carbon content. The BCA mechanism imposes a tax on emissions embodied in imported products to the EU in order to level the playing field between domestic and imported products with respect to carbon costs. With the right design, the BCA could prevent carbon leakage, incentivize foreign producers to shift toward lower emission technologies, and exert pressure on trade partners to strengthen their climate action. The BCA should comply with World Trade Organization rules and other international obligations of the EU. While BCA improves the global cost-efficiency of unilateral climate policies [7], it may cause adverse distributional impacts, as it may shift the economic burden of emission abatement from regulating to non-regulating countries [8], which may trigger increased international cooperation, but can also lead to detrimental trade conflicts, through retaliation measures.

The current study offers new insights on the risk of carbon leakage and industrial relocation in case of asymmetric climate policies (i.e., when climate policy measures have different ambition across countries) and provides an improved understanding of the sectoral and regional structure of leakage, as well as policy measures to mitigate the adverse impacts on industrial competitiveness. The paper goes beyond existing literature by: (1) assessing for the first time the competitiveness impacts of the ambitious EU Green Deal targets towards climate neutrality by mid-century, (2) using an advanced version of the leading GEM-E3-FIT model with an enhanced representation of energy system and technologies required for net zero transition, (3) exploring the impacts of first-mover coalitions conceptualising the most recent climate policy announcements of EU and China aiming to achieve carbon neutrality by mid-century, thus ensuring high policy relevance of the analysis and (4) assessing the cost-effectiveness of BCA mechanism in preventing carbon leakage and supporting the European EITE industries, as suggested by the EU Green Deal.

The study proceeds as follows. Section 2 presents the industrial and climate policy landscape in the EU, while Section 3 includes the study design and methodological approach. Section 4 presents the results of the model-based assessment on carbon leakage. Section 5 concludes the paper.

2. European Policy Landscape

2.1. The European Industrial Context

The industrial sector is a backbone of the European economy as it accounts for 68% of Europe's exports and about 18% of total EU value added with manufacturing activities providing about 32 million jobs in the EU in 2018. The EC acknowledges that the EU industry is undergoing a deep transformation, based on the uptake of new technologies, changing conditions in international markets, stronger energy and resource efficiency, new business models, bundling of manufacturing activities with services [9]. EC plans focus on boosting activity in key industry sectors, including steel, paper, construction, green technologies, renewable energies and maritime shipping [9]. The COVID-19 pandemic and the imposed restrictions heavily impacted the global and EU manufacturing activities and disrupted the international value chains, with potential for long-term economic impacts related to localisation of production, remote working and changes in travel patterns.

Several major trends shape the future development of the EU industrial sector, as European industries face growing international competition in traditional energy-intensive products like steel, cement and chemicals. European businesses will also encounter strong competition for new products entering the market, e.g., electric vehicles, renewable technologies, batteries, contributing to the UN Sustainable Development Goals [8]. Industrial policies and technological development are still too weakly coordinated, with large barriers of moving from R&D to market phase in most countries. The increasing carbon taxation would pose additional challenges for the European carbon- and material-intensive industries, which face increased risks of activity relocation to non-abating countries. The EU industrial policy should appropriately integrate considerations for key megatrends shaping future developments, including automation, digitization, globalisation, new business models and climate policies. Increasing globalisation would increase the competition of EU industries with their international competitors, as EU competitiveness in traditional industries depends crucially on labour costs and environmental regulation; however, increased innovation and R&D may increase European competitiveness in all high-tech branches. The increasingly globalised market requires the long-term transformation of the EU industry towards an "energy and resource efficiency" paradigm based on circular economy, uptake of low-carbon technologies (e.g., green hydrogen, electrification) and novel manufacturing principles, e.g., Industry 4.0 and ICT-based management. EU industrial policies should be aligned with overall economic, financing and education policies aiming to foster industrial competitiveness, through providing friendlier framework conditions for entrepreneurial activities and business venture formation, supporting firms' innovative capacities and developing better human resource reallocation and skills growth.

Energy-intensive industries have high capital and energy intensities, as the processing of raw materials requires high-temperature chemical conversions. Such processes have high efficiency potential for economies of scale, which results in large-scale manufacturing plants with high upfront costs. The recovery of high capital costs in highly volatile markets depends on high utilisation of manufacturing capacities, implying that plants may keep operating as long as product prices are higher than variable production costs, while the profitability of such plants is high during periods of high prices [10]. The high scale and capital intensity of energy-intensive activities pose barriers to market entry with new entrants often having to cooperate with or be absorbed by major established players.

2.2. European Policies on Carbon Leakage

Carbon leakage has been closely related with competitiveness in international markets and unilateral climate policies [11,12]. If a country or a group of countries adopt ambitious climate policies, while the rest of the world does not follow suit, energy-intensive manufacturing activities may relocate to non-abating regions aiming to minimise their carbon-related production costs. This issue is central in EU climate policy debate, especially as the EU Green Deal suggested the Border Carbon Adjustment (BCA) as a measure to minimise the risk of industrial relocation of European energy-intensive activities. It should be noted that there are limited indications that existing EU carbon pricing has led to carbon leakage [13], as the current EU ETS price remains at low levels and free allowances are given to energy-intensive and trade-exposed industries. However, the anticipated increase in EU ETS prices, due to the strengthening of GHG reduction targets, combined with changes in anti-leakage legislation would increase the risk of carbon leakage in the coming decades [9].

The transfer of manufacturing activities to countries with weaker environmental regulations due to cost increases related to carbon pricing is defined as carbon leakage, which results in increased emissions in non-regulating countries, thus reducing the climate policy effectiveness. Industries where energy costs account for a high share of their production costs face high risk of leakage. The EU ETS supports the competitiveness of industrial installations considered to be at significant risk of leakage by giving them a high share of free allowances. Phase 3 of EU ETS uses the below criteria to identify if a sector faces high risk of leakage:

- ETS pricing would increase its production cost, calculated as a share of gross value added, by at least 5%; and
- its trade intensity with countries outside the EU is above 10%.
- The value of at least one of the above indicators is higher than 30%

The amount of free allocation for each ETS installation is estimated based on its production quantity multiplied with the benchmark carbon intensity value for the specific product, which derives from the most efficient installations. Sectors facing high leakage risk (e.g., metals, chemicals, non-metallic minerals) receive 100% free allocation of emission allowances. However, the free allocation to other industrial sectors is gradually reduced from 80% in 2013 to 30% in 2020. In phase 4 of EU ETS, free allocation will focus on sectors facing the highest leakage risks. Figure 1 presents the carbon leakage exposure for EU industrial sectors as a function of their GHG emission intensity and trade openness. Basic metals, chemicals, non-metallic minerals, petroleum products, paper and pulp are the industries more exposed to carbon leakage, while sectors like motor vehicles, electrical equipment and electronic products face lower relocation risks due to their significantly lower GHG intensity (Figure 1).

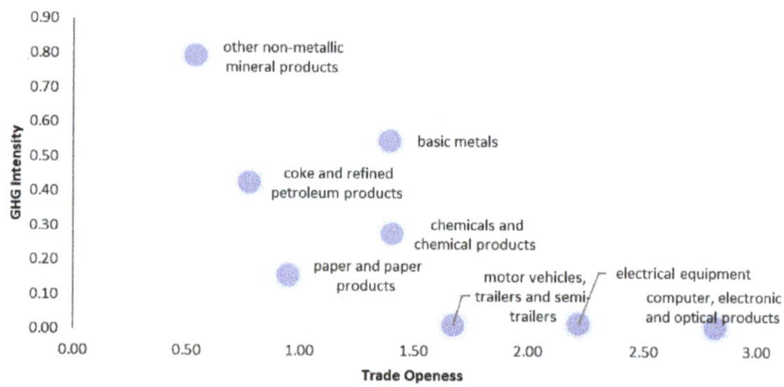

Figure 1. Carbon leakage exposure of EU sectors as a function of trade openness and GHG intensity.

The carbon leakage list of the European Commission includes those sectors with particularly high exposure to leakage, in which the contribution of energy to total production costs is about five times higher than in other industries [4]. Energy-intensive manufacturing accounts for about 7% of global GDP, but consumes 60% of global energy used by industries, mostly in the EU, USA, China, Korea and Russia. The European production of metals and chemicals is more exposed to foreign competition relative to other industries, due to their high trade intensity with non-EU countries. Non-energy intensive products (e.g., pharmaceuticals) do not face such high leakage risks as carbon pricing has limited impacts on their costs. Carbon leakage also depends on the easiness of industrial relocation (captured by transportation costs) and the degree of vertical integration of industrial activities (e.g., strong relation between car manufacturing and metals). The EU's main trading partners are China and the USA, followed by India, Russia, Turkey and Japan. Trade partners with good access and geographical proximity to the EU market and weak climate policies are the primary candidates for relocation of manufacturing activities.

Figure 2 shows that energy costs as a share of total production costs have declined from 2010 to 2017 for EU manufacturing sectors, with the largest reduction observed in the Pulp and Paper sector, in which costs fell from 11.4% in 2010 to a 6.7% share in 2017.

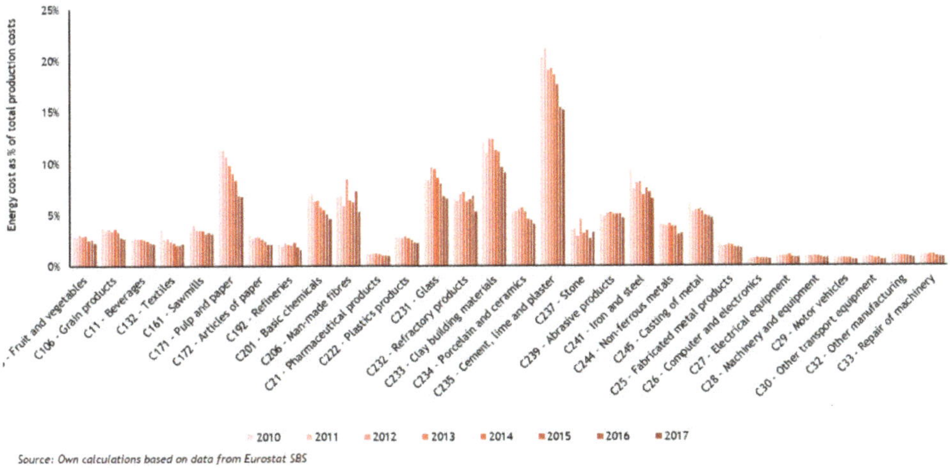

Figure 2. Energy costs as a share of production costs for EU manufacturing sectors over 2010–2017, source: [14].

Energy costs for EU manufacturing sectors typically account for 1–10% of total production costs, while costs exceed 10% for several sectors, including Cement, lime and plaster, Building materials, Glass, Pulp and paper. These sectors are most sensitive to changes in energy prices and carbon taxation. Among less energy intensive manufacturing sectors, energy costs typically constitute 0.5–3% of their production costs and are therefore a relatively small cost component for businesses in sectors like Computer and electronics, Machinery and equipment, Transport equipment and Motor vehicles. Energy cost shares have fallen across all manufacturing sectors over 2010–2017, but even more steeply after 2014 due to reduced energy consumption and lower prices for energy products. The EU has energy cost shares comparable to those of most international trade partners, although there are large differences across sectors and regions [14]. For example, the EU has a relatively high energy cost share in Iron and steel and Non-ferrous metals, but a low energy cost share in Refineries and Basic chemicals.

2.3. Policy Measures to Protect Domestic Industries

Several policy options to reduce the adverse impacts of high carbon pricing on industrial competitiveness have been proposed. These can be classified in three categories as in [12]:

Policy instruments adjusting the carbon costs at the border of the jurisdiction implementing the carbon pricing scheme. The EU Green Deal suggests to adopt a BCA mechanism in order to protect the competitiveness of the most venerable EU industries. This instrument aims to equalize the carbon costs on imports and exports in the jurisdiction implementing the carbon price through imposing the same carbon price on imports from non-regulated countries and rebating the carbon costs to exports to non-regulated countries. This can also take the form of an EU-wide horizontal tax on carbon content, through a carbon price levied on consumption of goods regardless of their origin. While the measure is effective in reducing carbon leakage of EITE industries, it may lead to negative socio-economic outcomes in EU countries.

Adjusting carbon costs upwards for non-domestic firms: These measures aim at increasing the carbon costs of firms outside the jurisdiction implementing the climate policy. This may take the form of sectoral agreements aiming to extend the participation of industries in climate action by offering options such as technology transfers, research and development collaboration, etc. However, a revision of existing multilateral trade rules with major trade partners is difficult, in particular agreeing on cutting industrial subsidies.

Policy measures to adjust carbon costs downwards for domestic firms by reducing their carbon-related costs but maintaining a marginal abatement incentive equivalent to that if such measures are not introduced. For example, free allocation of tradable permits, e.g., under an Emission Trading System like the EU ETS, can relieve industries from buying the emission permit but maintains the incentive to abate emissions. Carbon tax revenues can also be used to support investment (e.g., in low-carbon innovation) that will reduce the cost of low-carbon and efficient technologies.

Environmental tax reforms aim to shift taxation towards polluting activities, while output-based rebates return the revenues generated by a carbon tax to industries in proportion to their output. Carbon Border Adjustment has recently received increasing focus in policy debates after the EC suggestion to introduce BCA as part of the EU Green Deal. Given the high policy relevance of BCA and the drawbacks of other anti-leakage options (especially with regard to their non-compliance with WTO trade rules), the paper focuses on the assessment of BCA as a policy measure to reduce the cost burden on EITE sectors, by imposing the EU carbon price on imported products from non-regulating countries.

3. Methodology

3.1. The GEM-E3-FIT Modelling Framework

GEM-E3-FIT is an advanced and detailed applied computable general equilibrium (CGE) model, simultaneously representing 46 countries/regions, including all EU-28 countries individually and the largest economies globally (USA, Japan, Canada, Brazil, China, India, South Korea, Indonesia, Saudi Arabia, Russian federation, South Africa and others). GEM-E3-FIT covers the interactions between the economy, the energy system and the environment. It is a comprehensive model of the economy, covering the complex interlinkages between productive sectors, consumption, price formation of commodities, labour and capital, bilateral trade and investment dynamics. The model is dynamic, recursive over time, driven by accumulation of capital and equipment. The economic agents are assumed to exhibit optimising behaviour while market-derived prices are adjusted to clear all markets. The model features alternative market regimes, discrete representation of power producing technologies, equilibrium unemployment, energy efficiency standards and carbon pricing. GEM-E3-FIT formulates production technologies in an endogenous manner allowing for price-driven derivation of intermediate consumption and the services from capital and labour. The model can quantify the macroeconomic, trade, employment and distributional impacts of environmental and energy policies, both in the short and long term.

For alternative scenarios, GEM-E3-FIT provides for 51 countries: dynamic projections of national accounts; full Input–Output tables; distribution of income and transfers by agent; employment by economic activity and by skill; capital and investment by sector; CO_2 and GHG emissions by sector and fuel; consumption matrix by product and investment matrix by ownership branch; full bilateral trade matrices among countries and sectors; energy demand and supply by sector and fuel, energy efficiency investment and power generation mix. GEM-E3-FIT goes beyond a conventional CGE modelling approach, as it incorporates: a detailed representation of the financial sector, endogenous growth through learning-by-doing and R&D for low-carbon technologies [15], five labour skill levels and comprehensive energy system representation. GEM-E3-FIT has been extensively used for energy and climate policy assessment [16,17].

In GEM-E3-FIT, all regions and sectors are linked through endogenous bilateral trade flows. Total demand of each sector is optimally allocated between domestic and imported goods, under the hypothesis that they are imperfect substitutes (Armington assumption [18]). The supply mix is represented as a multi-level nested constant elasticity of substitution function: at the upper level, firms decide on the optimal mix between domestically produced and imported goods; at the next level the demand for imports is split by country of origin. Bilateral trade transactions are endogenous in GEM-E3-FIT and depend on relative production costs, transportation costs and consumer preferences, as captured by national account statistics on trade. The mathematical formulation adopted in GEM-E3-FIT is described in [4] and model details are presented in Appendix A. Table 1 includes the upper-level Armington elasticity values used in GEM-E3-FIT (σx) and the elasticity values driving the decision on imports by origin (σm), which differ among sectors, largely reflecting the degree of product homogeneity. The values of Armington elasticities are derived from the GTAP 10 database.

Table 1. Armington elasticities in GEM-E3-FIT sectors.

	σx	σm		σx	σm
Transport	1.90	3.80	Coal	3.05	6.10
Construction	1.90	3.80	Consumer Goods	3.21	6.43
Non Market Services	1.90	3.80	Chemical Products	3.30	6.60
Market Services	2.03	4.06	Transport equipment	3.55	7.10
Oil products	2.10	4.20	Electric Vehicles	3.55	7.10
Gas	2.80	5.60	Non-ferrous metals	3.98	7.95
Power Supply	2.80	5.60	Equipment goods	4.05	8.10
Ferrous metals	2.95	5.90	Batteries	4.05	8.10
Paper Products	2.95	5.90	Electronic Goods	4.08	8.15
Agriculture	3.03	6.07	Crude Oil	5.20	10.40

3.2. Study and Scenario Design

Alternative policy scenarios are modelled with GEM-E3-FIT to assess the macroeconomic and industrial implications of unilateral climate policy and evaluate possible measures to reduce the cost burden on EITE industries.

The Reference scenario represents a benchmark macroeconomic projection based on scientific expertise on growth patterns, technical progress, labour productivity and climate policies. The key macroeconomic and international energy price assumptions follow [19], while socio-economic assumptions for the EU are based on the Ageing Report [20]. This scenario assumes that already legislated climate policies, including Nationally Determined Contributions (NDCs), are implemented by 2030 (Table 2). After 2030, no additional emission reduction efforts are assumed. In modelling terms, this means that the carbon prices resulting from NDC targets in 2030 are kept constant until 2050, representing a lack of ambition in the international climate policy landscape. The costs of power generation technologies follow [21], while technology progress is included for low-carbon technologies.

Table 2. Nationally Determined Contributions (NDC) emission targets included in the Reference scenario.

Country	NDC Emission Targets	Energy-Related NDC Targets in 2030
EU	−40% GHG in 2030 relative to 1990	30% RES in gross final demand
China	−60% (−65%) CO_2 intensity in 2030 rel. 2005	20% Non-fossil in primary energy
India	−33% (−35%) CO_2 intensity from 2005	40% Non-fossil in power capacity
USA	−26% (−28%) GHG in 2025 from 2005	
Japan	−26% GHGs in 2030 from 2013	20–22% Nuclear and 22–24% RES share in electricity in 2030
Brazil	−43% GHGs in 2030 from 2005	
Russia	25–30% below 1990 levels by 2030	
S. Korea	37% below Business as Usual by 2030	
S. Africa	Peak GHG emissions in 2025 and plateau for a decade	

The 2DEG scenario is compatible with the 2 °C goal of the Paris Agreement, and as in [22], the global CO_2 budget of 1000 $GtCO_2$ by 2050 is used as proxy for the temperature target (Table 3). To ensure that the temperature increase relative to pre-industrial levels will stay below 2 °C, a carbon price is implemented in all regions and sectors over 2025–2050. The global carbon price is projected to increase from 75\$/$tnCO_2$ in 2030 to about 345\$/$tnCO_2$ in 2050 reflecting the increasing difficulty to reduce emission in hard-to-abate sectors including industries and freight transport. The scenario assumes equalization of marginal abatement costs across regions (i.e., common carbon price), leading to the cost-optimal achievement of the global constraint. As model results depend on the adopted carbon revenue recycling scheme, we assume that carbon revenues are recycled through the public budget.

Table 3. Scenario description.

	Scenario Description	EU Climate Target	Non-EU Climate Targets
REF	Reference scenario	Meets the EU NDC	All countries meet NDCs in 2030, no increase in policy ambition after 2030
2DEG	Decarbonisation to 2 °C with all options available	All countries adopt ambitious climate policies/universal carbon pricing to meet the 2 °C temperature target	
EUGD_Alone	EU meets the EU Green Deal Targets by 2030 and 2050	EU achieves 55/90% reduction in 2030/2050 from 1990	Non-EU countries meet their NDCs in 2030, policy ambition does not increase beyond 2030
EUGD_BCA	Green Deal Targets are met, BCA is implemented on EU imports	EU achieves 55/90% reduction in 2030/2050 from 1990	Non-EU countries meet their NDCs in 2030, policy ambition does not increase beyond 2030
EUGD_BCA_REC	As EUGD_BCA but BCA revenues are used to reduce social security contributions	EU achieves 55/90% reduction in 2030/2050 from 1990	Non-EU countries meet their NDCs in 2030, policy ambition does not increase beyond 2030
EUGD-CHN	EU and China adopt ambitious climate policies	EU achieves 55/90% reduction in 2030/2050 from 1990	Countries do not intensify policy ambition beyond 2030; China develops along a 2DEG trajectory

In the "EUGD-Alone" scenario, the EU unilaterally adopts ambitious policies to achieve the EU Green Deal targets of 55% emission reduction in 2030 from 1990 levels and the transition to climate neutrality by mid-century. As the EU Green Deal does not separately set a target for ETS and non-ETS, an EU-wide uniform carbon price is used from 2025 onwards. Energy system restructuring is mostly induced by high carbon pricing, while other instruments are also used, including ambitious technology standards, subsidies for buildings' insulation, lower risks for clean energy investment, electrification of energy services, subsidies for low-carbon innovation and uptake of disruptive mitigation options required for climate neutrality (e.g., Carbon Capture Use and Storage, green hydrogen, production of clean synthetic fuels from RES-based electricity). In contrast, non-EU countries follow their Reference climate policies.

As the EU intends to unilaterally adopt ambitious climate measures, the BCA instrument can complement domestic carbon pricing. The "EUGD-BCA" scenario explores the socio-economic and industrial implications of implementing an EU-wide BCA mechanism on imports building on "EUGD-Alone" scenario assumptions, but additionally assuming that the BCA is aligned with the EU ETS carbon pricing and applies to products subject to the ETS (e.g., energy-intensive industries). The BCA is implemented as a financial instrument (e.g., as a tax), based on the difference between EU ETS benchmarks and the carbon profiles of products originating from trading partners. Imported goods from non-EU regions are taxed according to their carbon content, which is calculated accounting for the direct GHG emissions. Tariff rates are differentiated by country, based on carbon flow data to determine country-specific coefficients. In particular, the BCA revenues of EU countries from non-EU regions for each sector s and year t can be calculated as follows:

$$BCA_{s,eu,noneu,t} = IMP_{s,eu,noneu,t} \cdot CI_{s,noneu,t} \cdot DIFF_{s,eu,\ t}$$

where $IMP_{s,eu,noneu,t}$ is the value of EU imports from non-EU countries.

CI represents the Carbon intensity of non-EU countries, calculated as the Scope 1 emissions divided by the value of production of sector s at year t.

DIFF: The EU carbon price difference between the EUGD-BCA and Reference scenario.

The BCA revenues are directed to the public budget of the regulating country. Export rebates can be considered as a subsidy under WTO's Agreement on Subsidies and Countervailing Measures [23] and are thus difficult to implement. The scenario assumes no retaliation measures from major trade partners. A variant of the BCA scenario assumes that BCA revenues are used to reduce social security contributions ("EUGD-BCA-REC" scenario). To explore if leakage depends on the size of regulating regions, we develop the "EUGD-CHN" scenario, where the EU and China jointly pursue an ambitious emission reduction effort. This scenario reflects the recent policy announcements by the EU and China, aiming for a carbon-neutral transition by 2050 and 2060 respectively. In particular, the joint emission reduction effort of the EU and China is equal to the aggregate reductions achieved by the EU (in the EUGD-Alone scenario) and China (in the 2DEG scenario). Decarbonisation is triggered by the imposition of a uniform carbon price, which constantly increases from 2025 onwards to achieve the ambitious EU-China emission reduction target. Other countries follow the same climate policies as in the Reference scenario.

The study focuses on industrial competitiveness impacts of unilateral climate policies and thus hydrocarbon supply is assumed to be elastic. Many reasons support the high elasticity of fossil fuel supply, as summarised in [5], including the small fraction of coal traded globally and the high oligopolistic rents of oil and gas whose prices are much higher than their production costs. This implies that changes in fossil fuel demand produce minor changes in international fossil fuel prices and limited leakage through the energy channel.

4. Scenario Results

The section explores the socio-economic and industrial impacts of asymmetric climate policies and possible protective measures for EITE industries.

4.1. Impacts of Unilateral Ambitious European Climate Policies

The EC recently proposed a new plan to increase its GHG emission reduction target for 2030, accompanied with the long-term climate neutrality goal by 2050. This proposal forms the basis of the EUGD-Alone Scenario, which meets the ambitious GHG reduction targets of 55% in 2030 and 90% in 2050 relative to 1990 levels. As the EU Green Deal does not set sectoral targets, an EU-wide carbon price is used, which reaches 75 $\$/tnCO_2$ in 2030 and 590 $\$/tnCO_2$ in 2050 indicating the increasing marginal abatement effort and difficulty in fully decarbonizing the European energy system, especially in hard-to-abate sectors with limited mitigation options, including energy-intensive industries and freight transport. The high carbon price drives a large-scale reduction in GHG emissions, which decline to about 575 Mt CO_2 in 2050, compared to 2350 Mt CO_2 in the Reference scenario, i.e., a reduction of 75%. This reduction is mostly driven by energy system decarbonisation, with extensive emission reductions projected in all sectors. Industrial CO_2 emissions decline significantly from Reference levels, as European industries are decarbonised through large-scale efficiency improvements, electrification of industrial processes and fuel switch towards energy carriers with low carbon intensity, e.g., biomass and green hydrogen, while domestic industrial activity also declines, as a part of energy-intensive manufacturing is relocated to countries without carbon pricing.

Figure 3 shows that all sectors and GHGs contribute to the ambitious EU emission reduction effort, but energy-related CO_2 emissions account for 73% of the overall mitigation effort, due to energy demand and supply restructuring. The European industrial sector accounts for about 25% of the overall EU mitigation effort relative to Reference, with large reduction of CO_2 emissions both from energy combustion and industrial processes.

Energy system decarbonisation is a capital-intensive process characterized by the transition towards low-carbon technologies, which require high upfront investment costs, but result in reduced operation and fuel costs in the longer term. A strict financial closure is adopted in the CGE modelling framework where investments are constrained by available savings and any additional investment plan needs to be financed by reallocating existing capital resources, leading to crowding-out of productive investment in other sectors. In this context, increased demand for low-carbon finance would increase the cost of capital with negative implications throughout the economy. The high carbon pricing increases the EU production costs of energy- and carbon-intensive processes, while the low-carbon technology costs decline due to their increased uptake (learning-by-doing). However, as non-EU countries do not intensify their policy ambition, the non-EU market demand for low-carbon technologies remains relatively small and thus the potential export benefits from a competitive EU low-carbon manufacturing are limited. In the EUGD-Alone Scenario, the driving factor for industrial relocation to non-regulating countries is the change in production costs of carbon-intensive sectors through higher carbon prices.

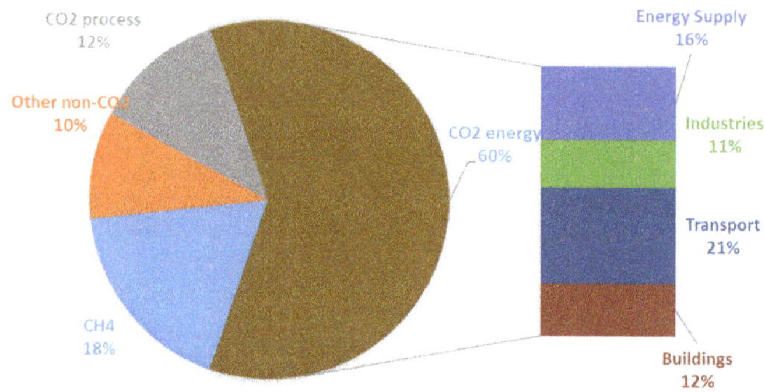

Figure 3. Distribution of the EU GHG mitigation effort by sector in the EUGD-Alone relative to the Reference scenario in 2050.

As the EU unilaterally adopts ambitious climate policies in the EUGD-Alone scenario, GHG emissions outside the EU increase as energy- and carbon-intensive industries are relocated to non-abating countries and domestically produced goods are substituted with imported goods. The cumulative carbon leakage is calculated as the ratio of increased emissions in non-abating countries over the amount of emission reductions in regulating countries compared to Reference scenario and is estimated at 24.6% with EU GHG emissions declining by about 24 Gt cumulatively over 2025–2050, while non-EU emissions increase by about 5.9 Gt relative to Reference scenario. Leakage rate increases over time following the increase in carbon price differential between countries compared to Reference scenario. Most carbon leakage occurs in Russia, Rest of World, China, India and the United States which increase emissions in the EU-GD-Alone scenario (Figure 4). Low transportation costs to the EU market favour Russia and Turkey, while China and India have sufficient production capacities at low cost and relatively high energy and carbon intensities inducing a higher increase in emissions (hence, higher leakage). It should be noted that the industrial relocation to the different countries is not proportional to the changes in GHG emissions as each country is characterised by different GHG intensities (e.g., one tonne of steel produced in the USA emits lower GHG emissions than in India).

The sectoral distribution of carbon leakage is presented in Figure 5. Given their high carbon intensities and openness to trade ratios, the sectors that are most vulnerable to carbon leakage are Chemicals, Metals, Non-metallic minerals and Air transport. The importance of sectoral leakage changes over time as the energy system is gradually decarbonised. An interesting finding is that carbon leakage occurs also indirectly through electricity-related emissions; the relocation of industrial activities to non-EU countries would lead to increased demand and production of electricity and thus higher carbon emissions in non-EU countries, especially as their power mix is still dominated by fossil fuels in the EUGD-Alone scenario. GHG emissions from refineries increase in most non-EU countries driven by the increased demand for oil products required to fuel industrial activities. However, in other countries, refinery emissions decline as the electrification of the EU energy services reduces the aggregate demand for refined oil products.

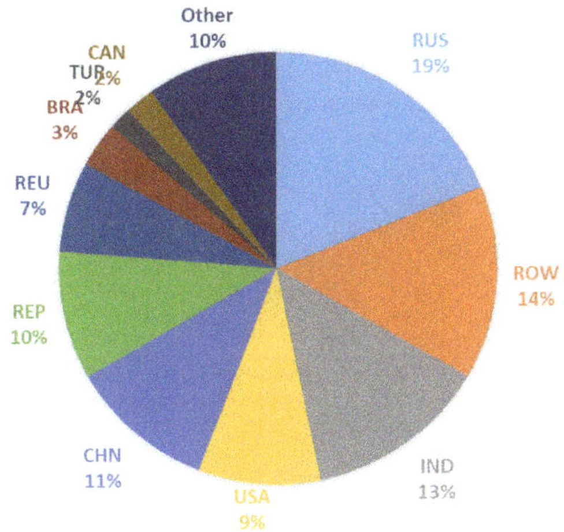

Figure 4. Regional decomposition of carbon leakage in EUGD-Alone over 2020–2050.

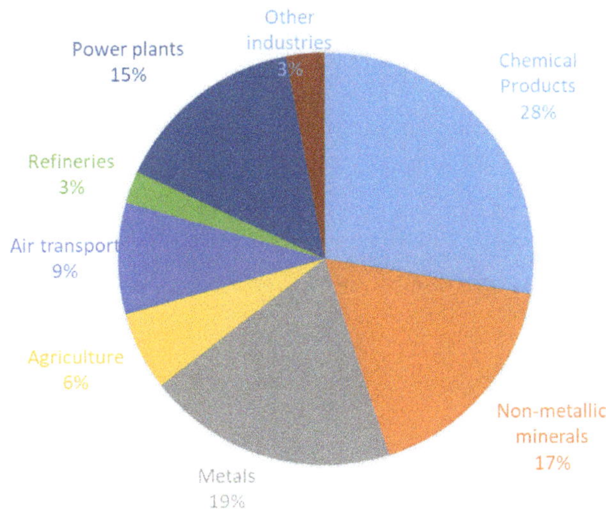

Figure 5. Sectoral decomposition of carbon leakage in EUGD-Alone over 2020–2050.

The model-based simulations show negative impact of high carbon pricing on economic activity, as higher costs for energy services increase production costs and depress demand, in the presence of crowding-out effects. The unilateral application of carbon pricing in an open economy weakens its international competitiveness leading to further decline in domestic activity, some parts of which are substituted by carbon-intensive production in non-EU countries emitting more GHGs. This reduced the effectiveness of unilateral climate policies through carbon leakage to non-regulating countries. The GEM-E3-FIT results confirm that the unilateral application of high carbon pricing would result in GDP and consumption losses in the EU compared to Reference scenario. The EU has a strong low-carbon innovation base and industrial know-how so as to build domestically a large part of the clean energy technologies [24], but the corresponding activity increase is not high enough to offset the depressing effects due to higher production costs and the relocation of industrial activity to non-abating countries.

GEM-E3-FIT results show that the cumulative EU GDP losses in the EUGD-Alone scenario over 2025–2050 are 1% below reference scenario GDP. As the carbon price differential between the two scenarios increases over time, EU GDP losses follow the same trend (Figure 6). Non-regulating countries benefit from their increased competitiveness, but also face losses as demand for their products declines in the EU, which experiences depressive effects on domestic demand due to carbon pricing. Therefore, the scenario has very small impacts on GDP of non-regulating regions, while GDP declines at the global level relative to Reference levels due to carbon pricing application.

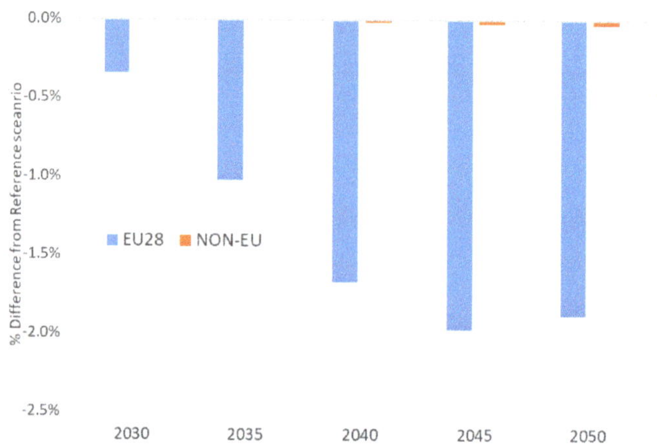

Figure 6. GDP implications of EUGD-Alone relative to Reference scenario.

In emission-abating countries, demand for energy-intensive products declines from Reference levels, because of increased production costs and despite their participation in building the low-carbon investment (e.g., metals contribute to wind turbine production). The reduction in domestic EU production of EITE industries is also driven by the worsening of their international competitiveness induced by unilateral application of carbon pricing. The degree of exposure of these industries to foreign trade is a critical factor influencing industrial relocation.

Figure 7 shows that the unilateral application of high carbon pricing impacts negatively the domestic EU industrial production. Energy system decarbonisation has profound negative impacts for the EU fossil fuel supply industries, which register a large activity reduction due to the reduced consumption of oil, coal and natural gas. A large activity decline is also projected for the domestic production of ferrous metals and chemicals (between 8–10%), while the decline is lower in other industrial sectors, as non-metallic minerals and paper are less traded. Part of the decreased activity is relocated to non-EU countries that do not adopt climate policies. The amount of industrial production increased in non-regulating regions is lower than the reduced EU production, because energy-intensive products face lower global demand relative to Reference scenario, as they are replaced by products with low carbon intensity induced by ambitious EU climate policies (Table 4). The sectors producing metals and chemicals are more exposed to foreign competition and bear higher relocation impacts than cement (and other building materials) which have to be located close to consumption due to high transportation costs.

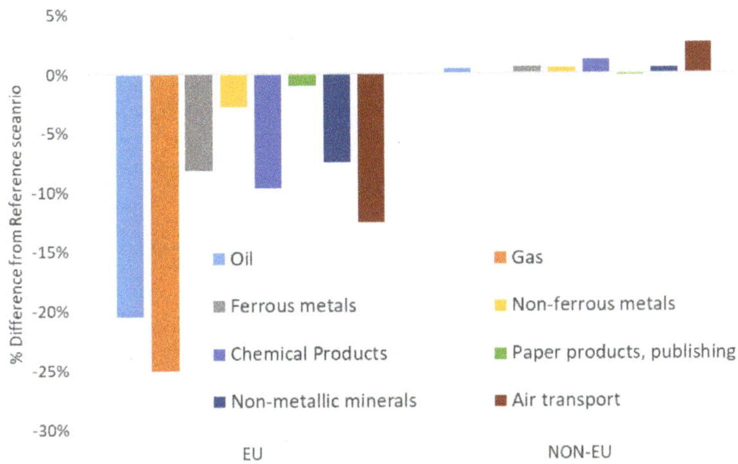

Figure 7. Cumulative changes in sectoral production in EUGD-Alone relative to the Reference scenario in EU and non-EU regions over 2025–2050.

Table 4. Impacts of EUGD_Alone on industrial production over 2025–2050.

Change from Reference over 2025–2050 (in bn Euro 2010)	EU Production	Non-EU Production
Oil	−3272	559
Gas	−142	5
Ferrous metals	−1103	795
Non-ferrous metals	−251	419
Chemicals	−2265	1803
Paper and pulp, publishing	−153	63
Non-metallic minerals	−1107	689
Air transport	−1206	914

Industrial leakage by sector is measured as the ratio of the amount of emission increases in non-abating countries over the amount of emissions reduced in countries pursuing climate action. The leakage rate is particularly high for energy-intensive industrial sectors, in particular for metals and chemicals, as a large part of the European production is relocated to non-abating regions, which have a considerably higher carbon intensity relative to the EU, as their energy mix is dominated by fossil fuels in the EUGD-Alone scenario. Industrial leakage rates as estimated by GEM-E3-FIT range from less than 5% for non-energy intensive production up to 70–80% for chemicals and metals, which are highly exposed to foreign competition and are relocated to countries with considerably higher carbon intensity than the EU (e.g., China, India, Russia). A large share of increased emissions in non-regulating regions comes from power generation, due to the increased electricity demand from industries and their fossil-based power supply system; thus the imposition of emission reduction measures on electricity production in non-abating countries can significantly reduce leakage rates.

4.2. Impacts of Joint EU–China Ambitious Climate Action

When China and the EU jointly apply high carbon pricing, leakage is significantly reduced. The EU and Chinese cumulative GHG emissions are reduced by about 129 Gt CO_2 relative to Reference scenario over 2020–2050. In the same period, GHG emissions in non-abating regions increase by 7.7 Gt CO_2 relative to Reference, indicating a carbon leakage rate of 6%, mostly in India, Russia and the Rest of world regions (Figure 8). The leakage rate is substantially lower relative to EUGD_Alone scenario, as Chinese emissions are much higher than the EU's and about 85% of the overall mitigation effort in the EUGD-China scenario is implemented in China, with EU accounting for only 15%. Therefore, a larger size of climate coalition would result in lower leakage rates. In addition, the imposition of common carbon price is more effective in abating emissions in countries with high carbon intensity, like China. Modelling outcomes suggest that carbon leakage can effectively decline in case that the climate coalition includes countries with high carbon intensity and with low industrial production costs, such as China.

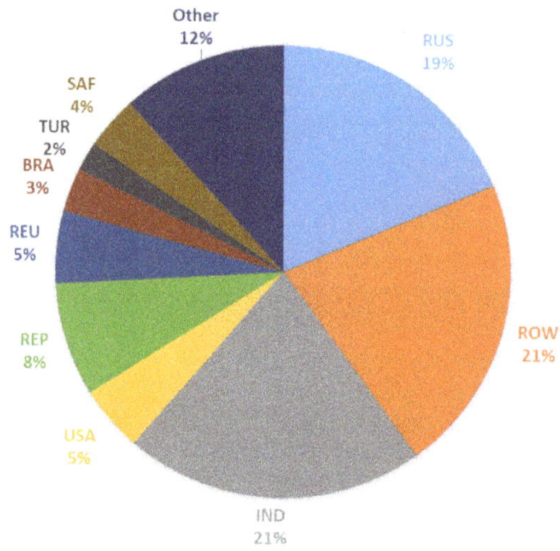

Figure 8. Leakage decomposition to regions in EUGD-China over 2020–2050.

The global demand for energy-intensive products (e.g., metals, cement) decreases in climate policy scenarios, due to the restructuring towards less carbon intensive products and services induced by carbon pricing. The higher relocation impacts are projected for the sectors producing metals and chemicals, due to their higher energy intensity and exposure to foreign competition.

Table 5 presents the EUGD-China scenario impacts on industrial activity with regulating countries experiencing activity losses in EITE industries relative to Reference. The losses are particularly high in the oil supply sector, but also in the production of metals and chemicals, as non-metallic minerals and paper are less traded. Part of the decreased activity in EU and China is relocated to non-regulating countries. The relative competitiveness and production shares in EITE industries of countries applying high carbon pricing change in the scenario. For example, the imposition of common carbon price would result in higher increase in Chinese production costs compared to EU, and thus the competitiveness of European EITE industries improves relative to Chinese industries. This means that energy-intensive industrial production declines more strongly in China (Table 5) and the Chinese economy bears higher GDP losses than the EU. It should be noted that in EUGD-China scenario, the global industrial production is lower from Reference in all EITE sectors. The activity impacts are larger in China with its GDP declining by 1.6% over 2020–2050 (with EU GDP declining by 0.7%), due to the higher carbon intensity of the Chinese economy and the large-scale reduction of Chinese energy-intensive industrial production, as carbon pricing impacts more negatively the low-cost industrial producers like China.

Table 5. Impacts of EUGD-China on industrial production over 2025–2050.

Production Change from Reference over 2025–2050 (in bn Euro 2010)	EU	China	Non-Abating Countries
Oil	−2620	−4642	701
Gas	−115	−30	14
Ferrous metals	−290	−4901	2710
Non-ferrous metals	52	−1907	1155
Chemicals	−1214	−3558	2401
Paper, publishing	−128	−239	87
Non-metallic minerals	−546	−4282	1748
Air transport	−794	−122	580

The relocation of manufacturing activities to non-abating countries leads to an increase in their energy and electricity consumption. As electricity trade across regions is limited, this would result in higher domestic production of electricity and higher CO_2 emissions in non-abating countries, especially as their power generation mix is dominated by fossil fuels. The increased emissions from electricity production in non-abating countries are not related to changes in energy prices, but are a direct consequence of industrial relocation, through the competitiveness channel.

4.3. How Effective Is the Border Carbon Adjustment?

Policy makers explore measures to protect the competitiveness of EU energy-intensive industries, which are vulnerable to relocation away from the EU if it unilaterally adopts high carbon pricing. The EC considers the implementation of Border Carbon Adjustment (BCA) mechanism as part of the EU Green Deal. The main principles of a BCA are relatively well-defined and are already part of the current policy debate in the EU. The EUGD-BCA scenario assumes that BCA is used to level the playing field between domestic and imported products with respect to carbon costs. With the right design, a BCA could prevent leakage, incentivise non-EU industries to shift toward lower emission technologies, and exert pressure on trade partners to strengthen environmental regulations. In the EUGD-BCA scenario, the BCA is aligned with the EU ETS pricing, entailing a similar coverage of products and the requirement for importers to purchase carbon allowances at prices equal to the EU ETS. This policy instrument targets the carbon content of imported goods that fall under the EU ETS sectoral classification. The EU ETS carbon price is applied to imported goods from non-EU countries whose carbon intensity exceeds a certain threshold, which is proxied with the notion of Best Available Technology (BAT). The scenario does not assume direct participation of non-EU industries in the EU ETS, but the ETS carbon price is simulated as a tax, which is paid by economic operators at the point of entry in European borders. Essentially, the mechanism imposes an additional cost on non-EU goods based on the difference between the EU carbon intensity benchmark and the intensity of the sector and country of origin. The benchmark by sector is calculated using the technology with the lowest carbon intensity across EU Member States. The BCA revenues are recycled through the public budget, while no retaliation is assumed by non-EU countries. If non-EU countries apply the EU ETS carbon price on EU exports (retaliation), the effect on EU production would be relatively small as the EU industries already produce goods with low carbon contents implying limited impacts on their production costs.

As BCA captures the regional differences in carbon intensities and the cost increases induced by the ETS price, this instrument is very effective in mitigating carbon leakage, which declines from 25% in the EUGD-Alone to less than 4% (with the USA and China accounting for most of this leakage). The imposition of BCA increases the cost of imported industrial products in the EU, thus resulting in a reduction of EU imports by 1.5% cumulatively over 2025–2050 (Figure 9). However, this instrument is not designed to support the competitiveness of European industries in international markets, as exported goods do not receive any compensation for their higher production cost due to ETS carbon pricing, as a direct intervention would be non-WTO compliant. Therefore, GEM-E3-FIT results show limited impacts on European exports.

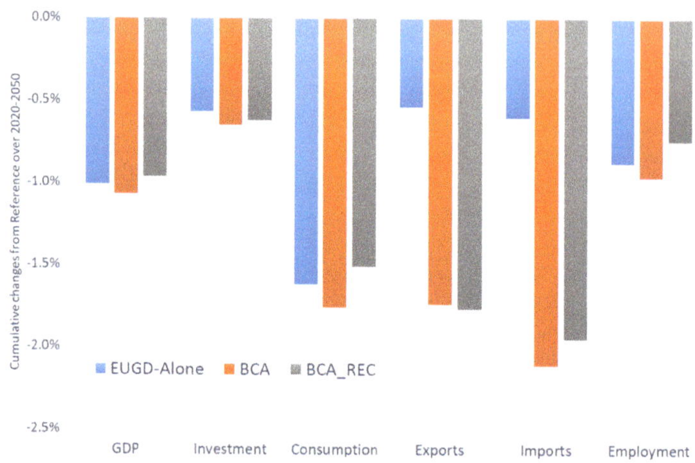

Figure 9. Macroeconomic impacts of alternative scenarios relative to Reference (cumulative changes over 2025–2050).

The recycling scheme used for the ETS and BCA revenues greatly affects the socio-economic impacts of climate policies. If these revenues are recycled through the public budget, the implementation of BCA has a slightly negative impact on EU activity (with GDP declining by 0.06% over 2025–2050 relative to EUGD-Alone); this is due to increased costs of imported products which is then diffused to domestic production and consumption through product value chains and complex inter-industrial relations. Additional taxes imposed on imported products further increase production costs of EU-based industries and reduce real disposable income for households. The negative impact on activity is found both in EU and non-EU countries as the tax imposition increases frictions in the economy.

The recycling of BCA revenues towards reducing social security contributions would reduce labour costs leading to the creation of additional jobs, with EU employment increasing by 0.3% relative to EUGD-BCA scenario. The increased labour income drives up private consumption and GDP, while also being beneficial for the EU trade balance. Our research shows that BCA is resilient to potential counteracting measures by non-EU competitors, as the low carbon intensity of EU products leaves little room for significant cost increases in its exported goods if a retaliation tax is applied.

The leakage-reduction effect of BCA is reflected in GHG emissions across regions, with global cumulative GHG emissions declining by 0.25% from EUGD-Alone scenario over 2025–2050. This means the same climate outcome and global emission budget can be achieved with lower ambition of domestic EU climate policies. The imposition of BCA results in reallocation of emissions among countries towards global cost-effectiveness, as global emissions decline, without impact on global GDP. However, BCA is only a second-best instrument as emissions in non-regulating countries are still higher than their Reference levels and considerably higher relative to emission reduction pathways compatible with the Paris Agreement goals.

Unilateral emission pricing increases the risks of relocation for EITE industries, thus putting these industries at a disadvantage relative to international competitors [2]. The EU maintains a competitive position in the global production of energy-intensive products, despite losing market share over 2015–2050 in the Reference scenario due to increasing competition from rapidly growing developing economies. The EU-based manufacturing activity declines in the EUGD_Alone scenario, due to high carbon pricing, reduced domestic demand and lower international competitiveness. The EU share in global production of energy-intensive products declines by about 0.5 percentage point relative to Reference scenario. The imposition of BCA minimises the risks of industrial relocation to non-EU countries and thus European industrial production returns close to the Reference scenario levels. Metals, Chemicals and Non-metallic minerals have the highest sectoral performance in the BCA scenario because of their higher carbon intensity and share of EU imports compared to the other EU ETS sectors (Figure 10).

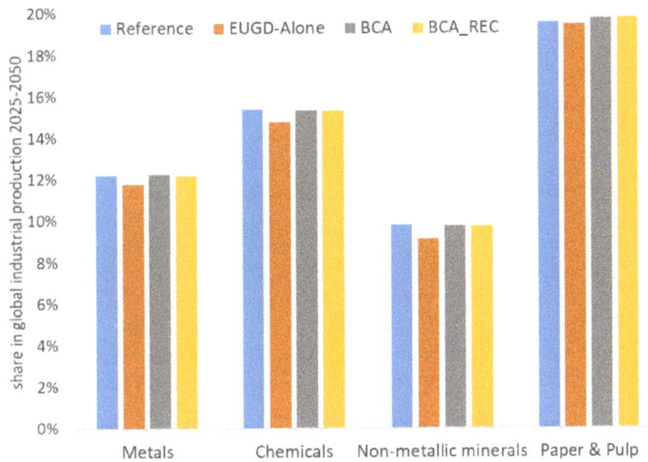

Figure 10. EU share in global industrial production in cumulative terms over 2025–2050.

5. Policy Recommendations and Conclusions

The ambition of European climate policies has increased in recent years, with the EC proposing to reduce EU GHG emissions by 55% in 2030 relative to 1990 levels, paving the way towards climate neutrality by mid-century. Unilateral climate action increases the risk of carbon leakage, i.e., the relocation of GHG emissions to countries with weaker environmental regulation. Here, we evaluate the leakage induced by the industrial competitiveness channel, through shifts in comparative advantage and activity relocation of EITE industries. The enhanced version of GEM-E3-FIT model is used to analyse highly policy relevant scenarios based on the most recent climate debate, including the EU and China's pledges towards carbon neutrality by 2050 and 2060, respectively and the potential introduction of Border Carbon Adjustment mechanism aiming to minimize industrial relocation to non-abating countries.

Model-based outcomes show that without appropriate preventative measures, unilateral climate policies could induce relocation of energy-intensive activities outside the EU. The redistribution of trade in commodities between countries due to changes in their relative competitiveness leads to a carbon leakage of 25% over 2025–2050, induced by the increased emissions in non-abating countries, which is close to the higher end projected by the literature [2,4,5,12,24]. The carbon leakage computed with the GEM-E3-FIT model lies at the upper bound—but well within the range—of similar model results from the EMF29 multi-model study [2]. In the current study, we find a leakage rate for the EU at 25% whereas the mean of EMF29 models is around 20%, but calculated in a limited horizon of 10 years and assuming less ambitious climate policies.

Most of the leakage occurs in China and India which have sufficient production capacities at low cost and relatively high energy and carbon intensities, inducing higher increase in emissions, while low transportation costs and proximity to the EU market favor Russia. The size and composition of countries participating in the climate coalition matters [25], as an EU–China coalition significantly reduces the leakage rate to about 6%

over 2025–2050. This is due to the high effectiveness of carbon pricing to mitigate emissions in countries such as China, which have significantly higher carbon intensity and lower industrial production costs compared to the EU and other developed economies. The sectors producing metals and chemicals experience the highest leakage rates, because of their high energy intensity and foreign competition, while leakage in non-metallic minerals is smaller due to high transportation cost preventing activity relocation.

In case that EU and China join forces to reduce GHG emissions, the activity impacts are larger in the Chinese economy, as it has higher carbon intensity than the EU and bears about 85% of the overall mitigation effort. In this context, the cost competitiveness of European industries vis-à-vis the Chinese improves considerably, and thus EU industrial activity losses are very small. This indicates that linking the carbon markets of EU and China reduces more the competitiveness of China's energy-intensive activity in international markets compared to EU's. The adverse competitiveness impacts on the Chinese economy can be reduced in case that the allocation of mitigation effort is different (e.g., limited linking of EU and China carbon markets as in [26] or alternative sectoral distribution of effort as in [27,28]) or strengthened mitigation efforts are applied in other jurisdictions [4,29,30]. Our analysis suggests that China and the EU may establish a linked carbon market to further increase their emission reduction efforts, while at the same time taking preventative measures to effectively reduce the negative impacts of linking the two markets.

Recently, the BCA has been suggested in the EU Green Deal as a potential instrument to protect domestic industrial activities. The study aims to inform European and international policy makers on the socio-economic and industrial impacts of imposing BCA complementing domestic carbon pricing. As there are inherent difficulties in implementing rebates of emission payments on EU exports, we assume that BCA is implemented through tariffs on embodied emissions of products imported to the EU from non-regulating counties, in order to level the playing field between domestic and imported products in the EU with respect to carbon costs. We find that the imposition of a BCA mechanism can reduce leakage through the competitiveness channel and minimize the negative policy impacts on European EITE industries; therefore, BCA can effectively complement domestic carbon pricing. The adopted recycling scheme is highly important, as using BCA and emission revenues to reduce social security contributions is highly beneficial for domestic employment.

The model-based results crucially depend on the assumptions made, especially on the values of specific elasticities. The values of Armington elasticities capture the ease of substitution between domestically produced and imported goods in GEM-E3-FIT. High elasticity values imply that countries can easily substitute the sources for commodities leading to strong leakage through industrial relocation. A comprehensive sensitivity analysis on the values of these elasticities is required to consistently evaluate the emission and economic impacts of unilateral policies.

The study highlights that while a BCA can be effective in reducing leakage and industrial relocation, careful consideration should be given to its design with particular emphasis on how the revenues are used while ensuring compliance with WTO rules so as to limit retaliation from trading partners. Overall, the BCA brings clear benefits related to cost-effectiveness and domestic acceptance, but legal and administrative barriers may reduce the efficiency gains of BCA, while its burden-shifting potential can be translated as a back-door trade policy against developing countries.

Author Contributions: Conceptualization, L.P. and P.F.; writing—original draft preparation P.F.; writing—review and editing, P.F. and K.F.; methodology, P.F., L.P. and K.F.; software, K.F.; validation, P.F., K.F., and L.P.; formal analysis, L.P. and P.F.; visualization, P.F. All authors have read and agreed to the published version of the manuscript.

Funding: The research leading to this study has received funding from the European Union Horizon 2020 research and innovation program under grant agreement No 730403 (INNOPATHS) and under grant agreement No 821124 (NAVIGATE).

Institutional Review Board Statement: Not applicable.

Informed Consent Statement: Not applicable.

Data Availability Statement: The data presented in this study are available on request from the corresponding author.

Conflicts of Interest: The authors declare no conflict of interest

Appendix A. GEM-E3-FIT Model Description

GEM-E3-FIT is a large-scale multi-sectoral CGE model that since the 1990s has been extensively used by governments and public institutions to assess the socio-economic implications of policies, mostly in the domains of energy and the environment. The development of GEM-E3 involved a series of modelling innovations that enabled its departure from the constraining framework of standard CGE models (where all resources are assumed to be fully used) to a modelling system that features a more realistic representation of the complex economic system. The key innovations of the model relate to the explicit representation of the financial sector, semi-endogenous dynamics based on R&D induced technical progress and knowledge spillovers, representation of multiple households, unemployment in the labour market and endogenous formation of labour skills. The model has detailed sectoral and geographical coverage, with 51 products and 46 countries/regions (global coverage), and is calibrated to a wide range of datasets consisting of Input–Output tables, financial accounting matrices, institutional transactions, energy balances, GHG inventories, bilateral trade matrices, investment matrices and household budget surveys. All countries in the model are linked through endogenous bilateral trade transactions identifying origin and destination. Particular focus is placed on the representation of the energy system where specialized bottom-up modules of the power generation, buildings and transport sectors have been developed. The model is recursive dynamic and produces projections of the economic and energy systems until 2100 in five-year time steps. The substitution elasticities of the model are derived from the general literature and are also econometrically estimated using the latest available datasets. The model is founded on rigorous and sound microeconomic theory allowing it to study in a consistent framework the inter-linkages of the economic sectors and to decompose the impacts of policies to their key driving factors. The model simulations are sensitive to a number of input parameters and modelling assumptions including capital costs of power producing technologies and associated learning rates, cost of capital and financing availability, easiness to substitute production factors, preferences over domestic and imported goods, etc. To address the uncertainty within, the model provides the option to make all its parameters stochastic according to user-defined probability distributions, and perform extensive sensitivity analysis.

The most important results provided by GEM-E3-FIT are: Full Input–Output tables for each country identified in the model, dynamic projections in constant values and deflators of national accounts by country, employment by economic activity and by skill and unemployment rates, capital, interest rates and investment by country and sector, private and public consumption, bilateral trade flows, consumption matrices by product and investment matrix by ownership branch, GHG emissions by country, sector and fuel and detailed energy system projections (energy demand by sector and fuel, power generation mix, deployment of transport technologies, energy efficiency improvements).

The representation of the financial sector in the GEM-E3-FIT model starts from the complete accounting of the financial flows-transactions among economic sectors. This accounting allows to determine the flow of funds, the debt profiles and the composition of agents' disposable income. The base year financial position of each agent is calculated using the institutional transactions statistics (full sequence of National Accounts that include all secondary transactions like property income, income from deposits, etc.) The net lending position of each agent is built from bottom-up data (all sources of income including dividend payments, interest rates, debt payments, bond interest rates, etc.) Data regarding the structure of the bilateral debt by agent are constructed according to current account statistics and proxies using cumulative bilateral trade transactions. All the financial transactions are arranged in a financial Social Accounting Matrix framework for each country that is represented in the model. From a modelling perspective, two additional economic sectors have been added (a world and a domestic bank) and six financial assets (deposits, time deposits, public bonds, corporate bonds, private loans and treasury bills). Banks collect savings from the economic agents in surplus and supply money at interest rates that clear the financial market (national or regional) while taking into account the risk premium and net credit position of each agent. The inclusion of the financial sector improves the simulation capabilities of the model in the following aspects:

(i) It moderates the short-term stress on capital markets by allocating capital requirements over a longer period (long-term financing schemes/loans). This effect is particularly visible in scenarios where the economy transits to a more capital-intensive structure and any limited availability of financing capital implies that capital costs will always rise.
(ii) It allows to simulate the role of carbon funds in implementation of ambitious climate policies.
(iii) It allows the assessment of socioeconomic impacts of investment projects characterized by different risk profiles performed by agents with different risk/debt profiles.
(iv) It allows for a detailed budgeting of debt by agent while it takes into account the impact of debt accumulation and debt sustainability in the ability of agents to borrow.
(v) Endogenous computation of interest rates for different financial assets (deposits, bonds, household and business financing, etc.) and direct link of nominal variables to the real economy.
(vi) Versatile financing options that correct market gaps (i.e., financing to low income households through energy-saving programs) and inclusion of financial repayment plans that allow to trace the interest payments in the future.

In GEM-E3, labour demand by firms depends on cost minimisation of their production function while labour supply is distinguished by skill and is modelled through an empirically estimated wage function (linking wages and labour supply) that allows for the existence of unemployment. The shift of labour demand to sectors requiring highly-skilled labour (i.e., a shift from agriculture to industrial manufacturing or financial services) can potentially cause a mismatch between demand and supply for specific skills and a potential skill shortage. The human capital module of GEM-E3-FIT allows households and firms to endogenously decide upon the optimal schooling-education years and on the optimal workforce training respectively. A household's decision to enter the labour market or acquire a skill (through additional education) depends on expected income (based on expectations on wages and unemployment rate by skill). The schooling decision of households concerns only certain age cohorts and allows to endogenously determine the participation rate and the supply of skills in the economy. The decision of firms to train their workers allows representing endogenous labour productivity growth through training.

In order to capture the "inequalities" within households that certain policies may imply, the new version of the GEM-E3-FIT model features for each country ten households that are distinguished by income class with different consumption patterns, different saving rates and sources of income according to the allocation of labour skills by type of household. This enhances the assessment of the social dimension of energy and climate policies enabling the assessment of income inequality within and across countries and the identification of vulnerable regions or agents. The inclusion of multiple households and human capital improves the simulation capabilities of the model by enabling the identification of potential bottlenecks due to skills scarcity and enabling productivity growth induced by R&I and knowledge spillovers.

GEM-E3-FIT adopts a bottom up approach for the electricity sector representation with different power producing technologies. Electricity-producing technologies are treated as separate production sectors with discrete investment decision. Electricity-producing technologies have different cost structures and conversion efficiencies. Total generation costs are conceived in three categories: (i) investment costs, (ii) operating and maintenance costs and (iii) fuel costs (including also potential carbon costs). Unit cost data and future projections for investment and operating costs were extracted from the PRIMES database, while the fuel costs depend on other variables of GEM-E3-FIT. At the first nesting level of electricity sector, production is split into two aggregates, one consisting of a bundle of power-producing technologies (TECH) and the other represents the transmission and distribution part (DIST). There is a zero elasticity of substitution between these two aggregates. At the second level, all power-producing technologies included in GEM-E3-FIT are in the same nest, whereas the (DIST) bundle is disaggregated to capital, skilled and unskilled labour and materials. With regards to data reconciliation, the electricity sector in IO tables is split by unbundling power generation (by technology) from electricity transmission and distribution based on PRIMES base year data and future projections.

To further improve the modelling of electricity supply, a new bottom-up module for electricity production (GEM-E3-Power) has been developed, aiming to fully endogenise the investment and operation of the electricity system. GEM-E3-Power is a technologically rich partial equilibrium model describing the development of the electricity generation mix under alternative policy assumptions. The module is hard-linked with the core GEM-E3-FIT model, through iterative exchange of their common variables (e.g., technology shares, electricity demand and supply). GEM-E3-Power decides the optimal investment and operation of the electricity system in order to minimize the intertemporal total costs to produce electricity, including capital costs, Operation and Maintenance expenditures, carbon costs and costs to purchase fuels (as inputs to power plants), while meeting system constraints (e.g., demand, technology potentials, resource availability, policy constraints, system reliability and flexibility). Thirteen power generation options are included (coal, oil, gas and biomass-fired, nuclear, hydro, PV, wind onshore, wind offshore, geothermal, CCS coal, CCS gas and CCS biomass) and compete based on their Levelised Cost of Electricity (LCOE) to meet the electricity requirements. The decision to invest in power technologies depends on their relative cost, barriers and potentials, while various policy instruments can influence the future development of the electricity system, e.g., ETS carbon prices, phase-out policies, renewable subsidies or feed-in tariffs, standards, etc.

In GEM-E3-FIT modelling, both passenger and freight are included, while the choice of transport modes and technologies and the way of using transport equipment is extensively simulated. GEM-E3-FIT distinguishes between public and private mobility. Private mobility is part of derivation of consumption by purpose of households from utility maximization under the income constraint. In the nested optimization, mobility is split between using private transport means and purchasing services from transport suppliers (public transport). To use private transport, the optimization involves purchasing of durable goods (e.g., cars) depending on stock turnover considering the choice of car types with different capital and fuel consumption features. In particular, three types of private cars are included: conventional Internal Combustion Engine (ICE) cars, plug-in hybrid vehicles and battery electric cars. Each car type uses a different mix of fuels, with conventional ICE cars using diesel, gasoline, gas and biofuels, electric cars using electricity, and plug-in hybrids using electricity, oil products and biofuels. The model separates between transport activity that can be covered by the existing fleet of private cars considering the annual car scrapping rate and new registrations of private cars. Private consumption projections in the transport sector are endogenously derived with GEM-E3-FIT model. The shares of three car types (r) in new car registration are calculated based on the Weibull discrete choice representation, as below:

$$xshcar_{r,t} = \frac{shcar_{r,t} \cdot \left(\frac{pcar_{r,t}}{pcar0_{r,t}}\right)^{sw_t}}{\sum_r shcar_{r,t} \cdot \left(\frac{pcar_{r,t}}{pcar0_{r,t}}\right)^{sw_t}}$$

where

$xshcar_{r,t}$ represents the share of car types in total new car registrations,
$shcar_r$ is the scale parameter used for the calibration of technology shares,
$pcar_{r,t}$ is the price by car type (reflecting total transport cost),
sw_t is the elasticity of substitution between alternative car types.

Mobility of private consumers is then translated into demand for specific car types, which in turn is related to demand for specific goods via the consumption matrix. The assumptions for fuel mix, technical efficiencies and other parameters (e.g., fuel use per passenger km) are based on PRIMES data. In ambitious climate policy scenarios, the technology and fuel mix in transport modes changes endogenously as a result of carbon pricing and other instruments and changes in technology costs, while the fuel shares in households' consumption matrix (than links consumption by purpose to demand for specific goods) can be modified. The supply of professional transport is represented as production sectors distinguishing between land, air and maritime transport. Each transport sector produces a homogenous service using inputs from capital, labour, materials and energy, based on endogenous choice of firms towards cost minimisation. The demand of other production sectors for transport services derives from cost minimization of their production input mix. Substitutions are possible between transport modes and between transport and non-transport inputs depending on relative prices of goods and services.

References

1. Fragkos, P. Global Energy System Transformations to 1.5 °C: The Impact of Revised Intergovernmental Panel on Climate Change Carbon Budgets. *Energy Technol.* **2020**, *8*, 2000395. [CrossRef]
2. Böhringer, C.; Balistreri, E.; Rutherford, T. The role of border carbon adjustment in unilateral climate policy: Overview of an Energy Modeling Forum study (EMF 29). *Energy Econ.* **2012**, *34*, S97–S110. [CrossRef]
3. Weyant, J.P. (Ed.) *The Costs of the Kyoto Protocol: A Multi-Model Evaluation*; The Energy Journal (Special Issue); Cleveland, OH, USA; pp. 1–398. Available online: https://web.stanford.edu/group/emf-research/docs/emf16/CostKyoto.pdf (accessed on 15 October 2020).
4. Paroussos, L.; Fragkos, P.; Capros, P.; Fragkiadakis, K. Assessment of carbon leakage through the industry channel: The EU perspective. *Technol. Forecast. Soc. Chang.* **2014**, *90*, 204–219. Available online: https://www.sciencedirect.com/science/article/abs/pii/S0040162514000602 (accessed on 15 October 2020). [CrossRef]
5. Carbone, J.; Rivers, N. The Impacts of Unilateral Climate Policy on Competitiveness: Evidence from Computable General Equilibrium Models. *Rev. Environ. Econ. Policy* **2017**, *11*, 24–42. [CrossRef]
6. European Commission. Communication from the Commission to the European Parliament, the European Council, the Council, the European Economic and Social Committee and the Committee of the Regions. In *The European Green Deal Brussels*; COM 640 final; European Commission: Brussels, Belgium, 2019.
7. Hoel, M. Global Environmental Problems: The Effects of Unilateral Actions Taken by One Country. *J. Environ. Econ. Manag.* **1991**, *20*, 55–70. [CrossRef]
8. Böhringer, C.; Carbone, J.; Rutherford, T.F. *Embodied Carbon Tariffs*; NBER working paper; National Bureau of Economic Research: Cambridge, UK, 2011; p. 17376.
9. European Commission. *State of the Union 2017–Industrial Policy Strategy: Investing in a Smart, Innovative and Sustainable Industry*; European Commission: Brussels, Belgium, 2017.
10. Wesseling, J.H.; Lechtenböhmer, S.; Åhman, M.; Nilsson, L.J.; Worrell, E.; Coenen, L. The transition of energy intensive processing industries towards deep decarbonization: Characteristics and implications for future research. *Renew. Sustain. Energy Rev.* **2017**, *79*, 1303–1313. [CrossRef]
11. Aldy, J.; Pizer, E. The Competitiveness Impacts of Climate Change Mitigation Policie. *J. Assoc. Environ. Resour. Econ.* **2015**, *2*, 565–595. Available online: http://www.nber.org/papers/w17705 (accessed on 15 October 2020).
12. DeCian, E.; Parrado, R.; Grubb, M.; Drummond, P.; Coindoz, L.; Mathy, S.; Stolyarova, E.; Georgiev, A.; Sniegoki, A.; Bukowski, M. *A Review of Competitiveness, Carbon Leakage and EU Policy Options in the Post-Paris Landscape*; Deliverable 3.1 of the COP21-RIPPLES H2020 project; European Commission: Brussels, Belgium, 2017.
13. Zachmann, G.; McWilliams, B. *A European Carbon Border Tax: Much Pain, Little Gain*; Policy Contribution 05/2020; Bruegel: Brussels, Belgium, 2020.
14. European Commission. *Study on Energy Prices, Costs and Their Impact on Industry and Households*; DG ENER.; European Commission: Brussels, Belgium, 2020.
15. Fragkiadakis, K.; Fragkos, P.; Paroussos, L. Low-Carbon R&D Can Boost EU Growth and Competitiveness. *Energies* **2020**, *13*, 5236.
16. Fragkos, P.; Tasios, N.; Paroussos, L.; Capros, P.; Tsani, S. Energy system impacts and policy implications of the European Intended Nationally Determined Contribution and low-Carbon pathway to 2050. *Energy Policy* **2017**, *100*, 216–226. [CrossRef]
17. Capros, P.; Vita, D.A.; Tasios, N.; Siskos, P.; Kannavou, M.; Petropoulos, A.; Evangelopoulou, S.; Zampara, Z.; Papadopoulos, D.; Nakos, C.; et al. *EU Reference Scenario 201–Energy, Transport. and GHG Emissions Trends to 2050*; European Commission Directorate General for Energy, Directorate General for Climate Action and Directorate General for Mobility and Transport: Brussels, Belgium, 2016.
18. Armington, P.S. Theory of Demand for Products Distinguished by Place of Production. *IMF Staff. Pap.* **1969**, *16*, 159–178. [CrossRef]

19. IEA. *World Energy Outlook;* International Energy Agency: Paris, France, 2019.
20. European Commission. *Ageing Report;* European Commission: Brussels, Belgium, 2018.
21. IRENA. *Renewable Power Generation Costs in 2019;* International Renewable Energy Agency: Abu Dhabi, UAE, 2020.
22. McCollum, D.L.; Zhou, W.; Bertram, C.; Boer, H.-S.d.; Bosetti, V.; Busch, S.; Després, J.; Drouet, L.; Emmerling, J.; Fay, M.; et al. Energy investment needs for fulfilling the Paris agreement and achieving the sustainable development goals. *Nat. Energy* **2018**, *3*, 589–599. [CrossRef]
23. Cosbey, A.; Droege, S.; Fischer, C.; Reinaud, J.; Stephenson, J.; Weischer, L.; Wooders, P. A Guide for the Concerned: Guidance on the elaboration and implementation of border carbon adjustment. *Int. Inst. Sustain. Dev.* **2012**, *2012*, 22. [CrossRef]
24. Karkatsoulis, P.; Capros, P.; Fragkos, P.; Paroussos, L.; Tsani, S. First-Mover advantages of the European Union's climate change mitigation strategy. *Int. J. Energy Res.* **2016**, *40*, 814–830. [CrossRef]
25. Paroussos, L.; MANDEL, A.; Fragkiadakis, K.; Fragkos, P.; Hinkel, J.; Vrontisi, Z. Climate clubs and the macro-economic benefits of international cooperation on climate policy. *Nat. Clim. Chang.* **2019**, *9*, 542–546. [CrossRef]
26. Li, M.; Weng, Y.; Duan, M. Emissions, energy and economic impacts of linking China's national ETS with the EU ETS. *Appl. Energy* **2019**, *235*, 1235–1244. [CrossRef]
27. Alexeeva-Talebi, V.; Böhringer, C.; Löschel, A.; Voigt, S. The value added of sectoral disaggregation: Implications on competitive consequences of climate change policies. *Energy Econ.* **2012**, *34* (Suppl. 2), S127–S142. [CrossRef]
28. Yumashev, A.; Ślusarczyk, B.; Kondrashev, S.; Mikhaylov, A. Global Indicators of Sustainable Development: Evaluation of the Influence of the Human Development Index on Consumption and Quality of Energy. *Energies* **2020**, *13*, 2768. [CrossRef]
29. Lisin, A. Prospects and Challenges of Energy Cooperation between Russia and South Korea. *Int. J. Energy Econ. Policy* **2020**, *10*, 130. [CrossRef]
30. Meynkhard, A. Long-Term prospects for the development energy complex of Russia. *Int. J. Energy Econ. Policy* **2020**, *10*, 224–232. [CrossRef]

energies

Article

Low-Carbon R&D Can Boost EU Growth and Competitiveness

Kostas Fragkiadakis *, Panagiotis Fragkos and Leonidas Paroussos

E3Modelling, 70–72 Panormou Street, PO 11523 Athens, Greece; fragkos@e3modelling.com (P.F.); paroussos@e3modelling.com (L.P.)
* Correspondence: fragkiadakis@e3modelling.com

Received: 8 September 2020; Accepted: 5 October 2020; Published: 8 October 2020

Abstract: Research and Innovation (R&I) are a key part of the EU strategy towards stronger growth and the creation of more and better jobs while respecting social and climate objectives. In the last decades, improvements in costs and performance of low-carbon technologies triggered by R&I expenditures and learning-by-doing effects have increased their competitiveness compared to fossil fuel options. So, in the context of ambitious climate policies as described in the EU Green Deal, increased R&I expenditures can increase productivity and boost EU economic growth and competitiveness, especially in countries with large innovation and low-carbon manufacturing base. The analysis captures the different nature of public and private R&I, with the latter having more positive economic implications and higher efficiency as it is closer to industrial activities. Public R&D commonly focuses on immature highly uncertain technologies, which are also needed to achieve the climate neutrality target of the EU. The model-based assessment shows that a policy portfolio using part of carbon revenues for public and private R&D and development of the required skills can effectively alleviate decarbonisation costs, while promoting high value-added products and exports (e.g., low-carbon technologies), creating more high-quality jobs and contributing to climate change mitigation.

Keywords: GEM-E3-FIT; low-carbon R&D; innovation-induced growth; endogenous technology progress

---

1. Introduction

Research and innovation (R&I) forms a key component of the EU's strategy calling for the delivery of stronger, lasting economic growth and the creation of more and higher quality jobs while respecting and promoting social and environmental objectives. The constant creation of new ideas and their transformation into technologies and products forms a powerful cornerstone for the 21st century society, with universities, research institutes and innovative companies cultivating this process. The EU R&I strategy ensures that innovative ideas are turned into new products and services in an effort to create growth, quality jobs and address societal challenges. R&I investment directed to research organisations, universities and innovative technology companies are key component of the EU's strategy, complemented with appropriate framework conditions, market regulation and the supply of the required skills and low-cost capital. Guided by the goals of the Innovation Union flagship initiative, the EU aims to radically change the way public and private R&I sectors cooperate, remove bottlenecks (like expensive patenting, market fragmentation, limited access to capital and skill shortages) that prevent the market development of innovative ideas and increase R&I investments to 3% of EU GDP. The central role of R&I in EU policy is demonstrated by the increased budget allocations to Horizon Europe program, the InvestEU fund (to mobilise private R&I) and the recent EU Green deal.

The "Clean Planet for all" long-term strategy (https://ec.europa.eu/clima/policies/strategies/2050_en) of the European Commission (EC) suggests that the acceleration of research, innovation, entrepreneurship and human capital investment are key for the achievement of a climate neutral Europe by mid-century while boosting the EU's industrial competitiveness. In this context, it is crucial to improve the understanding of how increased R&I spending in low-carbon technologies can induce innovation and affect firms' productivity, economic growth and societal transitions in order to better inform policy makers on the role of low-carbon R&D towards decarbonization. In turn, policymakers can deploy a series of policy instruments to promote low-carbon innovation directly and indirectly, including investment and subsidies in R&D and knowledge diffusion (e.g., open access to science results), investment in human capital upgrade (through education and training), developing a regulatory and policy environment that stimulates the entry and exit of new businesses, promoting venture capital and access to low-cost finance for businesses.

There is very little quantitative macro and socio-economic assessment of the impacts, costs and benefits associated to low-carbon R&D investment and how these are financed. Existing literature has analysed the R&D contribution to improve the costs and performance of low-carbon technologies [1,2], but not accounting for the R&D impacts at the macro-economy level. Energy system models are commonly used to explore low-emission transition pathways but do not capture the macro and socio-economic impacts of climate and innovation policies; conventional Computable General Equilibrium (CGE) models do not explicitly represent technological learning, innovation and diffusion. In order to bridge this research, the GEM-E3-FIT CGE model is further developed to consistently assess the interactions between energy system decarbonisation, low-carbon technology development and investment in public and private low-carbon R&D [3]. Due to its detailed treatment of energy technologies and sectors, technical progress, innovation and climate issues, GEM-E3-FIT model appears to be the most suitable modelling framework for exploring the socio-economic impacts of low-carbon innovation [4].

The improved macroeconomic modelling framework with enhanced representation of low-carbon R&I is used to analyse the synergies between climate and innovation policies, in particular how investment in low-carbon R&I and human capital upgrade would affect the European economy and industries within a decarbonisation context. The model-based assessment of the economic, trade and employment impacts of low-carbon R&I can be used to provide practical recommendations to EU policy makers on the potential allocation of ETS revenues towards R&D in low-carbon technologies to fully exploit the clean energy innovation benefits (i.e., enhanced competitiveness and productivity growth).

The main methodological novelty of the paper is the detailed modelling of technological change in a multi-sectoral CGE model combined with the representation of sectors manufacturing low-carbon technologies. In this way, the analysis provides novel insights on the activity growth and competitiveness impacts of innovation policies, capturing in detail the inter-industrial relations and the potential industry effects that being a global technology leader might bring about [5]. In addition, the distinction between public and private R&D in applied modelling sheds light on their differential socio-economic impacts, while the detailed representation of knowledge spillovers provides an enhanced framework to study the complex linkages between low-carbon innovation, human capital upgrade and decarbonisation.

This paper is organised as follows: In Section 2, we conduct a literature review on the role of low-carbon innovation and we provide an overview of how the modelling tools incorporate technical change. In Section 3, we provide a detailed description of the modelling approach and scenarios examined, focusing on the representation of low-carbon technology innovation and clean energy manufacturing markets. Section 4 includes the analysis of model-based results focusing on the macro-economic, employment and competitiveness impacts of low-carbon innovation. Section 5 includes a discussion and conclusions.

2. Literature and Context

2.1. The Role of Low-Carbon Innovation in the EU

The EU aims to increase the role of R&I encouraging EU countries to invest 3% of their GDP in R&D with specific national targets considering country differences, socio-economic priorities and current situation. The Horizon 2020 programme is designed to support transnational and multidisciplinary collaboration across the EU with public funds, while other initiatives, such as VentureEU and the European Innovation Council (EIC) pilot, aim to boost capital investment in innovative start-up companies across Europe. Governments deploy a series of instruments to promote innovation, including subsidies to R&D, human capital development and skills upgrade, provision of low-cost finance to innovative businesses and development of the required regulations stimulating innovation-based firms. Policies targeting innovation can take various forms, including instruments aimed at R&D (e.g., grants for R&D projects, tax incentives or direct subsidies for private R&D), options facilitating human capital development (e.g., subsidies to develop the required labour skills), R&D-specific finance instruments (e.g., low-cost loans to innovative firms) as well as measures that link R&D with innovation and industry, e.g., programmes of collaborative R&D and IPR-related policies.

Low-carbon innovation and development in the context of decarbonisation are key priorities for EU R&I activities, as demonstrated by their large share in H2020 and other funding mechanisms (e.g., the recent EIC €350 million programme calling for market-creating innovation contributing to the European Green Deal (https://ec.europa.eu/easme/en/section/sme-instrument/eic-accelerator-funding-opportunities). While the EU provides large amounts to R&I activities through various mechanisms (outlined above), the majority of public investment in R&I comes from national funding, with diverse performance among EU countries as measured in [6].

The Research and Innovation Observatory identifies two major challenges related to innovation, namely: low-levels of public and private R&I funding across European countries and barriers in industry collaboration, exploitation and commercialization of public research results. The EU needs to deepen its innovation capability to face global societal megatrends such as digitisation, an ageing society, sustainable development and climate action requiring innovation in a certain direction. Policy plays a critical role in shaping the R&I contribution to address these challenges, so that the EU would remain at the forefront.

2.2. Technology Progress

In the last decade, targeted innovation and uptake of low-carbon technologies has led to a rapid reduction in costs of key mitigation options. The costs of PV, wind onshore and offshore have declined significantly making renewable energy competitive with fossil fuel-based generation in many countries [7]. This reduction is driven both by technological innovations and breakthroughs, but most importantly by incremental learning, maturity and development of the industry and its supply chains at scale. However, the initial technology deployment was mainly induced by dedicated policies (e.g., German Energiewende), which were expensive but drove the PV revolution globally.

To capture these dynamics, most energy-economy models represent energy technology progress, in particular for new technologies. A major review of innovation modelling called for additional research in the empirical validation of technology learning due to the high prevailing uncertainty [8]. In the last decade, evidence about low-carbon innovation has further accumulated and the direction of causality between R&D and technology progress is clear [1]. Recent literature found that low-carbon innovation is influenced by market conditions, relative costs and the market size of clean energy technologies [9,10] adding further evidence and insights into their linkages. Innovation could massively reduce the cost of tackling climate change, but specific subsidies to innovation are required [11]. Pottier et al. criticized their study for the parameter assumptions and lack of path-dependence, being rooted in an 'incorporeal' world of sequential general equilibria [12]. Commonly, engineering-based models incorporate learning-by-doing, generating more complex behaviour and often indeterminate results, which are hard to communicate and implement computationally in applied studies. However, these issues should not be tackled using simplified exogenous technology assumptions and excluding dynamic realism in energy-economy models, but care should be taken to appropriately integrate an endogenous representation of innovation and technology learning.

Innovation is to a large degree a product of investment, in the form of public and private R&D or "spilling" from other sectors (like IT and digitization) into low-carbon technologies. Several empirical studies have confirmed the Hicks proposition that the direction of innovation is influenced by relative prices [13]. Popp reviewed numerous studies demonstrating that the share of private R&D devoted to energy increased after the 1970s oil price shocks [14]. Such evidence has been enhanced by more recent studies reviewed in [15], who documented a (statistically significant) link between energy prices and patent filing for energy innovations. Ley et al. find not only that higher energy prices increased patenting, but disproportionately enhance low-carbon patents [16], e.g., for solar PV [17] and electric vehicles [18]. Recent studies show a positive impact of carbon pricing and other environmental regulation on low-carbon innovation. Calel et al., found increased patents for low-carbon technologies across regulated firms in EU ETS [19], while Taghizadeh-Hesary et al. showed that energy price increases had a statistically significant impact in reducing PV module prices in USA, Japan and China [20]. Thus, evidence suggests that climate policies can enhance low-carbon innovation, support subsequent emission reductions and structural shifts.

In recent years, strong national leadership on key low-carbon technologies has played a critical role in driving technological innovation and cost reduction. Germany's efforts in driving PV, Denmark and Germany's early push for wind, and more recently the UK's efforts to drive offshore wind have reduced technology costs substantially. Aspiring to industrial leadership and seeking first-mover advantage in low-carbon technologies has been a major driver of climate policies, and this is expected to continue, especially in the decarbonisation context offering large export opportunities for low-carbon technology manufacturers [21].

The cost reduction of low-carbon technologies is induced by various forms of learning-by-doing, economies of scale, patentable innovations, along with the development of integrated supply chains, and growing confidence which reduces the perceived risks and cost of capital [22]. Energy-economy system analysis often uses "experience curves" which chart how much the cost of technologies decline with scale typically measured in terms of 'learning rates'—the decline of cost associated with a doubling of capacity. Rubin et al. reviewed numerous studies and found that learning rates vary among energy technologies and studies, but are uniformly positive [23]. Low-carbon technologies get cheaper as their markets grow, and learning rates have been remarkably high and stable for PV (at about 20%). Recent studies disentangle the effects of public R&D from private R&D and learning-by-doing; industrial economies-of-scale are particularly important for PV [24] and other low-carbon technologies [25], while the combination of all factors drives technology costs down and helps to build up new industries.

The empirical evidence suggests that investment in R&I is a key driver for the reductions in costs of low-carbon technologies [23]. Technological learning and innovation shape energy transitions through reduction of costs, increase of efficiency, the creation of new services or functionalities and promoting technology diffusion [26]. The combination of public and private R&D activities is a success factor for new technologies with private R&D being closer to commercialisation stage of technological development [27].

2.3. Modelling Innovation and Technology Change

Scientific findings confirm that learning effects in the form of learning-by-doing or learning-by-research are apparent in all energy-related technologies [28]. The relevance of learning-by-research and learning-by-doing and the relative importance of public vs. private sector R&D, vary greatly across technologies, depending on: technology maturity, inter-industrial linkages, and the geographical coverage of analysis. In all technology transitions, both supply-push and demand-pull drivers have played an important role. While a combination of both driver types seems to be optimal, supply push measures (i.e., R&D support) are more effective to support technologies at early development stages [15], whereas demand pull measures are better suited for technologies closer to market readiness. R&I incentivises cost reductions at early innovation stages, while economies of scale play an increasingly important role for technology development and diffusion as industry matures leading to further costs reductions and enhanced competitiveness with incumbent technologies [29], so the use of exogenous technology cost assumptions in energy-economy models is not the appropriate way to model technology learning, as it ignores the drivers of deployment, innovation and their complex interlinkages. The latter should be appropriately represented in models used for policy impact assessments, as these should capture all important policy-relevant elements, including cost-effective ways of stimulating innovation.

In addition to own R&D, spillover effects from other countries and sectors play a crucial role in technology learning and development, as they increase the stock of knowledge in the recipient industries and regions and contribute to improving productivity. Based on detailed statistical analysis on patent citations, Paroussos et al. illustrated the high importance of spillovers for the speed and magnitude of low-carbon technology innovation and knowledge diffusion [30]. Spillovers can be cross-sectoral (e.g., electronics industry spilling to battery manufacturing) and cross-regional, as spillovers are not uniformly distributed across countries; Japan and the US account for half of the weighted patent citations reflecting their overall size, advanced technological state, and well-developed system of patenting [30].

From a modelling perspective, assuming that productivity growth is driven purely by in-house R&I expenditures would be misleading, as this neglects the impacts of cross-sectoral spillovers and technology embodied in machinery, equipment and efficient IT structures. Griliches and Lichtenberg argued that not accounting for inter-industry technology flows leads to biased estimates of the contribution of R&I to productivity growth [31]. Parrado and De Cian argue that innovation is not limited to national R&I, but it further builds on knowledge spillovers and embodied technological progress from other regions through trade, foreign direct investment, research collaboration, technological similarities and cultural or institutional proximity [32]. Therefore, energy-economy models should represent both in-house innovation activities, but also knowledge spillovers across regions and sectors.

Representations of technological change and innovation dynamics are incomplete in energy-economy models, thus limiting their ability to assess climate policies. Missing or inadequate representation of endogenous technical change may lead to overestimation of mitigation costs, or findings of "optimal" mitigation strategies with too little short-term abatement [10]. The outcomes of models without endogenous technology learning were largely based on exogenous assumptions [33], e.g., on the presence of the so-called backstop technologies. Edenhofer et al. explored how leading modelling frameworks represent innovation and technological change and showed that model estimates for investment requirements for decarbonisation are lower if learning-by-doing progress is endogenised in models [34]. The ability of models to adequately capture endogenous technological change radically affects the conclusions drawn about e.g., costs of mitigation or the feasibility of rapid technology diffusion [10,35].

The induced innovation theory postulates that innovation can be heavily influenced by market conditions and policy priorities, which may drastically change the scale of new technologies. Energy-economy models require technology projections decades ahead, but a wide variety of evidence shows that innovation and associated cost reductions depend on investments in earlier periods, and hence should be endogenous to modelling. Ignoring the dynamic features of technology learning may produce misleading insights for policy-makers, potentially leading to a delay of ambitious decarbonisation measures.

The theory of endogenous innovation-driven growth has been developed in one-sector modelling frameworks, while its application in large-scale multi-sectoral models is sparse. The latter can realistically capture the high heterogeneity of sectors with respect to their cost structure, R&I intensities, labour skills and absorptive capacities. In contrast, IAMs do not account for the sectoral impacts of policies [36] and the potential domestic industry effects that being a global technology leader might bring about [5] and thus their results for specific regions can be misleading.

A more disaggregated representation of innovation and its linkages with the economy will contribute to improved assessment of policy interventions. Structural multi-sectoral macro-economic models can adequately capture the complexity of technological progress, integrating learning and innovation processes in a unique modelling framework and accounting for their complex interactions with other sectors (inter-industrial linkages) and countries. Studies to date offer quantifications of R&I impacts on economic performance at firm or sectoral level, but the comprehensive assessment of the economic effects at country level considering sector inter-dependencies remains scarce. CGE models can provide a robust quantification of macroeconomic impacts of R&I policies, but their application for innovation analysis remains limited due to the data intensity and their computationally demanding nature. Wing models induced technological change through the inclusion of knowledge capital in the production function [37]. Advanced macroeconomic models have been recently developed to assess the impacts of R&I policies [10,38]. Recent studies introduce international knowledge spillovers and R&I indicators including patent counts, technology flow and knowledge proximity matrixes [39], while spillovers are introduced in the CGE models via the trade flows of goods and services or via bilateral patent matrices [30,40].

Overall, most energy-economy models do not fully integrate modelling of R&D and its socio-economic impacts [4,10], while other models capture this rather simplistically, as they fail to represent real-world innovation dynamics, like knowledge spillovers across regions and sectors, the link to human capital and do not differentiate between public and private R&D [41]. The study overcomes the current limitations in modelling of low-carbon innovation by further improving the state-of-the-art GEM-E3-FIT model (see Section 3) aiming to offer novel, policy-relevant insights.

3. Methodology

In order to consistently assess the macro-economic impacts of low-carbon R&I, we further improve the GEM-E3-FIT model to represent low-carbon R&D expenditure—differentiated into public and private R&D—and its relation with productivity improvements, innovation spillovers, the workers' skill levels and the households' decision for schooling and education linked with human capital development and firms' absorptive capacity [30]. In addition to model extensions, key model parameters are updated with empirically validated evidence including the link between R&I expenditure and human capital with productivity growth [42].

3.1. Brief Description of GEM-E3-FIT

GEM-E3-FIT is a multi-sectoral, CGE model which analyses the complex interactions between the economy, the energy system and the environment. GEM-E3-FIT represents 46 regions (EU Member States are represented separately) and 51 production sectors linked through endogenous bilateral trade flows. It covers the interlinkages between productive sectors, consumption, labour and capital, bilateral trade, innovation and investment dynamics (Figure 1). GEM-E3-FIT is dynamic driven by accumulation of knowledge, capital and equipment until 2050. The model ensures that the economic system remains in general equilibrium in alternative scenario simulations.

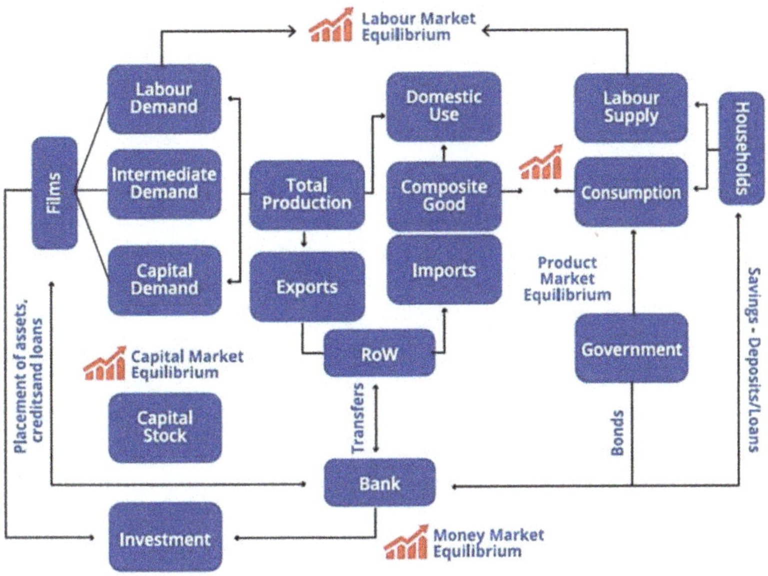

Figure 1. Overview of the GEM-E3-FIT model.

Each national economy is formed by a set of representative firms and a representative household whose interactions are governed by endogenously-derived prices. Firms maximize profits, considering the possibilities of substitution between capital, labour, materials and energy. Households maximize their intertemporal welfare under a budget constraint choosing the optimal level for current and future consumption levels. Households receive income from labour supply and from holding shares in companies. Firms' decision on investments is driven by the rate of return on capital and its replacement cost. Trade modelling is based on the Armington hypothesis that domestic and imported goods are imperfect substitutes. In GEM-E3-FIT, the supply mix of each good is represented as a multi-level nested constant elasticity of substitution (CES) function: at the upper level, firms decide on the optimal mix between domestically produced and imported goods; at the next level, the demand for imports is split by country of origin.

Technology constrains the set of paths upon which agents can make their investment and consumption choices. The standard technology representation in CGEs uses nested CES production functions that differentiate capital, labour, energy and material inputs, while the amount of economy-wide physical capital is considered fixed within a period. GEM-E3-FIT includes a detailed representation of energy system technologies as distinct sectors calibrated to recent EUROSTAT, GTAP and IEA datasets and covers fiscal elements, including taxes, subsidies and social security contributions.

GEM-E3-FIT includes several features that go beyond a conventional CGE approach, enabling an enhanced representation of socio-economic implications of climate and innovation policies. Its advanced features include: a detailed coverage of the financial sector [38], innovation-induced productivity growth through R&D, detailed modelling of the energy system [43], and disaggregated representation of employment by skill [42].

3.2. Endogenous Representation of Clean Energy Markets

Most global energy-economy models fail to represent the domestic industry effects that being a global technology leader might bring about and thus their results may be misleading [5]. To address this caveat, GEM-E3-FIT includes the manufacturing of low-carbon products and equipment as separate production sectors. The model consistently derives the future development of low-carbon manufacturing and trade patterns under alternative assumptions. GEM-E3-FIT database is extended to allow for a distinct representation of low-carbon technology producers in economic terms, namely for solar PV, wind, electric vehicles, batteries and biofuels. Data about the size, market shares, cost structure and trade flows of the above sectors are not included in GTAP and are derived from supplementary data sources, including Fraunhofer [44], Navigant [45], CEMAC [46] and IEA [47]. For example, IEA data [47] is used for the demand (sales) and manufacturing volumes of EVs and batteries which are combined with base year technology prices from [3] to estimate production and trade (in economic terms) of EVs in each country. The cost structure of low-carbon equipment is derived from [48,49] in terms of inputs required by other sectors (i.e., metals, electric equipment, machinery etc.) to produce clean energy products, which are different from the cost structure of fossil-fuel technologies. GEM-E3-FIT integrates differences in low-carbon technology production costs across countries as derived from [46]. Detailed data handling processes (i.e., RAS routines) are developed to ensure full consistency of GEM-E3-FIT Input-Output tables including low-carbon manufacturing with GTAP data, e.g., the production of conventional ICE and EVs sums up to GTAP sector 43 "Manufacture of motor vehicles, trailers and semi-trailers' [40]. This allows capturing consistently the potential growth effects driven by the uptake of low-carbon industries and innovation, as well as the changes in competitiveness and trade flows induced by ambitious decarbonisation and low-carbon innovation.

3.3. Modelling of Technological Change and R&D

GEM-E3-FIT represents public and private low-carbon R&I, knowledge spillovers and absorption linked to human capital and their impacts on technology costs. Modelling of technological change in GEM-E3-FIT draws on the endogenous growth theory [11,50]. In GEM-E3-FIT, technology progress is endogenous deriving from spending in R&I while productivity improvements are based on two-factor learning curves depending on learning-by-doing and R&D expenditure from the private and public sector. The learning-by-doing component represents the productivity gained through cumulative production (i.e., experience and economies of scale), while R&D learning describes the cost improvement for each doubling of cumulative R&D expenditure.

In conventional CGE modelling, total factor productivity (TFP) is determined exogenously in the baseline scenario. In GEM-E3-FIT, TFP development is endogenised as it includes an exogenous and an endogenous part, which represents innovation-induced growth and is composed of: (i) the learning by doing effect, (ii) the learning by research effect, (iii) the impact of knowledge spillovers, and (iv) the human capital stock measure. More precisely, TFP is decomposed into a part related to learning by doing, a part related to R&D expenditure and a part related to spillovers, i.e.,:

$$TFP_t = TFP_{t-1} \cdot GTFP_t{}^{lbd} \cdot GTFP_t{}^{R\&D} + GTFP_t{}^{spillover} \tag{1}$$

where $GTFP_t{}^{lbd}, GTFP_t{}^{R\&D}$, $GTFP_t{}^{spillover}$ denote the productivity growth due to learning by doing, learning by R&D and innovation spillovers, respectively.

Learning by doing is assumed to increase with cumulative production ("Wrights law"), so that

$$GTFP_t{}^{lbd} = \left(\frac{Q_t}{Q_{t-1}}\right)^{1+lbd} \tag{2}$$

where Q_t represents cumulative production up to period t and the parameter lbd denotes the percentage cost reduction induced by an increase in cumulative production given by learning rate LR. Namely:

$$lbd = -\frac{log(1-LR)}{\log(2)} \tag{3}$$

Learning rates for low-carbon technologies are based on a comprehensive literature review in [3] and are presented in Table 1.

Table 1. Learning rates used in GEM-E3-FIT.

	Learning by Doing	Learning by Research
Ethanol	0.10	0.11
Bio-diesel	0.10	0.11
Advanced Electric Appliances	0.10	0.10
wind technology	0.12	0.10
PV panels	0.23	0.12
CCS power technology	0.11	0.11
Other Advanced Heating and Cooking Appliances	0.10	0.10
Electric Vehicles	0.10	0.20

All parameters related to specification of endogenous TFP growth in GEM-E3-FIT are estimated using advanced panel data econometric techniques with cross country data for EU and non-EU countries over 2000–2016 [9].

GEM-E3-FIT separates public from private R&D expenditures. Public R&D is set exogenously, while private firms decide upon the optimal R&D spending so as to increase productivity and maximize their profits. Each sector optimizes allocation of resources in R&D simultaneously with decisions about acquiring capital, labour, materials and energy. R&D expenditures create demand for R&D services addressed to the R&D supply sector, which is represented as a separate sector in GEM-E3-FIT. Private R&D expenditure are undertaken by firms to develop product, process or other types of innovations, enhance firm productivity, reduce their production costs and improve their competitive position relative to other firms. R&D expenditures generate a stock of knowledge that in turn is linked to productivity increase, which is provided in the following Equations (4) and (5), where $RD_j^{Private}$ is the optimal demand of firms for R&D, θ_j^{rd} is the value share of R&D expenditures in production costs, Q_j represents the total sales of the firm, PQ_j is selling price and PRD is the unit cost of R&D:

$$RD_j^{Private} = \theta_j^{rd} \cdot Q_j \cdot \left(\frac{PQ_j}{PRD_j}\right)^{rho} \tag{4}$$

$$RD_j^{Public} = (Exogenous) \tag{5}$$

Public and private R&D expenditures accumulate over time increasing the stock of knowledge that leads to TFP growth:

$$GTFP_t^{R\&D} = \left(\frac{CRD_t}{CRD_{t-1}}\right)^{1+lbr} \tag{6}$$

where CRD_t represents the cumulative investment in R&D and lbr denotes the percentage cost reduction associated with a doubling in cumulative R&D, which is derived from [3]. TFP growth may also be induced by spillovers from R&D performed in other regions and sectors which may be driven by foreign direct investment, trade and transfers of intellectual property [51]. Kirchherr and Urban focus on international technology transfer schemes [52]. The impact of spillovers on TFP growth follows:

$$GTFP_{i,r,t}^{spillover} = \sum_{j,s} TFP_{i,j,r,s,t}^{spillover} \tag{7}$$

$$TFP_{i,j,r,s,t}^{spillover} = absorption_{i,r} \cdot spillover_{i,j,r,s} \cdot \left(GTFP_{j,s,t}^{R\&D} - GTFP_{j,s,t-1}^{R\&D}\right) \tag{8}$$

where $absorption_{i,r}$ represents the absorption capacity of sector i in region r and $spillover_{i,j,r,s,}$ denotes the rate of spillover from sector j in region s to sector i in region r. These spillover rates are estimated in [30] by using the patent citation methodology of Verspagen [53].

The availability of human capital is essential to enable productivity growth induced by R&D and knowledge spillovers. The index of human capital stock is constructed based on the shares of each skill type to the total labour force; these indicate that the respective skills embodied at high skill type are more productive relative to those in lower skill types [9]. The capacity of firms to absorb knowledge spillovers is linked to human capital availability, especially related to skilled labour. The accumulation of public R&D knowledge stock is global assuming perfect spillovers to other regions, while private R&D can be diffused through bilateral trade of goods and services and through knowledge spillovers based on a patent citations approach for low-carbon technologies. R&D expenditure has been widely used as an indicator of innovation, while patents are also widely used as output-based indicators. Generation of patents does not only have a direct positive effect on the industry that produces them, but also impacts positively other industries through knowledge spillovers. These spillovers benefit the country and industry that receives them, increase the income of innovator through royalties and reduce the monopoly rents of the innovator.

Productivity generated through R&D is diffused into other sectors and countries according to:

$$TFP\_SPILL_{i,j,r,s} = TFP_i \cdot spillover_{i,j,r,s} \quad (9)$$

3.4. Scenario Design

R&D investment is a key driver for cost reduction for low-carbon technologies, with clean energy R&D increasing in recent years and in 2017 amounted to $9.9 bn [54]; however, government R&D spending stayed flat at $5.1 billion despite the creation of Mission Innovation at the Paris conference in 2015. Europe maintained its lead in low-carbon R&D rising to $2.7 billion, with the US and China following closely; China leads the R&D in solar technologies. Among the leading economies, Japan registers the largest energy R&D expenditure as a share in GDP, ahead of China and Europe [55].

The EU strategy calls for increased R&D expenditure to stimulate growth and create quality jobs, with an increasing part of them directed to low-carbon technologies. To assess the socio-economic impacts of low-carbon R&D, a series of policy scenarios are modelled with GEM-E3-FIT (Table 2). These aim to explore the complex interactions between decarbonisation and low-carbon innovation, with EU carbon revenues used to finance R&D in low-carbon technologies. In all scenarios, public budget neutrality is ensured with the general equilibrium modelling framework. In the policy scenarios, the exogenous part of TFP does not change from Baseline levels, while changes in the endogenous part are driven by increased R&D expenditure and human capital upgrade in alternative scenarios. All other exogenous parameters (i.e., trade or substitution elasticities, value shares, price elasticities etc.) do not change from baseline levels.

To analyse the macro-economic and competitiveness effects of climate and innovation policies, we consider eight scenarios:

- A "baseline scenario" (BASE), where all regions implement their current energy, climate and innovation policies by 2030 and do not intensify their efforts beyond 2030. Low-carbon R&D remains constant as a percentage of GDP to 2015 levels. In this scenario, limited climate policies are adopted worldwide in line with the current fragmentation and lack of ambition in the international climate policy landscape.
- A "global well-below 2 °C scenario" (2DEG) assuming cost-efficient implementation of the 1000 Gt carbon budget over 2010–2050 (considered equivalent to "well-below 2 °C") based on the imposition of a global carbon price across all countries. In this scenario the EU achieves a GHG emission reduction of at least 80% over 1990–2050. Low-carbon R&D does not increase from Baseline levels.

Table 2. Scenarios examined with GEM-E3-FIT.

	Scenario	EU Climate Target	R&D Investment
BASE	Baseline (business-as-usual trends)	Continuation of trends	Current R&D intensities
2DEG	Decarbonisation to 2 °C with all options available	40% GHG emission reduction in 2030, at least 80% reduction in 2050 (relative to 1990 levels)	No additional R&D from Baseline
2DEG_GRD	Decarbonisation with increased EU public R&D		10% of EU carbon revenues are used for public low-carbon R&D
2DEG_GRDW	Decarbonisation with higher global public R&D		10% of global carbon revenues are used for public low-carbon R&D
2DEG_GRDH	Decarbonisation with increased EU public R&D & higher learning rates		10% of EU carbon revenues used for public low-carbon R&D, public R&D learning rates are same as private R&D
2DEG_PRD	Decarbonisation with increased EU private R&D		10% of EU carbon revenues are used for private low-carbon R&D
2DEG_SK	Decarbonisation with human capital upgrade		10% of EU carbon revenues used to subsidise wages and social security for high-skilled labour
2DEG_COMB	Decarbonisation with low-carbon innovation and education		30% of EU carbon revenues are used for private and public low-carbon R&D and to subsidise highly-skilled labour

We further consider six low-carbon innovation policy scenarios building on 2DEG specifications but assuming that EU countries implement additional investment in low-carbon R&I, either public or private, or/and increase expenditure to human capital development. The split of increased R&D spending in specific low-carbon technologies (PV, wind, biofuels, EVs, batteries and CCS) is determined by their base-year share in global clean energy market. Budget neutrality is ensured in all scenarios; increased low-carbon R&D is financed using a certain share of carbon revenues, while remaining carbon revenues are used to reduce the global interest rate, thus promoting investment required in the decarbonisation context. Innovation and human capital policy interventions can be simulated through various channels in GEM-E3-FIT, including direct government subsidisation to R&D and wages (i.e., for highly-skilled labour), tax credits, reduced barriers to buy patents, facilitation of spillovers and reduced risk premiums through provision of low-cost finance directed to R&D. In the current set-up, we use direct subsidisation of low-carbon R&D and high-skilled workforce as a policy driver, directly influencing productivity improvement and enhanced adoption of low-carbon technologies:

- In the "2DEG_GRD" scenario, EU countries use 10% of ETS carbon revenues to finance public low-carbon R&D, which leads to TFP growth in clean energy manufacturing and hence reduced low-carbon technology costs.
- In the "2DEG_GRDW" scenario, all countries use 10% of carbon revenues to finance public R&D in clean energy, which leads to an improved TFP and cost reduction globally, which is reinforced by spillovers across regions.

These scenarios reflect the major role of governments in energy innovation, often funding basic and higher-risk research as well as novel low-carbon technologies, which are costly and have uncertain market value.

- In the "2DEG_PRD" scenario, EU countries use 10% of carbon revenues for private low-carbon R&D, resulting in improved productivity in the country and industry performing the R&D. Non-EU countries benefit indirectly from EU low-carbon R&D spending through partial diffusion of knowledge spillovers and trade. This reflects intellectual property protection, obstacles for knowledge diffusion and patent spillovers, lack of human capital and costly replication of patents. Each country benefits from increased R&D spending only when its cumulative R&D stock increases beyond a certain threshold (set at 10% of the current global R&D stock). Countries with very limited knowledge stock cannot fully exploit innovation-induced productivity gains due to limitations in human capital, infrastructure, institutions, industrial and innovation base etc.

Public and private R&D can have different effects on the cost and/or performance of clean energy technologies [56,57]. Shinnosuke et al. developed a three-factor learning curve for PV costs as a function of cumulative capacity and the knowledge stock accumulated by public and private R&D spending [58]. They found that public R&D has a lower impact on PV cost reduction in Japan, implying a lower learning rate than private R&D (of about 30%). In addition, various empirical studies point to the different nature and impacts of public and private R&D [15,24,27]. Based on these empirical data, GEM-E3-FIT uses lower learning rates for public R&D relative to their default levels (Table 1). The 2DEG_GRDH variant is designed to explore this uncertainty and it is assumed that the values of learning rates for public and private R&D are set to be the same.

- In the "2DEG_SK" scenario, EU countries use 10% of ETS revenues to subsidise high-skilled jobs required for the clean energy transition in the form of direct wage subsidisation and a reduction of social security contributions for highly-skilled jobs. The subsidisation incentivises households to educate more in order to acquire additional skills needed for the low-carbon transition.
- The "2DEG_COMB" scenario combines all policy measures assessed in previous scenarios to quantify the socio-economic impacts of a variety of innovation and education policies. In particular, EU countries use 30% of carbon revenues for financing R&D in low-carbon technologies and for subsidising highly-skilled labour. In this scenario, the policy measures are simultaneously implemented in the model, thus simulating the adoption of a policy portfolio (based on low-carbon innovation and human capital upgrade) in order to boost EU's economic growth and employment in the decarbonisation context.

4. Results

Public budget neutrality is ensured in the general equilibrium framework and scenario results show the impact of resources shift within the economy (i.e., towards R&D and education), rather than just the impact of additional government spending on specific sectors.

4.1. Energy System Restructuring

The decarbonisation scenarios simulate a future consistent with the Paris Agreement goal to limit global warming to well-below 2 °C relative to pre-industrial levels, with a global CO_2 budget of 1000 Gt over 2010–2050; the EU meets its target to reduce domestic GHG emissions by at least 80% in the period 1990–2050. This is achieved through the imposition of universal carbon price reflecting the cost-optimal mitigation pathway equalizing marginal abatement costs across countries and sectors. To reflect increasing climate policy stringency, the global carbon price increases over time to about 80€/tn CO_2 in 2030 and 290€/tn in 2050, which is consistent with the results of multi-model comparison exercises on mitigation pathways [59].

Ambitious climate policies lead to structural transformation of global and EU energy systems with increased expansion of renewable energy, accelerated energy efficiency in end-uses and electrification of energy, mobility and heating services. Low-carbon technologies are massively deployed to substitute fossil fuel use, while their costs improve as a result of accelerated learning-by-doing induced by their increased deployment. GEM-E3-FIT incorporates several emission abatement technological options, including RES power generation technologies (wind onshore and offshore, PV, hydro, biomass), EVs, batteries, advanced biofuels, energy efficiency and electrification in end-uses, fuel substitution, and technologies to capture and store carbon dioxide (CCUS) emitted from power plants. Through its wide coverage, GEM-E3-FIT can provide a rigorous assessment of interlinkages between the various mitigation options, i.e., interplay between RES expansion and electrification, competition between advanced biofuels and EVs etc., and can assess the complex dynamics related to energy demand and supply, technology innovation and uptake of low-carbon technologies.

Global CO_2 emissions from fossil fuel combustion are projected to decline from 52 Gt CO_2 in Baseline to about 14 Gt CO_2 in 2DEG scenario in 2050 induced by increased carbon pricing (Figure 2). The power generation sector is the first to decarbonize and contributes to about 55% of emission reduction effort by 2050 through large-scale expansion of several low-carbon technologies, with are already cost-competitive with fossil-fired power plants in many countries. The electricity sector is rapidly transformed towards a low-emission paradigm with the share of fossil fuels declining from 68% in 2015 to only 8% in 2050 (Figure 2) accompanied by a rapid expansion of renewable energy and CCS (in specific countries) which is consistent with the projections of prior research using as suite of national energy-economy models [60]. The electrification of energy services combined with decarbonized power supply is a key mitigation option in hard-to-abate sectors (i.e., industries, heating, transport). Energy efficiency improves considerably relative to Baseline, while low-carbon and efficient energy forms (RES-based electricity, advanced biofuels) substitute fossil fuels in end-uses. The system transformation is even more pronounced in the EU, with a rapid coal phase-out by 2040, massive deployment of PV and wind (onshore and offshore) and a nearly emission-free electricity production by mid-century.

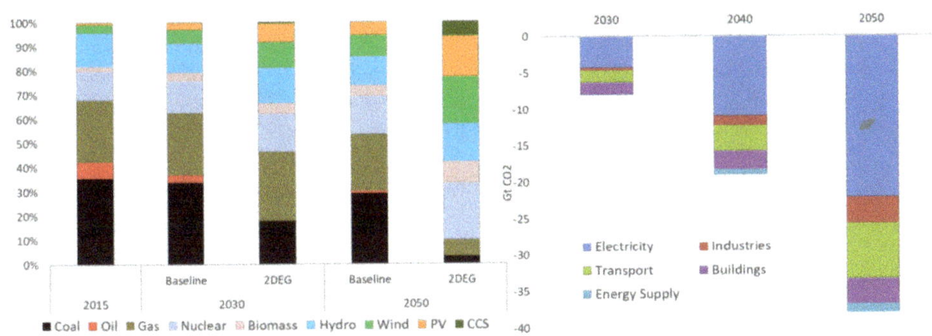

Figure 2. Global energy system restructuring in 2DEG: (**i**) Global power generation mix (in %), (**ii**) Emission reduction from Baseline by sector (Gt CO2).

The global market for low-carbon technologies is growing fast driven by technology uptake, innovation dynamics, technological improvements and climate policies and regulations. In 2015 the size of the global low-carbon technology market is estimated at about €250bn and is dominated by PV panels, biofuels and wind turbines. The global low-carbon market increases constantly in the baseline scenario and amounts to 18.7 trillion EUR over 2020–2050. In the 2DEG scenario, the high carbon pricing and accelerated uptake of low-carbon technologies induce an increase in the global low-carbon market, which amounts to €39 trillion cumulatively by 2050. EV and battery manufacturing are the largest sectors, accounting for 42% of the global market, with PV and wind representing about 35%.

The EU accounts for 15% of the global low-carbon market, but has very small shares in the production of PV and batteries (Figure 3).

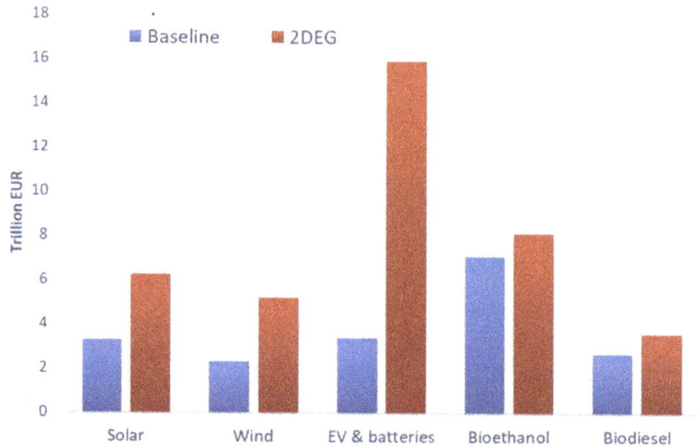

Figure 3. Global clean energy market (cumulative over 2020–2050).

The size of the low-carbon market is driven by the number of equipment units sold and technology costs that tend to decline as a result of learning by doing. Thus, the low-carbon market does not increase proportionally with technology uptake, as this is moderated by lower technology costs. In the baseline scenario, the deployment of PV and wind drives down their costs significantly, and hence their additional deployment in 2DEG leads to a relatively limited cost reduction from baseline. On the other hand, EVs and batteries will not reach maturity in terms of cost reductions in baseline scenario; thus they have high potential for cost improvements in the decarbonisation context.

Climate policies (and the imposition of carbon pricing) drive energy system restructuring towards a more capital-intensive structure, with increased investment to renewable energy, EVs and energy efficiency projects. High carbon prices increase the cost of energy services and hence production costs throughout the economy and have a depressing effect on private consumption and GDP, which is partly alleviated by increased investment in low-carbon and energy-efficient technologies. Ambitious climate policies in 2DEG lead to a 1.5% decline in cumulative global GDP over 2020–2050 (Figure 4) and 3.2% reduction in 2050.

Many macroeconomic models that have computed the costs of decarbonisation provide evidence of net mitigation costs; for example the modelling results reported in the IPCC WG3 AR5 estimated global consumption losses of 2–6% in 2050 associated with an emissions trajectory that limit global warming to less than 2 °C by 2100. Our model-based projection lies in the middle of these estimates and shows differential macro-economic impacts across countries:

- Major fossil fuel exporters, like Saudi Arabia and Russia, would face large negative economic impacts due to their high carbon intensity (per unit of GDP) and the reduced revenues from fossil fuel exports
- Mitigation costs in large developing countries (China and India) are generally higher than developed economies, as the former have higher carbon intensities and the imposition of universal carbon price results in higher relative mitigation effort for developing countries.

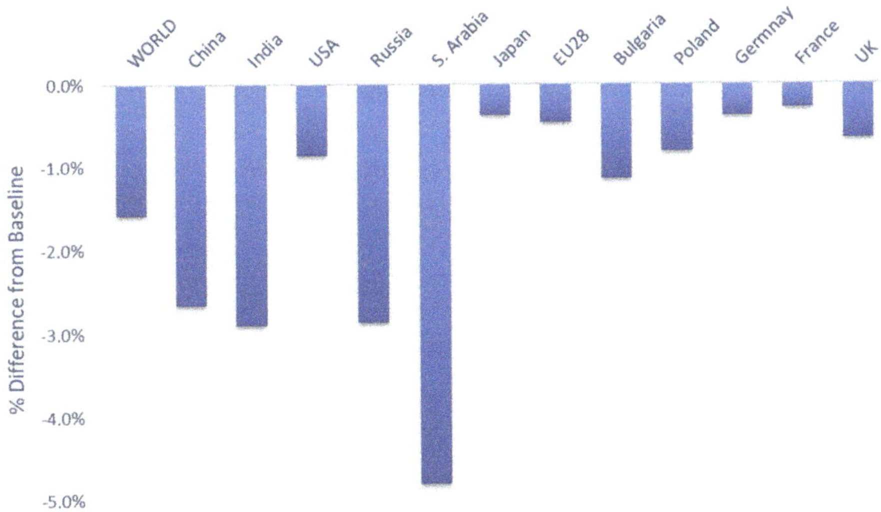

Figure 4. GDP impacts of 2DEG (% change from Baseline cumulatively over 2020–2050).

The macro-economic impacts across developed economies are limited, on average less than 1% of their cumulative GDP. Mitigation costs are higher in economies with relatively high carbon intensities (i.e., Bulgaria, Poland), while costs are very limited in countries with low carbon intensities (Japan, France, UK) that already implement climate policies and in countries that can benefit from increased low-carbon technology exports (e.g., Germany).

4.2. Impacts of Public Low-Carbon R&D

Energy innovation depends on public sector contribution, with governments funding basic research and novel low-carbon technologies, which are costly and have uncertain market value. This is reflected in the allocation of energy R&D [55], where public R&D represents more than 50% of R&D directed to renewable technologies, while its share in conventional energy is about 20%. In the last decade, energy-related R&D expenditure accounted for about 1–2% of total R&D expenditure in the EU-28 and US, with fossil fuels and nuclear having the biggest share till the early '80s, but innovation in renewables and energy efficiency is steadily gaining ground, enabling a more diverse and balanced energy R&D portfolio.

To construct base-year R&D stocks (to be used in GEM-E3-FIT modelling), R&D data from the IEA R&D database were used for low-carbon technologies, including PV, wind, batteries, biofuels and electric cars. In 2014, R&D on biofuels accounted for more than 50% of total low-carbon R&D in OECD economies, followed by PV and wind. This R&D structure changes considerably when China is added, as the Frankfurt School finds that solar technologies constitute 50% of global renewable energy R&D, followed by wind (20%) and biofuels (18%) [61]. Building on data from IEA R&D database, the R&D stock for low-carbon technologies in OECD countries is constructed following the methodology of [62] that consider R&D depreciation to develop consistent estimates for low-carbon R&D stock. For China, we have used estimates for public R&D in PV, wind and batteries from World Bank [63]. GEM-E3-FIT estimates for public R&D spending for PV, wind, biofuels and batteries are consistent with [54].

In Baseline and 2DEG scenarios, it is assumed that the R&D intensity of clean energy sectors (i.e., R&D expenditures as a percentage of value added) remains constant at 2015 levels in all countries implying that low-carbon R&D increases in line with the size of the clean market. Public R&D spending in low carbon technologies increases significantly in "2DEG_GRD" scenario as 10% of EU carbon revenues are directed to public low-carbon R&D, triggering increased innovation and learning in PV, wind and EVs. As EU carbon revenues increase to about 3% of GDP by 2050, the low-carbon R&D stock accumulates rapidly triggering cost reductions of 27% for wind, 18% for PV and 37% for batteries in 2050 relative to 2DEG. TFP growth induced by public low-carbon R&D is diffused across EU countries, with high amounts of R&D directed to wind, as the R&D split is determined by technological shares in global low-carbon market in 2015.

Increased public R&D expenditure would lead to improved productivity and cost reductions for low-carbon technologies resulting in positive macro-economic effects, with EU GDP increasing by 0.01% in 2030 and 0.05% in 2050 relative to 2DEG. The scenario impacts increase over time as low-carbon R&D stock accumulates and productivity improvements become more visible. Differential impacts are projected across EU countries depending (among others) on the country position in low-carbon manufacturing and trade and the amount of carbon revenues directed to public R&D. Energy system decarbonisation would create market opportunities for countries and industries that manufacture clean energy products, as already manifested by large export surpluses in China and the EU (major PV and wind manufacturers respectively). The explicit modelling of low-carbon products and equipment in GEM-E3-FIT allows to consistently capture the trade, competitiveness and inter-industrial production effects of decarbonisation. The largest GDP increase is projected for Denmark (0.11% in 2050), which is the leading wind turbine producer and highly benefits from productivity improvements and increased turbine exports. Leading producers of PV (China), wind turbines (Germany, Denmark), batteries (Japan, Korea) and biofuels (Brazil) register larger GDP impacts than the global average. As public R&D diffuses globally, all countries benefit and register positive GDP growth relative to 2DEG (global average of 0.05%), while leading EU low-carbon technology producers benefit the most from increased public R&D in terms of exports, production and employment. For example the implicit multiplier of low-carbon R&D (ratio of GDP gains to public R&D spending) is particularly high in key wind manufacturers (Denmark, Germany); note that most of low-carbon R&D is directed to wind turbine manufacturing as technology shares in R&D investment are determined by their shares in 2015 global low-carbon market. Technology importers register lower economic benefits, as financial resources are limited in the general equilibrium framework and R&D expenditures can exert a crowding out effect in investment in other sectors at least temporarily (Figure 5). However, productivity gains induced by R&D enlarge the market prospects and can induce higher investment and activity in the long term. Our findings reinforce prior research on the topic, as Edenhofer et al. illustrated positive GDP impacts from low-carbon R&D spending [34], while Dechezleprêtre et al. argue that increased public low-carbon R&D is required for the transition to a low-emission economy while providing socio-economic benefits [64].

In case that all countries use 10% of their carbon revenues to finance low-carbon public R&D (2DEG_GRDW), the accumulation of knowledge and innovation accelerates leading to increased productivity growth via leaning by research. Productivity improvements are diffused to other countries and industries based on knowledge spillovers of public R&D. As global carbon revenues amount to about 7 trillion EUR in 2050 (i.e., about 3.5% of global GDP), the public low-carbon R&D stock increases significantly leading to cost reductions of 52% for wind, 43% for PV and 60% for batteries in 2050 relative to 2DEG. As productivity improvements induced by public low-carbon R&D diffuse globally, higher public R&D expenditures have positive economic and employment impacts for EU and non-EU countries. Global GDP increase is significantly higher than 2DEG_GRD (Figure 6), as the additional low-carbon R&D spending is 10 times higher than 2DEG_GRD. The economic impacts differ across countries depending on their low-carbon technology innovation potential. Leading producers and exporters of PV (China), wind (Denmark, Germany), biofuels (Brazil, USA) and batteries (Japan, Korea) register the largest GDP gains. On the other hand, GDP gains are low in countries with limited innovation dynamics and low clean energy manufacturing base (as observed in [39]).

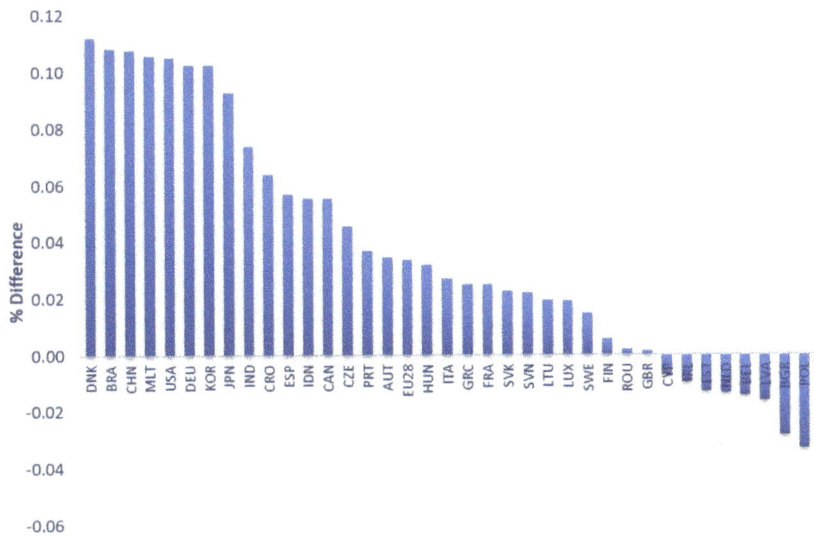

Figure 5. GDP Changes in 2DEG_GRD in 2050 (% difference from 2DEG).

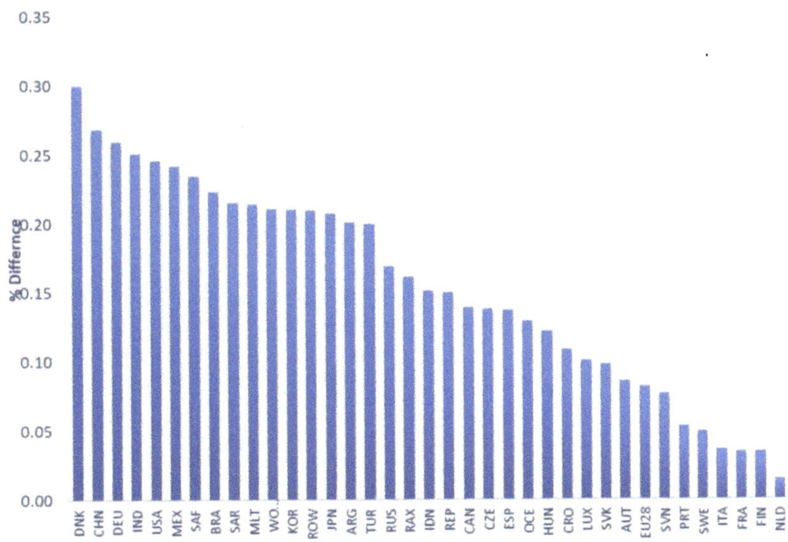

Figure 6. GDP Changes in 2DEG_GRDW in 2050 (% difference from 2DEG).

To explore the uncertainty surrounding R&D learning rates, the variant 2DEG_GRDH is developed where the learning rates of public low-carbon R&D are assumed to be the same as private R&D (Table 2). This variant shows larger macro-economic benefits, with global GDP increasing by 0.3% relative to 2DEG, with even higher growth in large low-carbon technology manufacturers (Denmark, China, Germany, Brazil, Japan etc.). The uncertainty regarding the link between R&D expenditures and cost reduction is more relevant to new, currently immature technologies like batteries. Paroussos et al. implemented a full-scale sensitivity analysis to quantify the macro-economic impacts of uncertain low-carbon R&D learning rates [30].

4.3. Impacts of Private Low-Carbon R&D

Several low-carbon technologies do not have global system boundaries, as e.g., the learning system of photovoltaics is national or regional with spillovers across regions [65], while wind turbines were analysed both at the national and global basis [66]. Clas-Otto argues that public and private R&D have different objectives, as the former focuses on basic higher-risk research and novel low-carbon technologies with uncertain market value [56]. Private R&D focuses is commonly directed to mature technologies with limited risk and high market value. As government and private R&D have different nature and impacts on technology costs and performance [56], our analysis differentiates between public and private low-carbon R&D.

Private R&D expenditure can increase knowledge stocks and is linked with positive externalities of technological progress. To construct the private R&D stock for low-carbon technologies, we combine R&D data from [54] with estimates on the share of corporate to government R&D by technology for major economies. For China, estimates on private low-carbon R&D are used [63], which match well with [54] estimates. Based on these sources, the private R&D stock for low-carbon technologies is constructed following [62] considering R&D depreciation over time. In China low-carbon R&D is largely based on government spending, while Japan and Korean R&D efforts are driven by the private sector; in the EU and USA, private and public low-carbon R&D expenditures are comparable.

In the 2DEG_PRD scenario, 10% of EU carbon revenues are assumed to be directed to private low-carbon R&D (in the form of subsidies), triggering increased innovation and learning in low-carbon technologies. The R&D split across technologies is determined by their shares in global clean energy market in 2015. The private EU low-carbon R&D stock increases significantly, triggering significant cost reductions of 26% for wind, 20% for PV and 43% for batteries in 2050 (average of EU countries) relative to 2DEG scenario. The private low-carbon R&D expenditures lead to TFP growth in the country and industry performing the R&D, while non-EU countries benefit indirectly through bilateral trade and spillovers with partial knowledge diffusion to other countries, reflecting Intellectual Property Protection, obstacles for knowledge diffusion and costly replication of patents. In GEM-E3-FIT, countries with low knowledge stock cannot fully exploit innovation-induced productivity gains due to limitations in human capital, infrastructure, institutions, regulation, industrial and innovation base [3]; countries benefit from increased R&D spending only when their R&D stock increases beyond a certain threshold (set at 10% of the 2015 global low-carbon R&D stock).

R&D expenditures reduce the costs of low-carbon technologies. As resources are limited in the general equilibrium framework, additional R&D can exert a crowding out effect on investment in other sectors, but temporarily, because productivity gains induced by R&D enlarge the market prospects inducing higher growth and investment in medium-term. Thus, private R&D expenditures may induce positive economic growth with EU GDP increasing by 0.9% in 2050 compared to 2DEG scenario (Figure 7). The positive impact is higher relative to public R&D scenario (2DEG_GRD) indicating the higher efficiency of corporate R&D, which is closer to industrial activities that can benefit directly from innovation activities in contrast to public R&D activities that commonly focus on basic research and immature, highly uncertain technologies, as argued in [67]. Major EU manufacturers of low-carbon technologies (Germany, Spain, Denmark) register higher GDP gains relative to EU average, as they benefit from their competitive advantage and increase their production and exports. High GDP gains are also projected in countries with large amounts of carbon revenues as a percentage of GDP, i.e., Poland, Hungary, Spain, as they benefit from higher investment in private low-carbon R&D and overtake the "efficiency" threshold of private R&D. In contrast, impacts are smaller in countries where increased low-carbon R&D does not suffice to reach the "efficiency threshold" and thus they fail to benefit massively from low-carbon innovation.

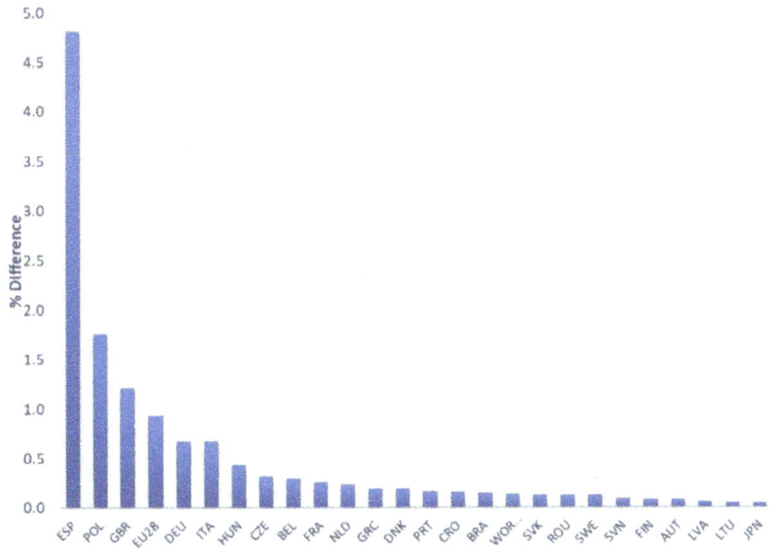

Figure 7. GDP Changes in 2DEG_PRD in 2050 (% difference from 2DEG).

Higher private low-carbon R&D expenditures improve the competitiveness of EU manufacturers in global markets, reflecting that "*To become competitive, consolidated R&I action at a European level is required in the short term to secure Europe's strong industrial position in clean energy manufacturing in the future*" [68]. Based on technology assessments [46], the EU produces PV and batteries at a cost 30% higher relative to the main supplier (China). Increased private R&D can eliminate the price gap, as EU technology production costs decline induced by R&D expenditure, and leading EU manufacturers achieve price-parity with global main suppliers by 2040 (Figure 8).

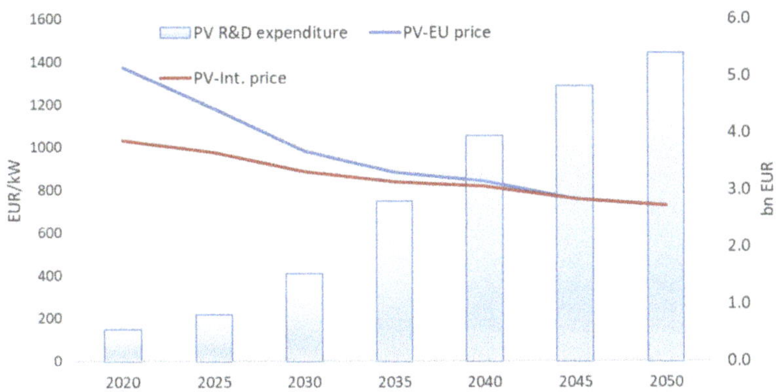

Figure 8. EU R&D expenditure in PV (bn EUR) and development of PV costs (EUR/kW).

Increased low-carbon R&D reduces costs for EU manufacturers, leading to an increase in EU share in global low-carbon production from 15% in 2DEG to 25% 2DEG_PRD (Figure 9), provided that knowledge advantage gained from low-carbon R&D is not quickly diffused to foreign competitors. Leading European manufacturers would enhance their competitive advantage in wind turbine and EV manufacturing, while R&D expenditure directed to batteries and PV would significantly reduce EU production costs allowing EU producers become competitive in global markets and achieve price-parity with their main competitors (China, Japan, Korea) by 2050. The development of a competitive battery and PV manufacturing capacity boosts domestic activity through growth of industrial activities, increased employment and positive impacts on household income.

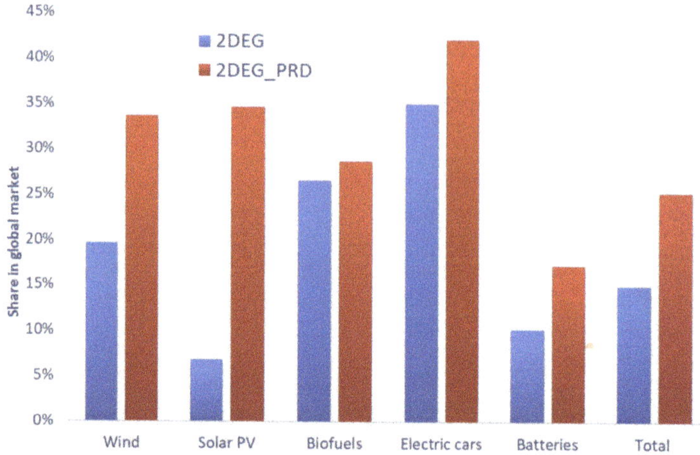

Figure 9. EU share in global low-carbon market (cumulative over 2015–2050).

4.4. *Upgrade of Human Capital*

Energy system decarbonisation leads to structural changes in the economy and labour market, requiring a different set of skills relative to conventional technologies [36]. The enhanced labour skills can be provided either by higher attainment in tertiary education (e.g., advanced skills for high-tech manufacturing activities), public training programmes (aiming to develop new skills required by the market) and private on-job training inducing higher labour productivity [69]. The 2DEG_SK scenario assumes that 10% of ETS revenues are used to subsidise high-skilled jobs required for the low-carbon economy, with half of them directed to firms to reduce social security contributions for highly-skilled jobs, and the other half subsidising the wage of highly-skilled labour, i.e., those completed tertiary education. This makes it attractive for young age cohorts to attain tertiary education and develop skills required for the transition, i.e., engineers, IT, technicians, STEM skills, managers [9]; thus inducing an upgrade in human capital and increased productivity of workforce entering the labour market.

In GEM-E3-FIT modelling, the wage rate is higher for high-skilled relative to low-skilled labour reflecting their higher labour productivity. The subsidisation of high-skilled workers incentivises firms to employ high-skilled workforce through the reduction of social security contributions. Households are incentivised to attain tertiary education with the prospect of increased wage in the future (i.e., for engineering or IT jobs) due to the subsidisation of wages of high-skilled labour.

In the short-term, as younger cohorts increasingly decide to attain tertiary education and delay entering the labour market, labour supply declines inducing small negative impacts on EU GDP. However, in the longer-term, EU GDP would increase by 0.1% relative to 2DEG induced by higher employment (+0.16% in 2050), increased labour productivity and higher wage rate (+0.15% in 2050), which result in increased private consumption of households (+0.22% in 2050). The improved matching of labour demand and supply leads to reduced EU unemployment rate from 7.4% in 2DEG to 7.2% in 2DEG_SK in 2050. The differentiation of impacts across countries largely reflects the amount of carbon revenues as share of GDP used to subsidise highly-skilled labour.

The 2DEG_SK scenario leads to a 0.5% increase in highly-skilled jobs relative to 2DEG in 2050 (Figure 10). Impacts on other labour skills are relatively limited, driven by indirect mechanisms: increased GDP –through higher labour productivity- would positively impact labour demand for all skills (as positive effects cascade in all productive sectors); however, firms have incentives to choose high-skilled employees that offer higher productivity and can take advantage of the subsidisation. As an increasing number of young cohorts chooses to attain tertiary education, the supply of low-skilled labour declines exerting a limited downward pressure on their employment.

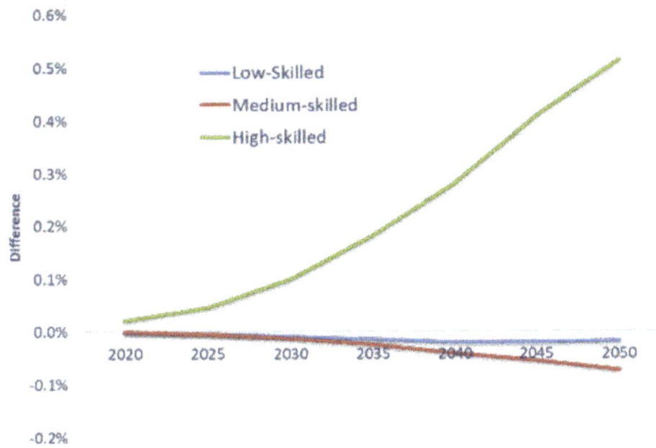

Figure 10. EU employment impacts by skill of 2DEG_SK scenario (% changes from 2DEG).

4.5. Combined Policy Scenario

The "2DEG_COMB" scenario simulates the impacts of a policy package to facilitate clean energy transition, combining low-carbon innovation and development of the required skills. It aims to boost low-carbon R&I with an ambitious allocation of resources, as 20% of ETS carbon revenues are used to finance public and private low-carbon R&D, while an additional 10% is used to subsidise the wages and social security contributions of highly-skilled labour required for the transition. The interactions and synergies between policies promoting public and private low-carbon R&D and labour skills are explored through a combination of increased expenditure for low-carbon R&D and human capital upgrade.

The "2DEG_COMB" scenario has positive macro-economic impacts driven by: (1) high productivity improvements and cost reductions in low-carbon technologies and (2) higher labour productivity, as a higher percentage of the workforce would attain tertiary education and develop advanced skills. The European GDP would increase by 1.1% relative to 2DEG in 2050, triggered by impacts from increased public (0.08%) and private low-carbon R&D (0.9%) and human capital upgrade (0.11%). The positive socio-economic impacts of "2DEG_COMB" are higher in the longer term with EU GDP gains increasing from 24 bn EUR in 2030 to 311 bn EUR in 2050 (Figure 11). This is driven by the increasing amounts of ETS revenues directed to low-carbon R&D combined with the accumulated effect of innovation and human capital policies on productivity growth. The major driver of GDP growth is the increase of low-carbon technology exports triggered by improved EU competitiveness in international markets, while investment and consumption also contribute to EU GDP increase. The results of our analysis confirm prior research on the benefits of endogenous technical change, as Edenhofer et al used 10 energy-economy models and showed that low-carbon technology improvements result in a significant reduction of mitigation costs [34].

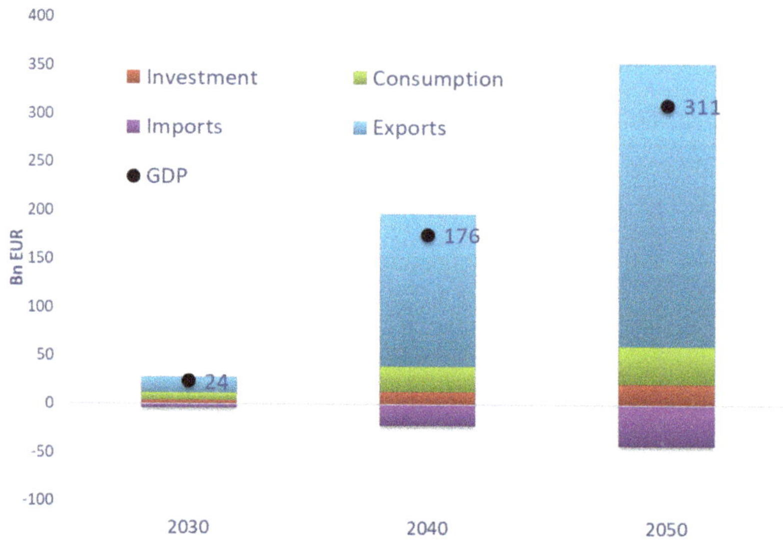

Figure 11. Changes in GDP components in 2DEG_COMB relative to 2DEG.

The macro-economic impacts of the combined policy scenario are significant in countries with:

- Large low-carbon manufacturing base (Germany, Denmark, Spain) as they benefit from higher low-carbon R&D to enhance their competitive advantage and technology exports.
- High amounts of carbon revenues as a percentage of GDP, i.e., Poland, Hungary, Spain, that benefit through higher low-carbon R&D leading to higher gains in innovation-induced productivity growth.

The impacts are smaller in magnitude in EU countries with limited low-carbon innovation base, as they do not massively benefit from low-carbon innovation, as they do not reach the "efficiency threshold" imposed for low-carbon innovation, and receive limited knowledge spillovers. Using ETS revenues to finance low-carbon R&D has positive impacts for EU GDP growth as R&D expenditure induces productivity improvements through innovation, with positive modest impacts on employment triggered by improved industrial competitiveness (Figure 12). In case that the policy focus lies on job creation, carbon revenues should be directed towards subsidization of skills required for the transition (engineers, IT, technicians, managers). Private low-carbon R&D induces higher productivity and economic growth than public R&D (as industrial activities directly benefit from corporate innovation) with EU-based companies improving their competitiveness and increasing low-carbon technology exports. The Combined Policy scenario leads to an even higher GDP growth as a larger part of ETS carbon revenues are directed towards low-carbon R&D and human capital upgrade. GDP growth is mostly driven by increased low-carbon technology exports triggered by improved EU competitiveness in international markets.

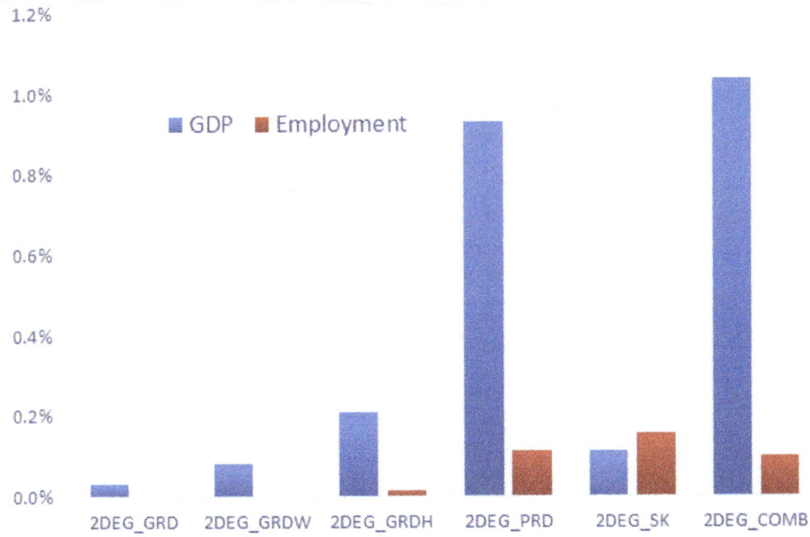

Figure 12. GDP and Employment impacts of policy scenarios (% difference from 2DEG in 2050).

5. Discussion and Conclusions

The EU strategy calls for increased R&I investment in an effort to deliver stronger and inclusive growth, create more and better jobs and promote social and environmental objectives. The ambitious EU emission reduction targets included in the EU Green Deal would drive a massive deployment of renewable and energy efficient technologies. In this context, the EU can exploit its large R&I and manufacturing base to further innovate and develop these technologies domestically creating new jobs and growth. The study investigates the socio-economic impacts of increased public and private R&D in low-carbon technologies and assesses its contribution towards a cost efficient low-emission transition while enhancing the EU competitiveness in global markets.

Using the state-of-the-art multi-sectoral CGE model GEM-E3-FIT enhanced with a novel representation of learning and innovation dynamics, we find that extensive R&I is required to support private businesses in low-carbon technology innovation and manufacturing and the development of domestic value chains, if the EU wants to be a competitive technology supplier. The policy scenarios examine different ways to support EU low-carbon innovation using a part of ETS carbon revenues, e.g., increased public R&D, subsidies to support private R&D or subsidisation to develop labour skills required for the transition. The complex interactions between energy system decarbonisation, technology development and low-carbon R&D are analysed, in the ambitious climate policy context consistent with Paris Agreement goals. Based on the general equilibrium framework, all scenarios ensure public budget neutrality where government expenditures need always to be backed up by the generation of respective revenues. Model results thus show the impact of specific policies leading to a reallocation of resources and innovation-induced productivity growth, rather than just the impact of additional spending on R&D or on education.

The decarbonisation of the energy system is based on large-scale uptake of renewable energy, improved energy efficiency and accelerated electrification of energy services. Ambitious climate policies induce a transition towards a more capital-intensive structure of the economy, while carbon prices increase the cost of energy services throughout the economy with negative impacts on GDP and consumption. The low-emission transition results in increased development of low-carbon technologies that substitute for fossil fuel use in energy demand and supply, leading to a large-scale growth of the global low-carbon market amounting to €39 trillion cumulatively by 2050, dominated by EVs, solar PV, wind and batteries. Empirical scientific findings show that R&D expenditure can reduce the costs of low-carbon technologies, a mechanism that is incorporated in GEM-E3-FIT. So, what is the impact of policies using carbon revenues to enhance low-carbon R&D and develop the required labour skills?

The analysis shows that increased public and private EU investment in low-carbon R&D (funded by carbon revenues) leads to lower technology costs, improved productivity and GDP growth fuelled by innovation and increased EU competitiveness and exports in the international market. The public R&D scenario shows macro-economic benefits for all countries as productivity improvements are diffused worldwide but the highest GDP gains are registered in large low-carbon manufacturers, as they exploit their competitive advantage in a large-growing market; these include Denmark and Germany (wind turbines), China (PV), Japan (batteries) and Brazil (biofuels). In case all countries use 10% of their carbon revenues to finance low-carbon R&D, the technology cost reductions will be significant resulting in large GDP gains (0.2% at the global level in 2050). As resources are limited in the general equilibrium framework, additional low-carbon R&D can crowd-out investment in other sectors, but only temporarily, because innovation-induced productivity gains enlarge market prospects, leading to more efficient use of economic resources triggering accelerated activity growth in the medium and longer term.

The different nature of private R&D is reflected in the policy scenarios. Private R&D investment leads to improved productivity in the country and industry performing R&D, while other countries benefit indirectly through bilateral trade and knowledge spillovers but with limitations in knowledge diffusion. At the EU level, GDP gains are higher relative to public R&D, indicating the higher efficiency of corporate R&D that is closer to industrial activities, while public R&D commonly focuses on basic research and immature highly uncertain technologies. Major EU low-carbon manufacturers (Germany, Spain, Denmark) register even higher GDP gains, as they benefit from their competitive advantage and increase technology exports. In addition, countries with large amounts of carbon revenues as a percentage of GDP benefit from higher private R&D investment and overtake the innovation efficiency threshold. In contrast, impacts are smaller in countries with limited innovation and low-carbon manufacturing base that fail to massively benefit from low-carbon R&D and their competitiveness vis-à-vis other EU countries worsens. The model-based results show that the EU should increase expenditure for private low-carbon R&D relative to its major competitors, in order to boost activity growth, create more and high-quality jobs and expand its first mover advantage in low-carbon manufacturing exploiting new market opportunities with high export potential.

The low-carbon transition leads to structural economic and labour market changes and requires different labour skills relative to conventional technologies. In case that carbon revenues are used to subsidise high-skilled jobs required for the transition (e.g., engineers, IT, STEM skills, managers, technicians), the mismatch between labour demand and supply is reduced leading to human capital upgrade and increased productivity of the workforce. The subsidisation of high-skilled jobs leads to increased attainment in tertiary education, so that younger cohorts delay entering the labour market with slightly negative short-term GDP impacts. However, increased labour productivity and household income boost long-term GDP and employment.

These findings can provide robust evidence to policy makers, especially related to potential allocation of carbon revenues in the decarbonisation context. Using ETS revenues to subsidise public or private low-carbon R&D will reduce technology costs and improve productivity leading to increased EU competitiveness and GDP. If the policy focus lies on the creation of jobs, carbon revenues can be directed towards subsidisation of high-skilled labour required for the low-carbon transition. A holistic strategy aiming to boost low-carbon R&D and develop the required skills can fully alleviate the cost burden of decarbonisation for the European economy and create more and better jobs, especially in transition-related sectors. Investment and consumption contribute to GDP growth, but the major driver is the increase in low-carbon technology exports triggered by the improved EU competitiveness in international markets. Therefore, innovation-related policies –including deployment– should be integrated in climate policy impact assessments, as they can provide socio-economic co-benefits, while ensuring the cost-efficient transition to a climate neutral and resource efficient economy.

The analysis estimates the potential socio-economic and trade impacts of supporting low-carbon R&I to facilitate EU decarbonisation. The EU currently has a small share in PV and battery value chain and thus large support to low-carbon R&D is required to assure the competitiveness of EU-produced low-carbon products. The relocation of low-carbon manufacturing to the EU brings benefits in terms of reduced imports of PV and batteries and lower dependence on non-EU manufacturers. The design of well-planned climate, innovation and education policies would lead to enhanced development of cost-efficient clean energy manufacturing capability in the EU boosting domestic activity and employment through increased production of low-carbon technologies. The study informs policy makers on the need for a technology-smart policy strategy, integrating a portfolio of measures facilitating private low-carbon R&D (to boost domestic activity), public R&D (to enhance research for currently immature clean technologies) and labour market policies (to ensure increased employment, labour productivity and development of the required skills). The incentivization of private R&D along the supply chain of low-carbon technologies can enhance the international competitiveness of EU-based manufacturing. A clear, well-predictable and ambitious strategy for the development of low-carbon technologies is needed to create an attractive European market and provide well-anticipated price signals and planning security to European investors, industries and innovators [40]. This strategy should integrate demand (i.e., though ambitious climate policies, carbon pricing, emission standards, subsidies or mandates) and manufacturing aspects—largely focused on innovation and skills development—related to clean energy transition.

There is no doubt that despite the significant methodological improvements presented above, this kind of modelling has limitations and requires additional research. The modelling results greatly depend on the trade and price elasticities and learning rates used in the model. The study has benefited from the most recent estimates on elasticities and technology learning rates available in the literature but future research can develop a sensitivity analysis around the core model parameters specified by country and sector. Further research is also required on the distinction between public and private R&D, as empirical literature provides evidence on their different nature, but stronger empirical foundation is needed.

Author Contributions: L.P., P.F. and K.F. conceived and designed the experiments, K.F. and P.F. performed the experiments, P.F., K.F. and L.P. analysed the data, P.F., K.F. and L.P. wrote the paper. All authors have read and agreed to the published version of the manuscript.

Funding: This research was funded by the European Union's Horizon 2020 research and innovation programmes under grant agreement No 727114 ("MONROE" project) and No 730403 ("INNOPATHS" project).

Acknowledgments: The information and views set out in this paper are those of the authors and do not reflect the official opinion of the European Commission.

Conflicts of Interest: The authors declare no conflict of interest.

References

1. Verdolini, E.; Anadón, L.D.; Baker, E.; Bosetti, V.; Reis, L.A. Future Prospects for Energy Technologies: Insights from Expert Elicitations. *Rev. Environ. Econ. Policy* **2018**, *12*, 133–153. [CrossRef]
2. Bosetti, V.; Catenacci, M.; Fiorese, G.; Verdolini, E. The future prospect of PV and CSP solar technologies: An expert elicitation survey. *Energy Policy* **2012**, *49*, 308–317. [CrossRef]
3. Paroussos, L.; Mandel, A.; Fragkiadakis, K.; Fragkos, P.; Hinkel, J.; Vrontisi, Z. Climate clubs and the macro-economic benefits of international cooperation on climate policy. *Nat. Clim. Chang.* **2019**, *9*, 542–546. [CrossRef]
4. Di Comite, F.; Kancs, D. Macro-Economic Models for R&D and Innovation Policies. In *IPTS Working Papers on Corporate R&D and Innovation—No 02/2015*; Publications Office of the European Union: Luxembourg, 2015. [CrossRef]

5. De Cian, E.; Keppo, I.; Bollen, J.; Carrara, S.; Förster, H.; Hübler, M.; Kanudia, A.; Paltsev, S.; Sands, R.D.; Schumacher, K. European-led climate policy versus global mitigation action. Implications on trade, technology, and energy. *Clim. Chang. Econ.* **2013**, *4*, 1340002. [CrossRef]
6. European Commission—Joint Research Centre. *Current Challenges in Fostering the European Innovation Ecosystem*; EUR 28796 EN; Publications Office of the European Union: Luxembourg, 2017; ISBN 9789279738623.
7. IRENA. *Renewable Power Generation Costs in 2019*; International Renewable Energy Agency: Abu Dhabi, UAE, 2020.
8. Gillingham, K.; Newell, R.G.; Pizer, W.A. Modeling endogenous technological change for climate policy analysis. *Energy Econ.* **2008**, *30*, 2734–2753. [CrossRef]
9. Fragkiadakis, K.; Paroussos, L.; Capros, P. Technical description of the R&I module of GEM-E3-RD model, MONROE Deliverable 4.3.2. 2019. Available online: https://www.monroeproject.eu/wp-content/uploads/2019/05/D4.3.2-Technical-description-of-the-RI-module-of-GEM-E3-RD-model.pdf (accessed on 1 September 2020).
10. Mercure, J.-F.; Knobloch, F.; Pollitt, H.; Paroussos, L.; Scrieciu, S.S.; Lewney, R. Modelling innovation and the macroeconomics of low-carbon transitions: Theory, perspectives and practical use. *Clim. Policy* **2019**, *19*, 1019–1037. [CrossRef]
11. Acemoglu, D.; Aghion, P.; Bursztyn, L.; Hemous, D. The Environment and Directed Technical Change. *Am. Econ. Rev.* **2012**, *102*, 131–166. [CrossRef]
12. Pottier, A.; Hourcade, J.-C.; Espagne, E. Modelling the redirection of technical change: The pitfalls of incorporeal visions of the economy. *Energy Econ.* **2014**, *42*, 213–218. [CrossRef]
13. Hicks, J. *A Theory of Wages*; MacMillan: New York, NY, USA, 1932.
14. Popp, D. Induced Innovation and Energy Prices. *Am. Econ. Rev.* **2002**, *92*, 160–180. [CrossRef]
15. Popp, D. Environmental Policy and Innovation: A Decade of Research. Available online: https://www.ifo.de/DocDL/cesifo1_wp7544.pdf (accessed on 1 September 2020).
16. Ley, M.; Stucki, T.; Woerter, M. The Impact of Energy Prices on Green Innovation. *Energy J.* **2016**, *37*, 41–75. [CrossRef]
17. Vincenzi, M.; Ozabaci, D. The Effect of Public Policies on Inducing Technological Change in Solar Energy. *Agric. Resour. Econ. Rev.* **2017**, *46*, 44–72. [CrossRef]
18. Aghion, P.; Dechezleprêtre, A.; Hémous, D.; Martin, R.; Van Reenen, J. Carbon Taxes, Path Dependency, and Directed Technical Change: Evidence from the Auto Industry. *J. Political Econ.* **2016**, *124*, 1–51. [CrossRef]
19. Calel, R.; Dechezleprêtre, A. Environmental Policy and Directed Technological Change: Evidence from the European Carbon Market. *Rev. Econ. Stat.* **2016**, *98*, 173–191. [CrossRef]
20. Taghizadeh-Hesary, F.; Yoshino, N.; Inagaki, Y. Empirical analysis of factors influencing the price of solar modules. *Int. J. Energy Sect. Manag.* **2019**, *13*, 77–97. [CrossRef]
21. Karkatsoulis, P.; Capros, P.; Fragkos, P.; Paroussos, L.; Tsani, S. First-mover advantages of the European Union's climate change mitigation strategy. *Int. J. Energy Res.* **2016**, *40*, 814–830. [CrossRef]
22. Egli, F.; Steffen, B.; Schmidt, T.S. A dynamic analysis of financing conditions for renewable energy technologies. *Nat. Energy* **2018**, *3*, 1084–1092. [CrossRef]
23. Rubin, E.S.; Azevedo, I.M.; Jaramillo, P.; Yeh, S. A review of learning rates for electricity supply technologies. *Energy Policy* **2015**, *86*, 198–218. [CrossRef]
24. Kavlak, G.; McNerney, J.; Trancik, J.E. Evaluating the causes of cost reduction in photovoltaic modules. *Energy Policy* **2018**, *123*, 700–710. [CrossRef]
25. Nemet, G.F. *How Solar Energy Became Cheap*; Informa UK Limited: London, UK, 2019.
26. Creutzig, F.; Agoston, P.; Goldschmidt, J.C.; Luderer, G.; Nemet, G.; Pietzcker, R.C. The underestimated potential of solar energy to mitigate climate change. *Nat. Energy* **2017**, *2*. [CrossRef]
27. Boulatoff, C.; Boyer, C. What is the impact of private and public R&D on clean technology firms' performance? An international perspective. *J. Sustain. Financ. Invest.* **2016**, *7*, 147–168. [CrossRef]

28. European Commission. Cost Development of Low-Carbon Energy Technologies: Scenario-Based Cost Trajectories to 2050, 2017 Edition. 2018. Available online: https://ec.europa.eu/jrc/en/publication/cost-development-low-carbon-energy-technologies-scenario-based-cost-trajectories-2050-2017-edition (accessed on 1 September 2020).
29. Fouquet, R.; Pearson, P.J. Past and prospective energy transitions: Insights from history. *Energy Policy* **2012**, *50*, 1–7. [CrossRef]
30. Paroussos, L.; Fragkos, P.; Vrontisi, Z.; Fragkiadakis, K.; Pollitt, H.; Lewney, R.; Chewpreecha, U. A Technical Case Study on R&D and Technology Spillovers of Clean Energy Technologies. 2017. Available online: https://ec.europa.eu/energy/sites/ener/files/documents/case_study_3_technical_analysis_spillovers.pdf (accessed on 1 September 2020).
31. Griliches, Z.; Lichtenberg, F. R&D and Productivity Growth at the Industry Level: Is There Still a Relationship? In *R&D, Patents, and Productivity*; University of Chicago Press: Chicago, IL, USA, 1984; Chapter 21; pp. 465–502. ISBN 0-226-30884-7.
32. Parrado, R.; De Cian, E. Technology spillovers embodied in international trade: Intertemporal, regional and sectoral effects in a global CGE framework. *Energy Econ.* **2014**, *41*, 76–89. [CrossRef]
33. Grubb, M.; Köhler, J.; Anderson, D. Induced Technical Change in Energy and Environmental Modeling: Analytic Approaches and Policy Implications. *Annu. Rev. Energy Environ.* **2002**, *27*, 271–308. [CrossRef]
34. Edenhofer, O.; Lessmann, K.; Kemfert, C.; Grubb, M.; Köhler, J. Induced Technological Change: Exploring its Implications for the Economics of Atmospheric Stabilization: Synthesis Report from the Innovation Modeling Comparison Project. *Energy J.* **2006**, *35*. [CrossRef]
35. Vartiainen, E.; Masson, G.; Breyer, C.; Moser, D.; Román Medina, E. Impact of weighted average cost of capital, capital expenditure, and other parameters on future utility-scale PV levelised cost of electricity. *Prog. Photovoltaics Res. Appl.* **2019**, 1–15. [CrossRef]
36. Fragkos, P.; Paroussos, L. Employment creation in EU related to renewables expansion. *Appl. Energy* **2018**, *230*, 935–945. [CrossRef]
37. Wing, I.S. Representing induced technological change in models for climate policy analysis. *Energy Econ.* **2006**, *28*, 539–562. [CrossRef]
38. Paroussos, L.; Fragkiadakis, K.; Fragkos, P. Macro-economic analysis of green growth policies: The role of finance and technical progress in Italian green growth. *Clim. Chang.* **2019**, 1–18. [CrossRef]
39. MONROE Project. D6.4.1: Technical Description of the Modelling Results, Including Interpretation of the Differences between the Outcomes of the Models. 2019. Available online: https://www.monroeproject.eu/wp-content/uploads/2019/05/D6.4.1-Technical-description-of-the-modelling-results-including-interpretation-of-the-differences-between-the-outcomes-from-the-three-models.pdf (accessed on 1 September 2020).
40. Fragkiadakis, K.; Charalampidis, I.; Fragkos, P.; Paroussos, L. Economic, Trade and Employment Implications from EVs Deployment and Policies to Support Domestic Battery Manufacturing in the EU. *Foreign Trade Rev.* **2020**, *55*, 298. [CrossRef]
41. Criqui, P.; Mima, S.; Menanteau, P.; Kitous, A. Mitigation strategies and energy technology learning: An assessment with the POLES model. *Technol. Forecast. Soc. Chang.* **2015**, *90*, 119–136. [CrossRef]
42. MONROE Project. MD3.4.2: The Role of Human Capital in Creating Knowledge, Absorbing Spillovers and the Importance of Skilled Labour Migration. 2019. Available online: https://www.monroeproject.eu/wp-content/uploads/2019/05/D3.4.2-The-role-of-Human-Capital-in-creating-Knowledge-absorbing-spillovers-and-the-importance-of-skilled-labour-migration-E3M.pdf (accessed on 1 September 2020).
43. Fragkos, P.; Tasios, N.; Paroussos, L.; Capros, P.; Tsani, S.Z. Energy system impacts and policy implications of the European Intended Nationally Determined Contribution and low-carbon pathway to 2050. *Energy Policy* **2017**, *100*, 216–226. [CrossRef]

44. Fraunhofer ISE. *Photovoltaics Report*; Fraunhofer Institute for Solar Energy Systems (ISE): Freiburg im Breisgau, Germany, 2018.
45. Navigant Research. In *World Wind Energy Market Update 2018*; Navigant Consulting Inc.: Burlington, MA, USA, 2018.
46. Clean Energy Manufacturing Analysis Center (CEMAC). *Benchmarks of Global Clean Energy Manufacturing*; National Renewable Energy Laboratory: Golden, CO, USA, 2016.
47. International Energy Agency. *Global EV Outlook 2019*; International Energy Agency: Paris, France, 2019.
48. IRENA. *Renewable Energy and Jobs Annual Review 2016*; International Renewable Energy Agency: Abu Dhabi, UAE, 2016.
49. Fries, M.; Kerler, M.; Rohr, S.; Schickram, S.; Sinning, M.; Lienkamp, M. An Overview of Costs for Vehicle Components, Fuels, Greenhouse Gas Emissions and Total Cost of Ownership, Update 2017. Available online: https://steps.ucdavis.edu/wp-content/uploads/2018/02/FRIES-MICHAEL-An-Overview-of-Costs-for-Vehicle-Components-Fuels-Greenhouse-Gas-Emissions-and-Total-Cost-of-Ownership-Update-2017-.pdf (accessed on 1 September 2020).
50. Romer, P.M. Endogenous Technological Change. *J. Polit. Econ.* **1990**, *98*, S71–S102. [CrossRef]
51. Glachant, M.; Ménière, Y. Technology Diffusion with Learning Spillovers: Patent Versus Free Access. *Manch. Sch.* **2012**, *81*, 683–711. [CrossRef]
52. Kirchherr, J.; Urban, F. Technology transfer and cooperation for low carbon energy technology: Analysing 30 years of scholarship and proposing a research agenda. *Energy Policy* **2018**, *119*, 600–609. [CrossRef]
53. Verspagen, B. Estimating international technology spillovers using technology flow matrices. *Rev. World Econ.* **1997**, *133*, 226–248. [CrossRef]
54. United Nations and Bloomberg New Energy Finance. Global Trends in Renewable Energy Investment 2018. 2018. Available online: http://www.iberglobal.com/files/2018/renewable_trends.pdf (accessed on 1 September 2020).
55. International Energy Agency. *World Energy Outlook 2018*; International Energy Agency: Paris, France, 2018.
56. Clas-Otto, W. Energy Technology Learning through Deployment in Competitive Markets. *Eng. Econ.* **2008**, *53*, 340–364.
57. International Energy Agency. *World Energy Investment 2018*; IEA/OECD: Paris, France, 2018.
58. Hayamizu, S.; Furubayashi, T.; Nakata, T. Quantification of technological learning by R&D and its application for renewable energy technologies. *Trans. JSME* **2014**, *80*. [CrossRef]
59. Kriegler, E.; Riahi, K.; Bauer, N.; Schwanitz, V.J.; Petermann, N.; Bosetti, V.; Marcucci, A.; Otto, S.; Paroussos, L.; Rao, S.; et al. Making or breaking climate targets: The AMPERE study on staged accession scenarios for climate policy. *Technol. Forecast. Soc. Chang.* **2015**, *90*, 24–44. [CrossRef]
60. Fragkos, P.; Fragkiadakis, K.; Paroussos, L.; Pierfederici, R.; Vishwanathan, S.S.; Köberle, A.C.; Iyer, G.; He, C.-M.; Oshiro, K. Coupling national and global models to explore policy impacts of NDCs. *Energy Policy* **2018**, *118*, 462–473. [CrossRef]
61. Frankfurt School-UNEP Centre/BNEF. Global Trends in Renewable Energy Investment 2016. 2016. Available online: https://www.actu-environnement.com/media/pdf/news-26477-rapport-pnue-enr.pdf (accessed on 1 September 2020).
62. Bointner, R.; Pezzutto, S.; Grilli, G.; Sparber, W. Financing Innovations for the Renewable Energy Transition in Europe. *Energies* **2016**, *9*, 990. [CrossRef]
63. Kuriakose, S. *Accelerating Innovation in China's Solar, Wind and Energy Storage Sectors*; World Bank: Washington, DC, USA, 2017.
64. Dechezleprêtre, A.; Martin, R.; Bass, S. Climate Change Policy, Innovation and Growth: Policy Brief. In *Handbook on Green Growth*; Grantham Research Institute and Global Green Growth Institute: London, UK, 2016.
65. Schaeffer, G.J.; Seebregts, A.J.; Beurskens, L.; De Moor, H.; Van Sark, W. *Learning from the Sun; Analysis of the Use of Experience Curves for Energy Policy Purposes: The Case of Photovoltaic Power*; ECN: Petten, The Netherlands, 2004.
66. Junginger, M.; Faaij, A.; Turkenburg, W. Global experience curves for wind farms. *Energy Policy* **2005**, *33*, 133–150. [CrossRef]

67. Reinaud, J.; Clinckx, N.; Ronzeau, K.; Faraggi, P. "Scaling up Innovation in the Energy Union to Meet New Climate, Competitiveness and Societal Goals: Scoping the Future in Light of the Past", i24c and Capgemini consulting manuscript. 2016. Available online: https://www.capgemini.com/wp-content/uploads/2017/07/scaling_up_innovation_in_energy_union_capgemini_and_i24c_report_1.pdf (accessed on 1 September 2020).
68. Lebedeva, N.; Di Persio, F.; Boon-Brett, L. *Lithium Ion Battery Value Chain and Related Opportunities for Europe*; EUR 28534 EN; JRC105010; Publications Office of the European Union: Luxembourg, 2017; ISBN 978-92-79-66948-4. [CrossRef]
69. Canton, E. Social Returns to Education: Macro-Evidence. *De Econ.* **2007**, *155*, 449–468. [CrossRef]

MDPI
St. Alban-Anlage 66
4052 Basel
Switzerland
Tel. +41 61 683 77 34
Fax +41 61 302 89 18
www.mdpi.com

Energies Editorial Office
E-mail: energies@mdpi.com
www.mdpi.com/journal/energies

www.ingramcontent.com/pod-product-compliance
Lightning Source LLC
Chambersburg PA
CBHW040825250926
51850CB00007BA/1262
9783036538853